BIODIVERSITY ENDANGERED

INDIA'S THREATENED WILDLIFE AND MEDICINAL PLANTS

A.B. Chaudhuri

Ex-Indian Forest Service
Ex-Director Forest Survey of India

&

D.D. Sarkar

Ex-State Forest Service
Ex-Divisional Forest Officer

SCIENTIFIC PUBLISHERS (INDIA)

P.O. BOX 91 JODHPUR

Published by
PAWAN KUMAR
Scientific Publishers (India)
5-A, New Pali Road, P.O. Box 91,
JODHPUR - 342 001 (Raj.)
Tel.: +91-291- 433323
Fax : +91-291-512580
E-mail: scienti@sancharnet.in
artrans@satyam.net.in
www.scientificpub.com

ISBN : 81-7233-312-9

Lasertype Set : Rajesh Ojha
Printed in India

DEDICATION

In Memory of the Foresters of Yore

....you braved the severest hazards to protect an invaluable wealth of flora and fauna. You left them to our care. Have we been worthy of this priceless heritage ?

CONTENTS

FOREWORD

The present book under review has emanated from the pen of Shri A.B. Chaudhuri, who is an eminent ecologist. Earlier he had turned-out a number of books on forest resources and allied subjects, which provide basic data meticulously collected by him, when he was serving as a member of I.F.S. and the department of Forest Survey of India.

In this book he has attempted an in-depth analysis of data regarding biodiversity. Although, there are some excellent treatises on this subject, these lack quantitative data. Shri Chaudhuri has made sincere efforts to bridge this gap. These vital inputs are very necessary for better conservation of forest resources and propagation of plant species. What is the real picture of forest resources is difficult to know from most of the available publications.

An euphoria has been created by people, that India is the home of many medicinal plants. It might have been true for situation in ancient India. Scenario has now drastically changed since the colonial times. By wanton destruction of forest wealth of India,. many medicinal plants have been lost. As a result, Ayurved, which has a glorious past is also suffering. It is possible to retrieve the situation if proper methods are adopted for conservation of forest resources as suggested by Shri Chaudhuri.

Shri D.D. Sarkar, an eminent Forester, has worked with Shri Chaudhuri throughout the study and compilation.

Dated at : Salt Lake,
19th of September, 2001

(A.B. Biswas)
Hony. Secretary
Centre for Study of Man & Environment,
Parivesh Kendra,
CK-11, Sector -2, Salt Lake,
Kolkata - 700 091
&
Former Director,
Geological Survey of India

PREFACE

Several purposes dominated the writing of this book, the first and foremost being to collate the various concepts of 'Biodiversity' and to bring them together in a simple equation for easy perception of common readers.

It is also an exercise in attempting comprehensive synthesis of the various ideas and conception about biological diversity. Even in the early 90's of the last century, literature on the subject, in so far, as India is concerned was scanty. Of late, however, a good number of studies have been published on biological diversity.

In the present work the concept of genera, species and individual has been applied to the medicinal plant species of India, indicating that the quantitative aspect in determining the status of biodiversity of such plants has not received attention in the past and the belief that India has inexhaustible resources of medicinal plants is no longer true. The authors have marked the important plant species which are common, or sporadic or are threatened. Quoting the figures of analysis of tree crop in various parts of India, compiled by the Forest Survey of India, they have calculated the density of each species. For herbs and shrubs and factual report on their status has also been presented.

Many authors have written books of excellence on the plant species resources growing in India. But till now little effort has been made to make on analysis of the qualitative and factual quantitative status of plants, which is absolutely necessary for research and development, effective conservation, propagation and for determining the status of individual plant or a group of plants. Apart from their use as timber, fuel and pulpwood etc., plants possess tremendous potential as regards medicinal value. Irrespective of their life-form that is whether they are trees, shrubs, climbers, herbs, grasses or reeds, thousands of plants contain properties of immense medicinal value.

India was a pioneer in discovering the medicinal properties

of scores of plants and also in preparing both prophylactic and curative drugs for the benefit of mankind. Ayurveda was perhaps the oldest amongst all such compilations of the world. It was compilation par excellence regarding preparation of different drugs from plants and its therapeutic use amongst people suffering from different ailments. This was followed by many more compilations over long period of time by a galaxy of Ayurvedic scholars. Nevertheless, the status of the various medicinal plants did not receive proper attention.

Though, Allopathy has reached a very high stage of advancement due to outstanding technological achievements during the twentieth century, particularly in the surgical field, Ayurveda is gaining ground slowly but steadily in respect of the preparation of prophylactic and curative medicines. In this situation it is imperative that proper assessment of the status of various plant species, facing heavy biotic and abiotic impact be conducted.

The authors have made the utmost effort to analyse the factual qualitative and quantitative status of plants, growing in India. India being a very large country with high bio-diversity resources, the authors have made an assessment of the status of only selected plant species and have kept their work restricted to a few areas of vital importance for such diversity study. The following areas have been taken-up for the study of tree species, which are the plains of Assam (Nawgaon), West Bengal (Jalpaiguri and Purulia), Bihar, Meswasi and Melghat of Maharastra, Dang, Surat, Bulsar and Bharuch of Gujarat, Little Andamans, the hills of West Bengal, North Cachar, West Nepal, Bhutan, Sikkim, Kameng, Lohit and Tirap of Arunachal Pradesh and Meghalaya.

However, to keep the book to size observations on faunal bio- diversity have been confined to some selected faunal groups only namely, avifauna, ichthyofauna, amphibia, reptiles and mammals of India.

Regarding avifauna, observations have been made on various facts of the fauna but account of number of individual species is not available in any work which leaves a big-gap in biodiversity studies of this group. A similar situation exists in the study of ichthyofauna. It has been observed that most of the species are in decline due to biotic and abiotic impact. Moreover, a research break-through in breeding carps in the laboratory and sizeable success in introducing Tilapia, Grass-carp, Silver-carps etc. have brought about revolutionary changes in pisciculture. Such fishes have now reached the common man at a low price and therefore, fishermen are becoming increasingly averse to protect, preserve and breed most of the other species which need labour, protraction and rigorous time schedules and also sizeable funds.

As for mammals, amphibias and reptiles, biotic and abiotic impact have been the prime cause of depletion of these groups of fauna. It has been pointed out that without a reliable statistical design for measuring density, frequency and abundance of biotal species, one cannot properly assess the status of various groups of biota.

Productivity (primary productivity) is the basic resource that fuels Life on earth. Productivity is correlated with diversity, but sometimes it is also negatively correlated with species diversity. These factors have received little attention.

The authors expect that most readers will find some useful data and exciting ideas in this publication. They are aware that many pioneers in the field of bio-diversity and ecology of flora and fauna have contributed to a broad and clear concept of basic environmental and biotic processes. Yet, they humbly, believe that the present book will provide some basic facts and insights. The authors hope that they have been successful in hinting at a new dimension of the issue and that various government and non- governmental agencies will find the work useful.

131 N S C Bose Road
Block 10, Flat 4
Kolkata - 700 040

A.B. Chaudhuri
D.D. Sarkar

ACKNOWLEDGEMENTS

The authors are grateful to many individuals and organisations who have helped them directly or indirectly in the preparation of this manuscript. They apologize for being unable to mention their names individually. They are also grateful to those pioneer workers who permitted them to use their diagrams, charts and their reproduction on due acknowledgements.

We are indebted to those persons who encouraged us to work for the success of this endeavour as it deals with various implications of biodiversity and the status of the medicinal plants including some discussions on Ayurveda, which are of prime importance of the day.

Thanks are also due to Dipyaman Sarkar for helping in secretarial and computerised typographical works of the manuscript.

A.B. Chaudhuri
D.D. Sarkar

CHARTS

Page No.

Chapter 1.

Chapter 2.

TABLES

Page No.

Chapter 2.

MAPS OF INDIA AND STATES SHOWING VEGETATION AND MARKING VARIOUS BIOTIC AND ABIOTIC IMPACTS

MAPS SHOWING PRESENT AND PAST DISTRIBUTION OF SOME ANIMALS/BIRDS.

MAP 1. Vegetation of India, marking various biotic and abiotic impact zones

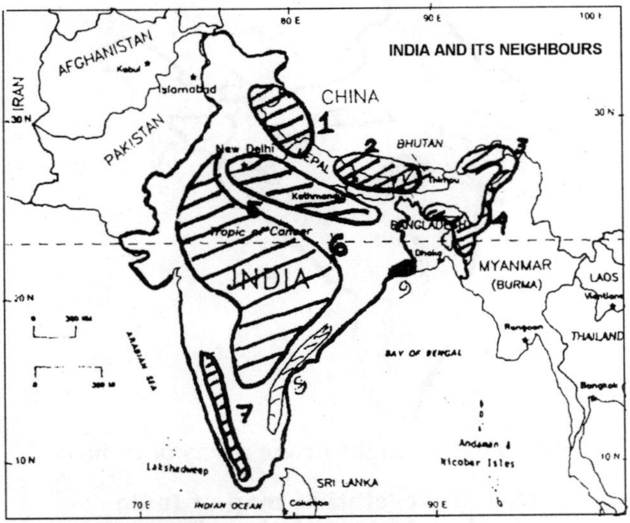

MAP 1A. Marking zones as per detailed legends.

LEGENDS

1. Western Himalayas i.e. Jammu and Kashmir and Himachal Pradesh. Rich zone, biotically affected.
2. Uttar Pradesh hills and Nepal (Central Himalayas). Rich zone, biotically affected.
3. Eastern Himalayas, i.e. West Bengal, Sikkim, Bhutan, Arunachal Pradesh. Rich zone, biotically affected.
4. Assam — Meghalaya Hills (North Eastern Himalayas) i.e. Nagaland, Manipur, Mizoram, North-Cachar, Tripura and Meghalaya. Rich zone, affected by shifting cultivation.
5. Western Plains i.e. Haryana of Punjab and Rajasthan.
   Central Plains of Uttar Pradesh and Bihar and
   Eastern Plains of West Bengal and Assam. Extension of agriculture and human population exploitation.
6. *Central zone* —
   Aravalli-Satpura Upland i.e. Rajasthan, Madhya Pradesh and Maharastra.
   South-eastern uplands i.e. Madhya Pradesh, Andhra Pradesh, Bihar and Orissa.
   Deccan upland, i.e. South Maharastra, Karnataka and Andhra Pradesh.
   Disturbed area —
   Western Ghats, i.e. Maharastra, Karnataka, Tamilnadu and Kerala. Rich but biotically affected.
   Eastern Ghats i.e. Andhra Pradesh, Tamilnadu and Orissa. Biotically affected.
7. *Coastal area (Plains)* —
   Coastal area of Sunderbans. Biotically affected.

MAP 1B. Drought prone areas of India

MAP 2. Vegetation map of India (as in mid-twentieth century)

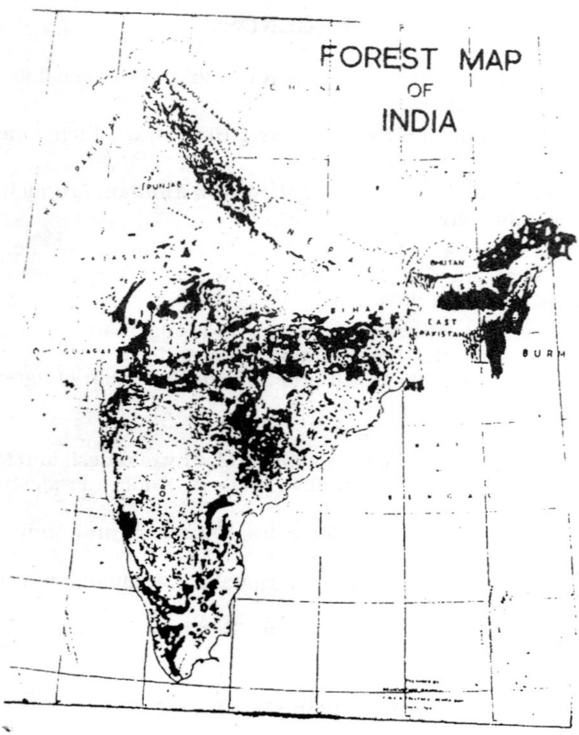

MAP 3. Forest Cover Profile

MAP 3A. Forest cover profile of Western Himalayas, Jammu and Kashmir, Himachal Pradesh and Uttar Pradesh (Hill Areas).

These hills are sources of medicinal plants and timber. Grazing, fire and pilferage are burning problems of vegetation conservation. Corbett Tiger Reserve is a protected site in U.P.

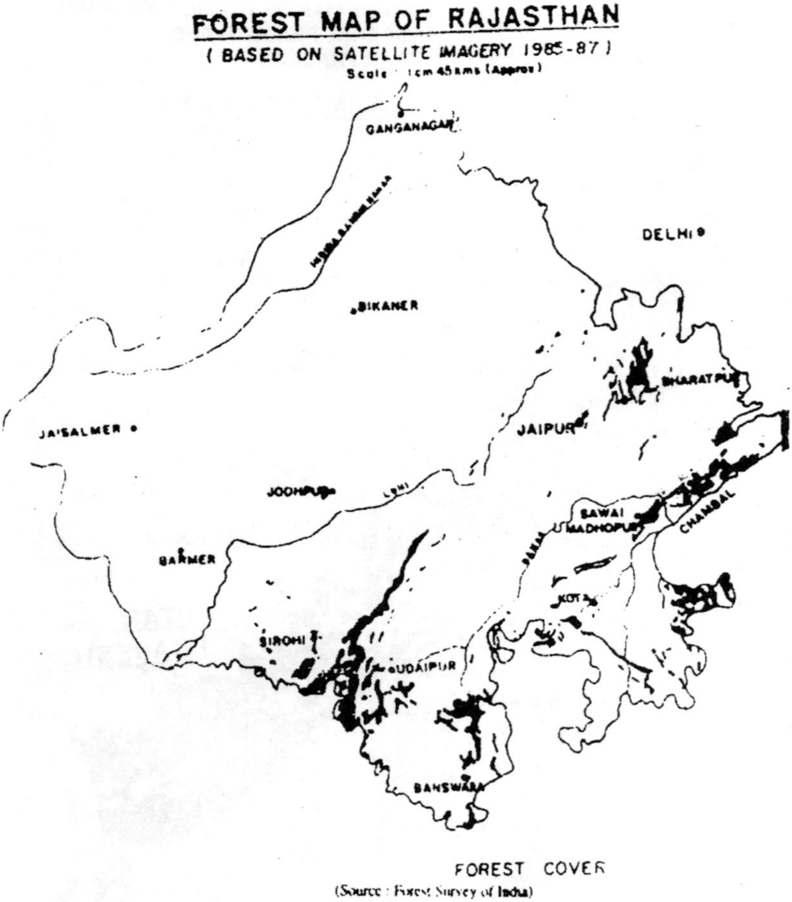

MAP 3B. Forest cover profile of Rajasthan

Dry deciduous and Scrub forests are located in north-east and south-east. Other areas are arid, and desert and have a very very scanty tree cover.

Vegetation with low diversity and low resource of medicinal plants. Seriska and Ranthambhore Project Tiger areas are protected.

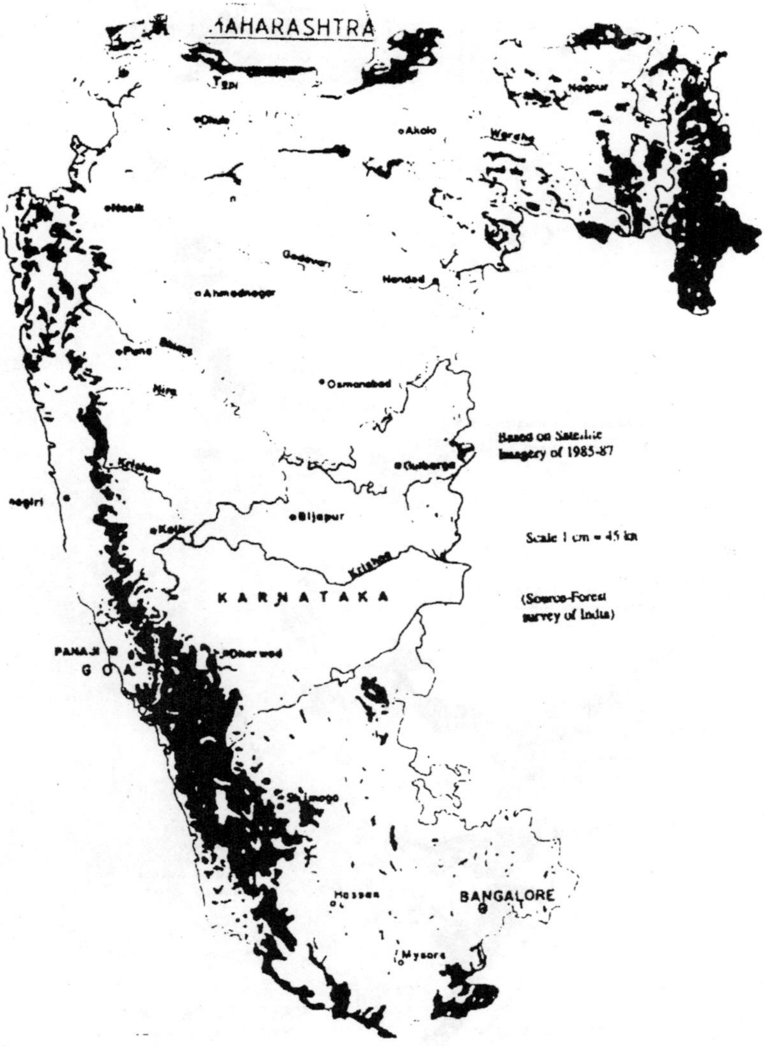

MAP 3C. Forest cover profile of Karnataka, Maharashtra, Gujarat, Kerala and Tamilnadu

Rich biodiversity areas have been found at Melghat Tiger Reserve, and the forest of Western Ghat, in Agumbe Evergreen and Bandipur Tiger Reserve.

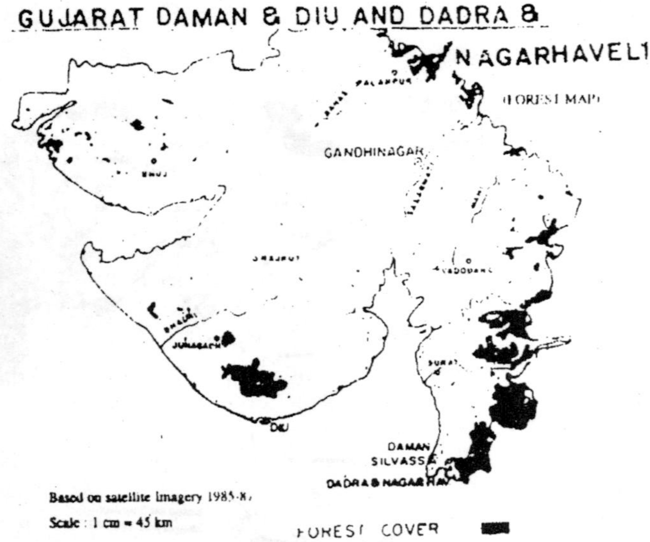

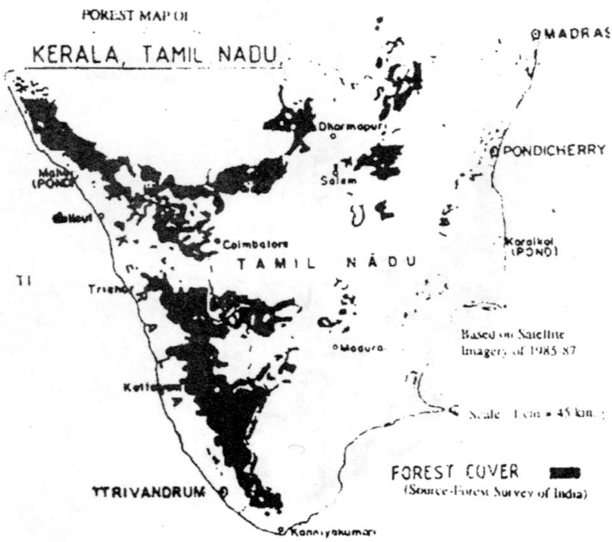

MAP 3D. Dry deciduous forests of Gujarat (Open forests)

Kerala, South western Tamil Nadu and Karnataka Forests and rich in Biodiversity and Medicinal Herbs.

Periyar Project Tiger area is a protected area.

MAP 3E. Forest cover profile of North-Eastern India

Once rich in Biodiversity and Store - House of Medicinal Plants; But now have depleted.

Plant Resources in North-Eastern India

There are many protected sites with high biodiversity e.g. Manas, Namdhapa and Dampa reserves, besides Kaziranga, Loktak National Parks.

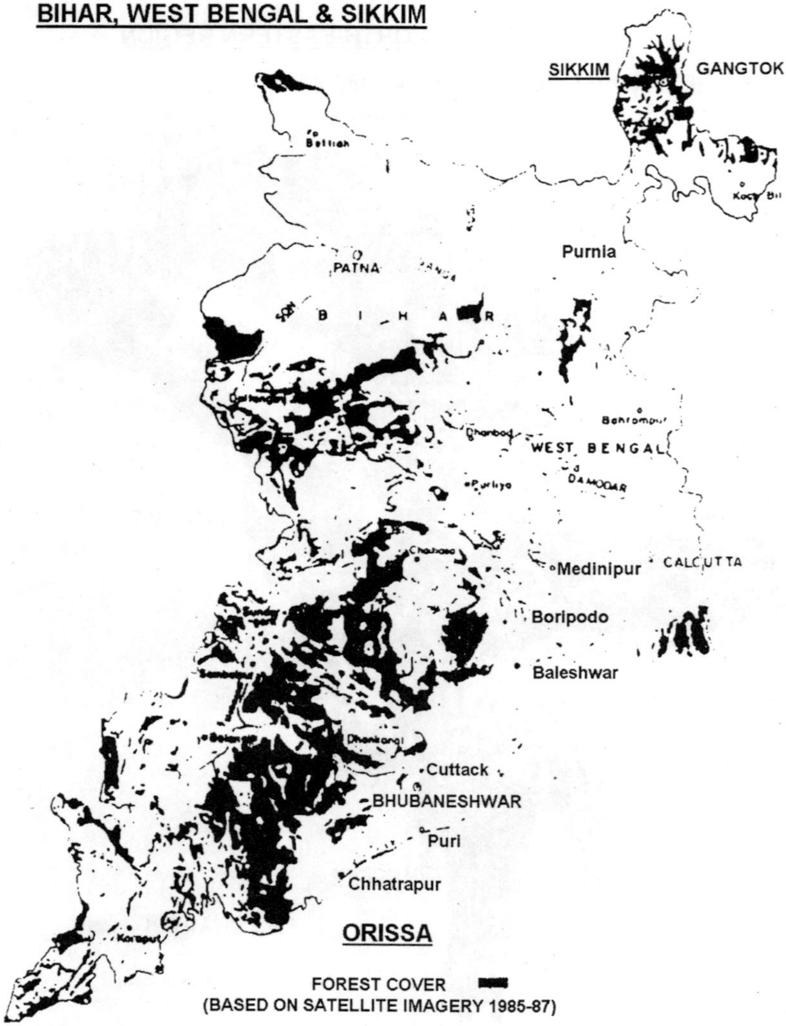

(Source: Forest Survey of India)

MAP 3F. Forest cover profile of Sikkim, West Bengal and Bihar

Rich Biodiversity and Medicinal plant resources were available in Sikkim, Darjeeling and Duars Sal forests of West Bengal; In Bihar Saranda forests and other forests of Tholkabad, Chaibasha & Palamu.

Palamu Tiger Reserve is a protected area with high biodiversity.

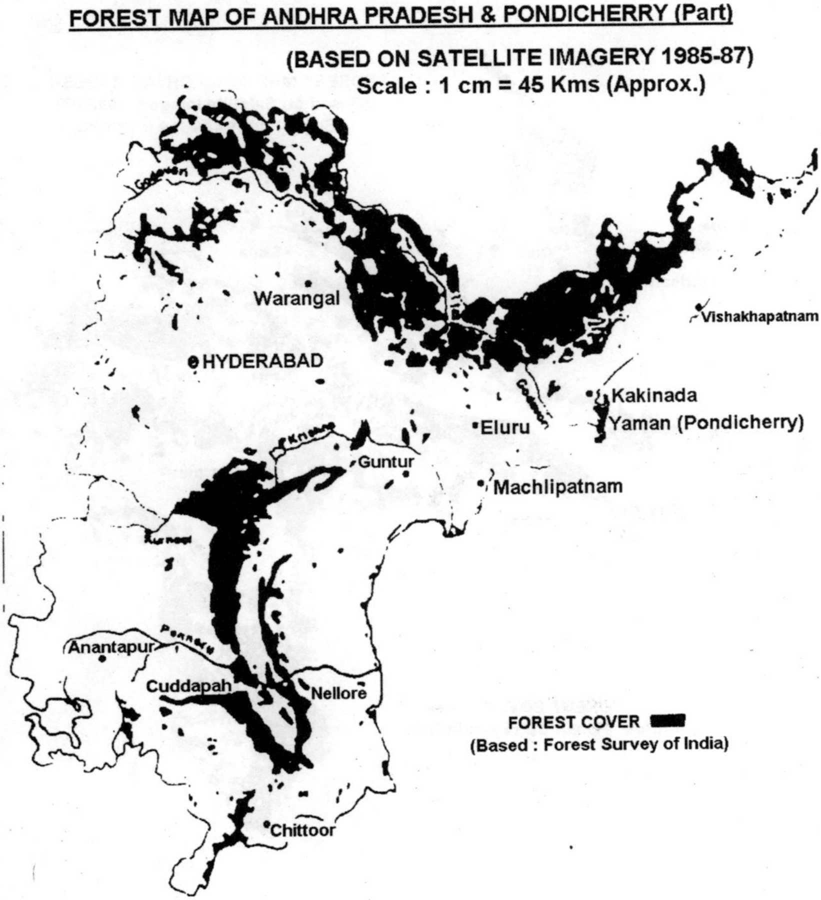

MAP 3G. Forest cover profile of Andhra Pradesh Forests

Rich resources have been found over the areas in the northern border with Madhya Pradesh. Nagarjun Sagar Project and Tiger Reserve is a protected areas.

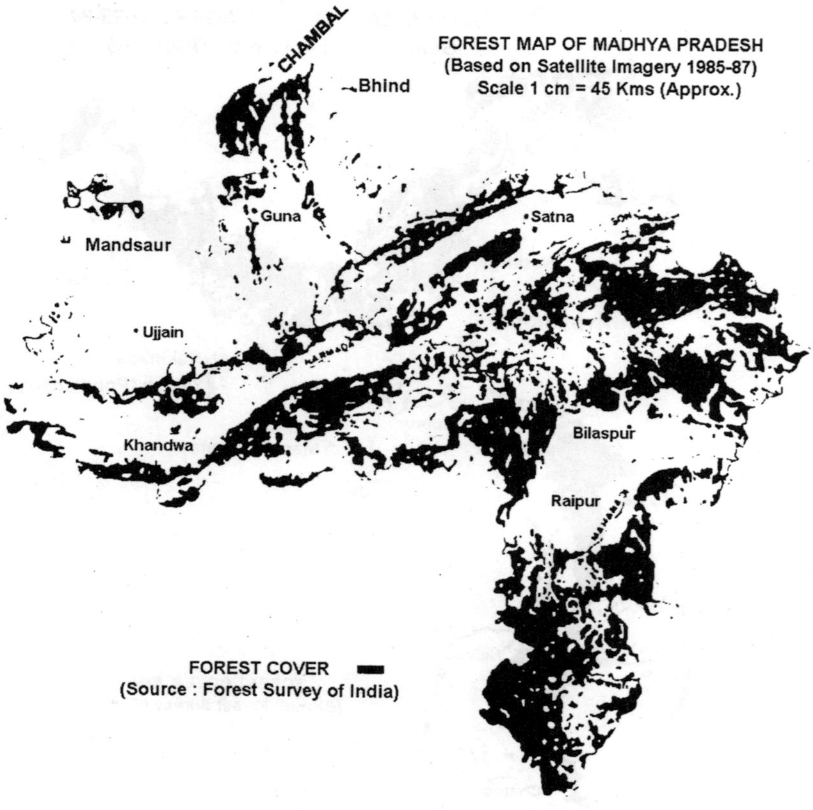

MAP 3H. Forest cover profile of Madhya Pradesh.

Extensive forested tracts of Madhya Pradesh were the home of faunal richness; the remnants now lie in the protected areas of Bandhabgarh Kanha, Taroba, Pench and Indrawati Tiger Reserve. Forests of Baster, north of Indrawati Tiger Reserve at Abhujmak are rich in plant diversity.

MAP 4(A&B). Coal mines, Oil fields and Industrial Sites of India

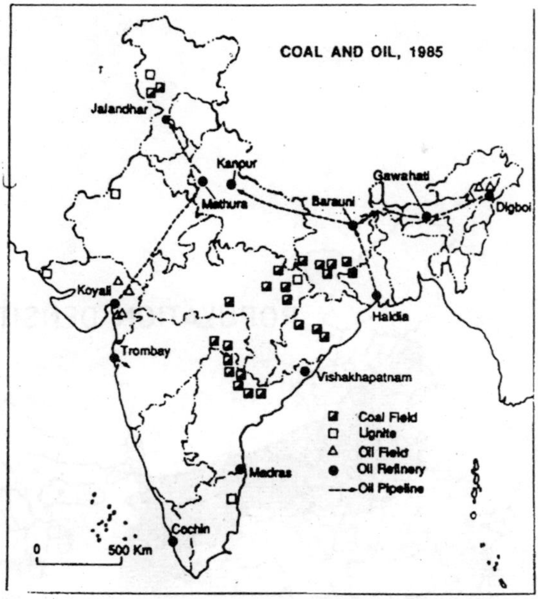

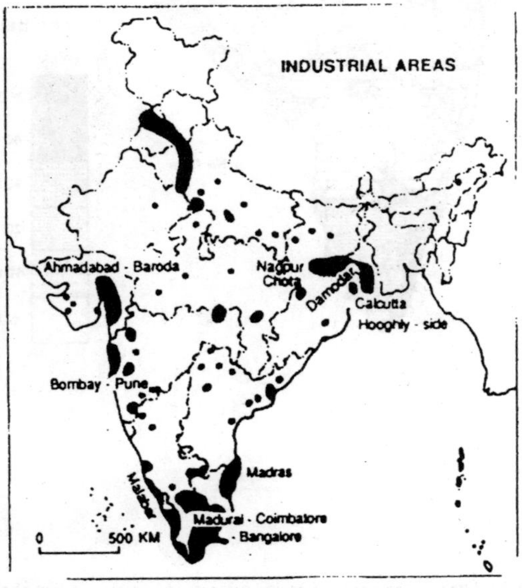

Two maps — one showing Coal and Oil producing areas and another Industrial areas, show the extent of impact, these would make on the conservation and reproduction of the existing vegetations.

MAP 5 (A & B). Population

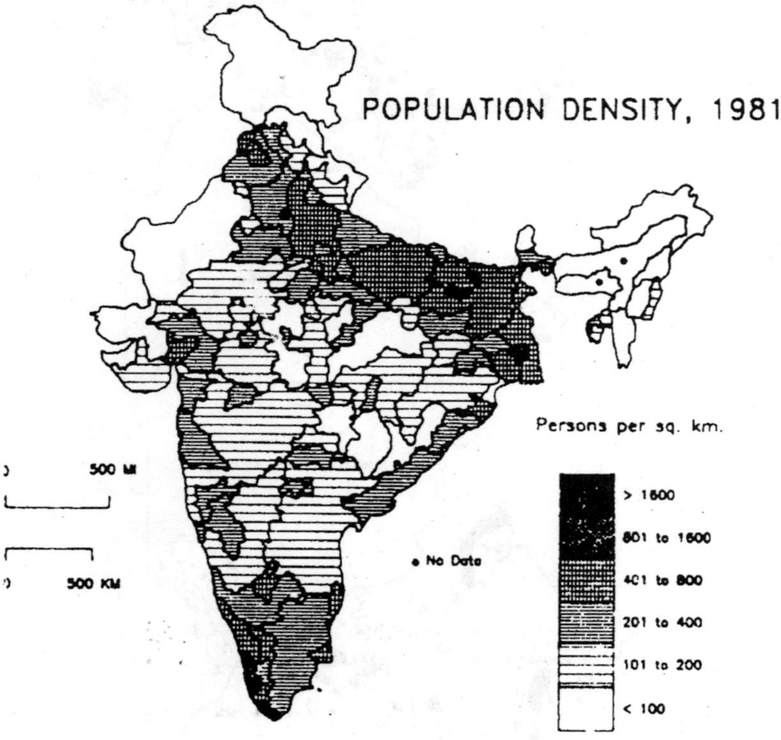

MAP 5A. India's Population vis-a-vis impact on vegetation

Population density is shown on the map. This gives an idea, how the forests and scattered plant resources are facing immense pressure.

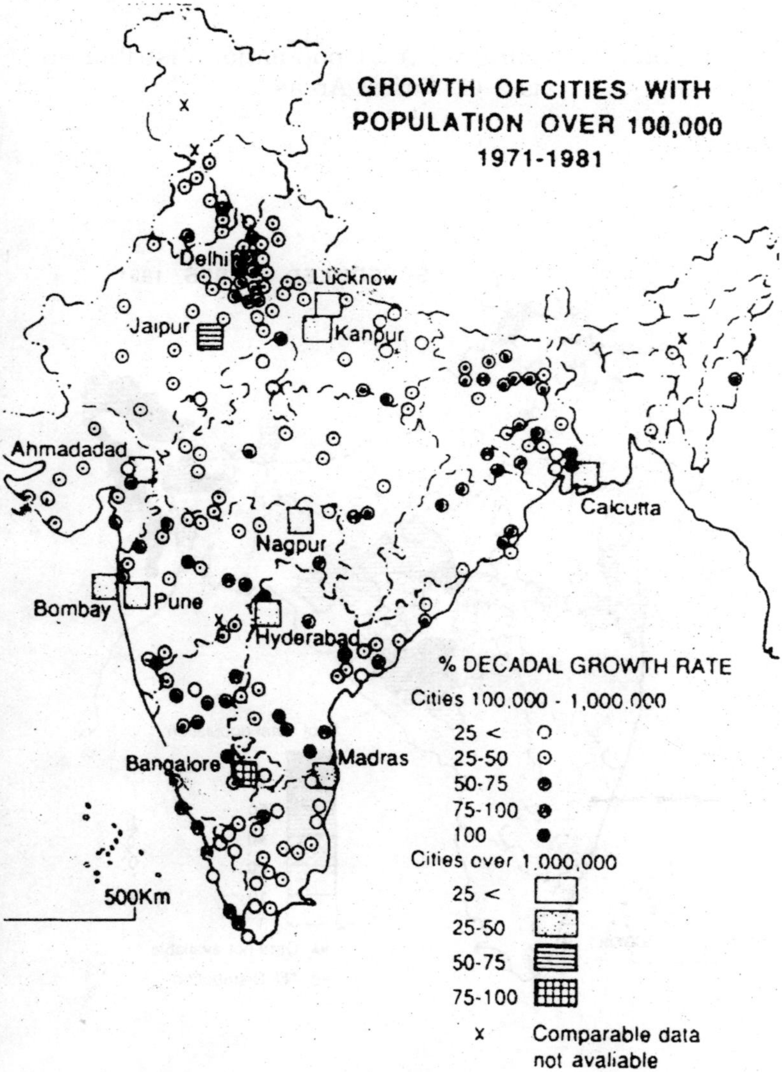

MAP 5B. Growth of City Population

The population increase has resulted in the influx of people towards the cities; new cities are growing-up and the vegetation is going poorer.

MAP 6 (A & B)

Map of India, depicting SC & ST population distribution vis-a-vis Forest Areas.

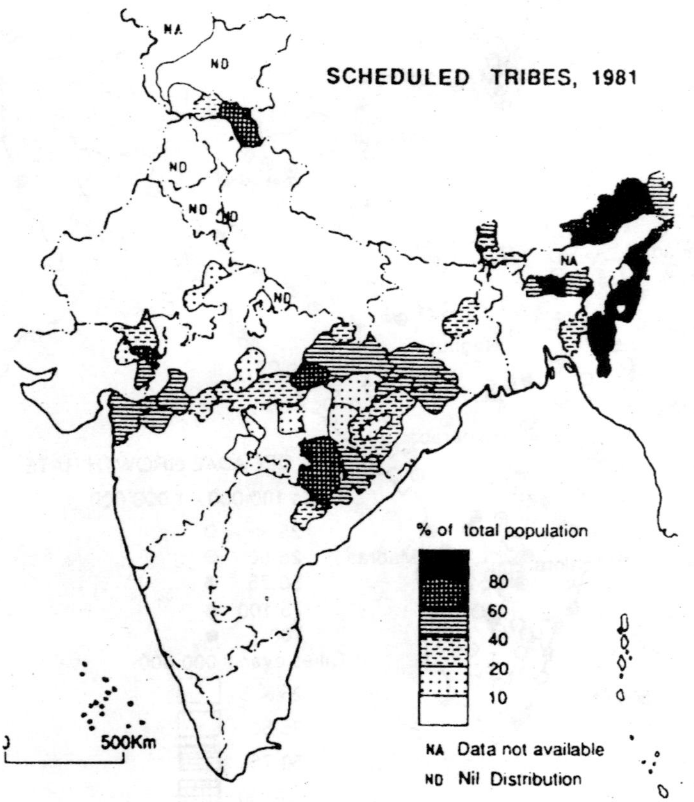

MAP 6A. Areas occupied by Sheduled tribes and Non-forest members.

If super imposed on the forest map which reveal that tribal population have their home in dense forested tracts.

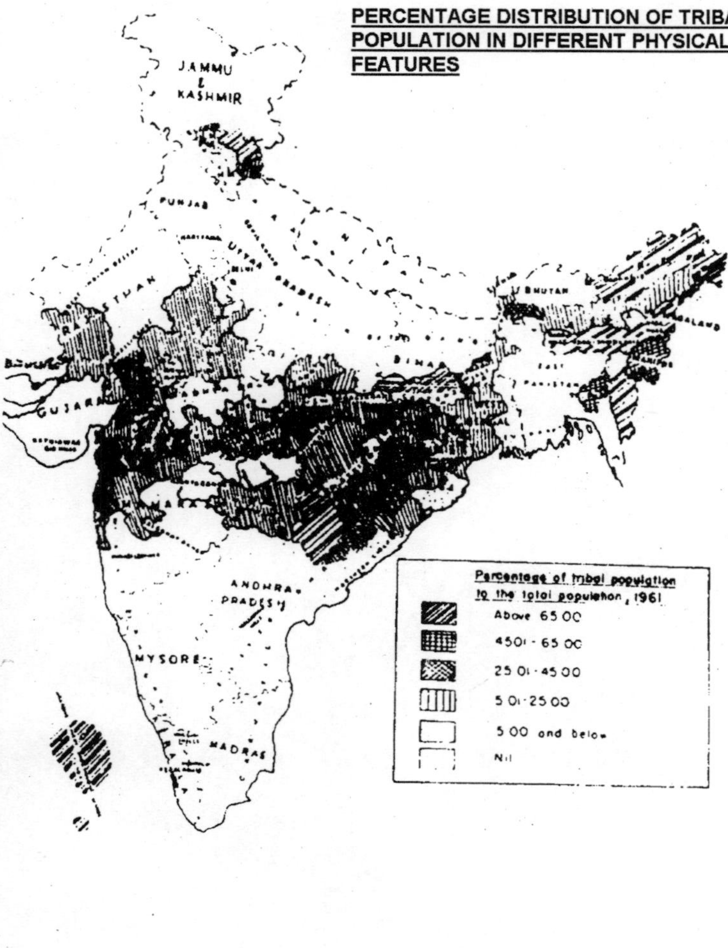

MAP 6B. Percentage distribution of Tribal population to the total population, 1961.

CHAPTER - 1

BIODIVERSITY AND CONSERVATION

1.1. INTRODUCTION

The concept of Bio-diversity had its origin in mid 1980, though the earth and evolution processes are very ancient phenomena. It has become a popular culture of the Scientists, more so with various non-government organisations to use this term sometimes inappropriately. The concept is outwardly simple when bio- diversity is used for the richness and variety of life on earth. In fact it represents the very foundation of human existence as many conceive.

All types of flora and fauna (including insects and bacteria, coral reefs etc.) are elements of bio-diversity. In the present days this term has been increasingly used as a conceptual focus for conservation policy to evaluate and qualify biological diversity, species loss and extinction, genetic variation, the number of species, the extent of eco-system etc. and measure ecological and evolutionary processes.

However, some of the definitions of bio-diversity are placed below.

1.2. DEFINITIONS

The most acceptable definition of Bio-diversity is the one held by the Convention of Biological Diversity (C.B.D.), which was signed by more than 150 nations on June 5, 1992 at Rio-De- Janeiro. But there are at least twelve formal definitions.

The C.B.D. states that 'Biological diversity' means the variability among living organisms from all sources, inter alia, terrestrial, marine and other aquatic ecosystems and the ecological complexes of which they are part, this includes diversity within species, between species and of ecosystems.

To put more colloquially, bio-diversity is the variety of life, in all its many manifestations. It encompasses all forms, levels and combinations of natural variation and thus serves as a broad unifying concept.

According to convention the actual definition of bio-diversity, as given above, is neutral with regard to any importance it may be perceived to have. The convention, in contrast far from a neutral document, is amply revealed by its objectives as below :

"...... the conservation of biological diversity, the sustainable use of its components and the fair and equitable sharing of the benefits arising out of the utilization of genetic resources, including by appropriate access to genetic resources and by appropriate transfer of relevant technologies, taking into account all rights over those resources and technologies and by appropriate funding."

The term 'biodiversity' carries with it connotations that biodiversity is a good thing per se that its loss is bad, and that something should be done to maintain it. It is important to recognize that there is rather more to use of the term than a formal definition in the Convention (or for that matter else where) and its application reveals just as much about the values of the person using it. This should always be kept in mind when interpreting what is being said about biodiversity.

However, the World Conservation Monitoring Centre (1992) in their edited work viz., "Global Biodiversity" hold the following definition :

Diversity is a concept which refers to the range of variation or differences among some set of entities; biological diversity thus refers to variety within the living world. The term 'biodiversity' is indeed commonly used to describe the number and variability of living organisms. This very broad usage, embracing many different, parameters, is essentially a synonym of 'Life' on Earth.

The Scientific definition looks at biodiversity in terms of genes, species and ecosystems, corresponding to three fundamental and hierarchically-related levels of biological organisation.

Several definitions framed by distinguished Scientists are also laid below :

— D.L. Perlman and G. Adelson defines:

"Biological diversity refers to the variety and variability among living organisms and the ecological complexes in which they occur. Diversity can be defined as the number of different items and their relative frequency. For biological diversity, these items are organised at many levels, ranging from complete eco-systems to the chemical structures that are the molecular basis of heredity. Thus the term encompasses different ecosystems, species, genes and their relative abundance".

— Fidler and Jain (1992) defines Biological Diversity (=Biodiversity) as full range of variety and variability within and among living organisms, their associations and habitat oriented ecological complexes. The term encompasses ecosystem, species and landscape as well as intra-specific (genetic) levels of diversity.

— International Council for Bird Preservation 1992 defines - "Biodiversity is the total variety of life on earth. It includes all genes, species and ecosystems and the ecological processes of which that are part..... "

— Meffe and Carnll (Glossary entry) defines

"*Biodiversity* : The variety of living organisms considered at all levels, from genetics through species, to higher taxonomic levels and including the variety of habitats and ecosystems".

— Hunter 1996 states -"Definitions of biodiversity usually go one step beyond the obvious - the diversity of life - and define biodiversity as the diversity of life in all its forms, and at all levels of organization ..... 'All levels of Organization' indicates that biodiversity refers to the diversity of genes and ecosystems, as well as species diversity."

— U.S. Office of Technological Assessment, 1987 have the following view:

Ecosystem diversity : A landscape interspersed with croplands, grass lands and wood lands has more diversity than a landscape with most of the woodlands converted to grass lands and crop lands.

Species diversity : A rangeland with 100 species of annual and perennial grasses and shrubs has more diversity than the same range land after heavy grazing has eliminated or greatly reduced the frequency of the perennial grass species. Species are regarded as : populations within which gene flow occurs under natural conditions; members of one species do not breed freely with members of other species.

Genetic diversity : Economically useful crops are developed from wild plants by selecting valuable inheritable characteristics. Thus, many wild ancestor plants contain genes not found in today's crop plants. An environment that includes the domestic varieties of a crop (Such as corn) and the crop's wild ancestors has more diversity than an environment with wild ancestors eliminated to make way for domestic crops.

— Walter G. Rosen (1986) defines [refer Forum in Biodiversity, washington D.C., September, 86), that the term biodiversity was probably first used in this forum. Rosen has stated that he felt that the definition from the 1987 O.T.A. report captured what he had in mind when he came up with the term biodiversity. The O.T.A. articulates some of the various levels (ecosystem, species and genes) at which biological diversity is easily understood.

— Mc. Neely *et al.*, (1990) define bio-diversity as "an umbrella term for the degree of nature's victory. It 'encompasses all species of plants, animals, and micro-organisms and the ecosystems and ecological processes of which they are part'. Biodiversity can be seen as a measure of nature and its diversity, rather than an entity in itself, and is usually measured at three levels- genes, species and ecosystems.

Genetic diversity is the variety of genes within species. This form of diversity can be between populations of the same species, or within distinct population. Species diversity is the variety of different species found in an area - the number of different species is often used as a measure, but in some cases taxonomic diversity is used, as it considers the relationship of species to each other".

The schematic presentation in the following figure of the different levels of biodiversity is a useful way of illustrating the different levels at which biodiversity is important. It must not be taken to imply a hierarchy of importance where, the bottom level is more important than the upper levels, because biodiversity is concerned with inter relationships, all of these types of diversity are of equal importance.

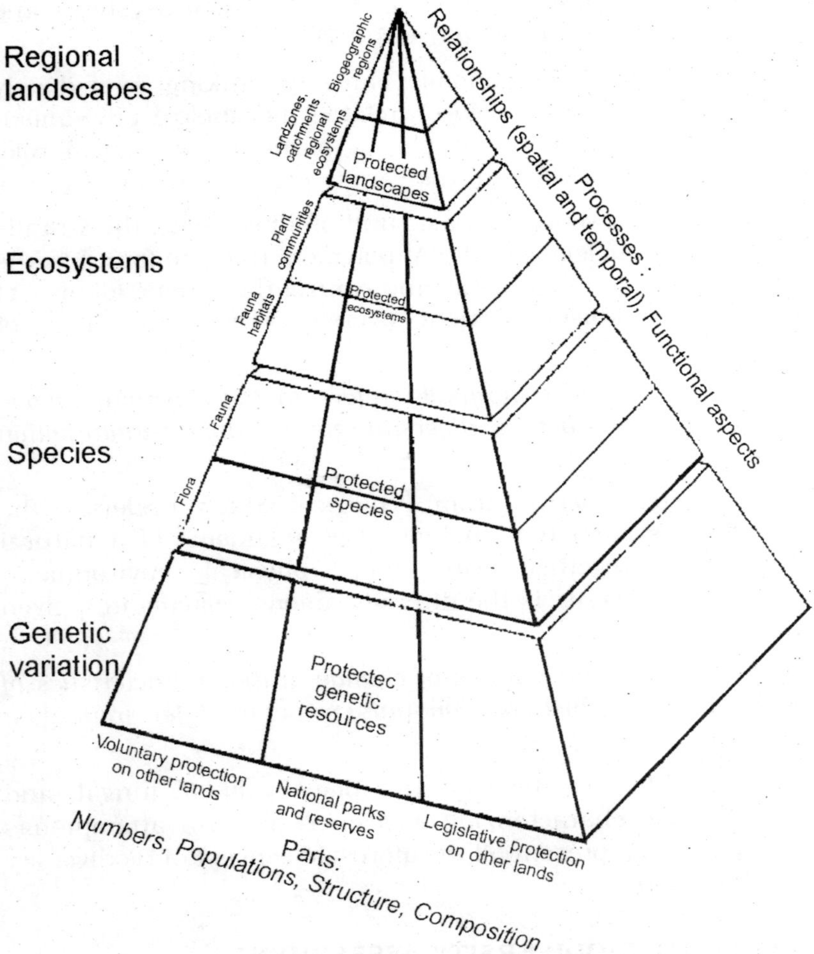

A conceptual hierarchy of biodiversity, its parts and processes, at four levels: regional landscape, ecosystem, species and genetic variation (adapted from Sattler, 1991)

— U.S. National Research Council (1992). defines : Biological diversity (or biodiversity, as it has come to be called) refers to the variety and variability among living organisms and the ecological complexes in which they occur. Diversity can be defined as the number of different items and their relative frequency. For biological diversity, these items are organized at many levels, ranging from chemical structures that are the molecular basis of heredity to complete ecosystems. Thus, the term encompasses different genes, species, ecosystems and their relative abundance (OTA, 1987).

Species is the taxonomic category ranking immediately below genus; it includes closely related, morphologically similar, individual organisms that play a particular ecological role. Species diversity refers to the variety of different species.

Genes represents the basic unit of inheritance, the strands of deoxyribonucleic acid (DNA) polymers that are found in the Chromosomes in Cell nuclei and control the genetic identity of individual organisms. Genetic diversity refers to the variety of genes.

Species diversity normally refers to the diversity among species, whereas genetic diversity refers to the diversity within species.

Ecosystem (Derived from "ecological system") refers to the functional system that includes the organisms of a natural community together with their physical environment. Ecosystem diversity is the diversity among systems in a given area.

Evolution is the process of change in the characteristics of organisms by which descendants come to differ from their ancestors.

Biota refers to the collective plant, animal, fungal, and microbial life characterizing a given region. An attempt has been made to represent all the aforesaid concept of Biodiversity in Charts 1-9.

1.3 GLOBAL BIODIVERSITY ASSESSMENT

UNEP (United Nations Environment Programme) has brought out a volume entitled 'Global Biodiversity Assessment' 1995 which has received contributions from more than 15000 experts. Various chapters deal with magnitude and distribution, loss, ecosystem functioning, inventorying and monitoring, resource base, data and information management, biotechnology, human influence, economic values, conservation and sustainable use of biodiversity (Charts 10-15).

BIODIVERSITY UMBRELLA

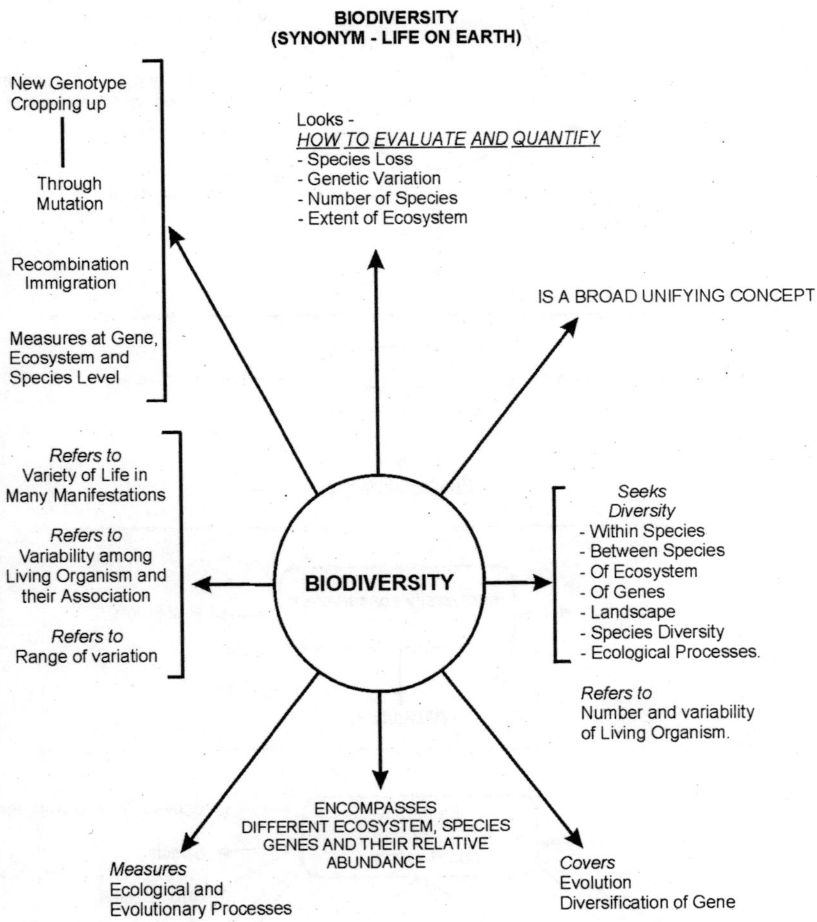

Chart Showing Various Facets of Biodiversity

CHART – 1

BIODIVERSITY
FURTHER CONCEPTS

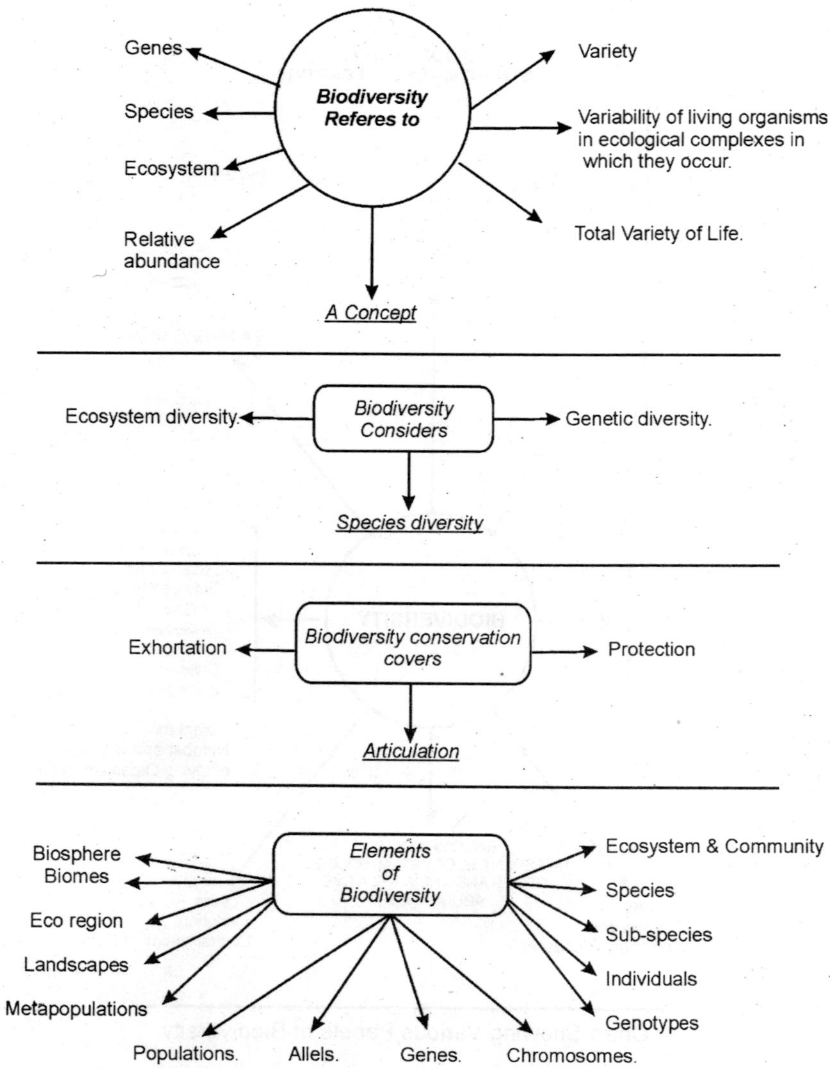

Source : Modified from Biodiversity by D. Perlman & Glenn Adelson, 1997.

CHART – 2

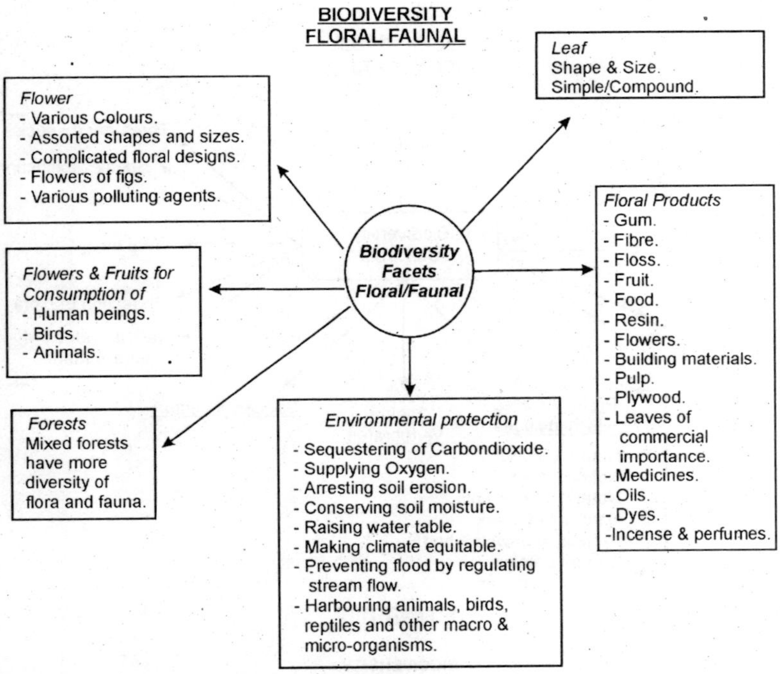

CHART 3

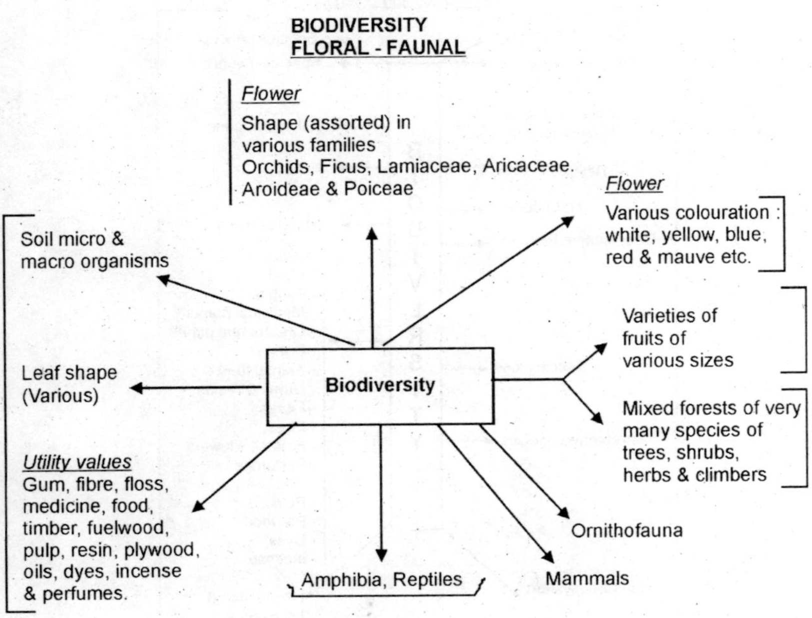

CHART 4

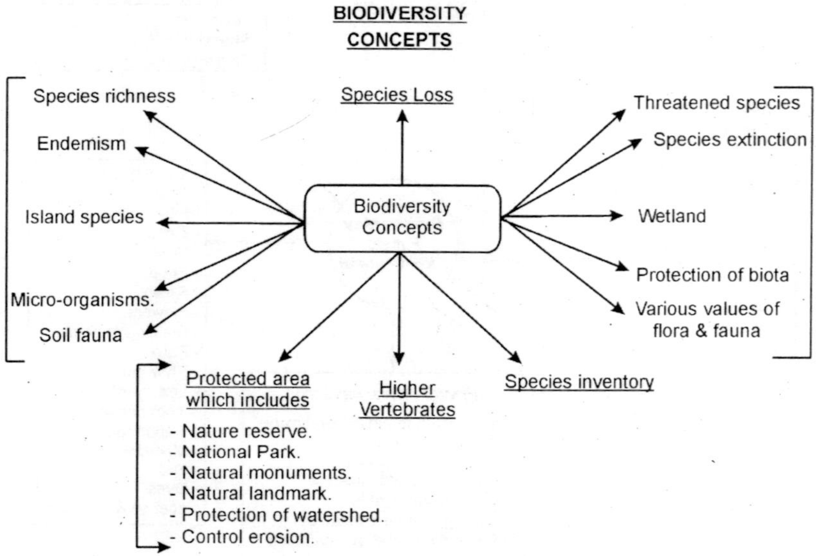

CHART 5

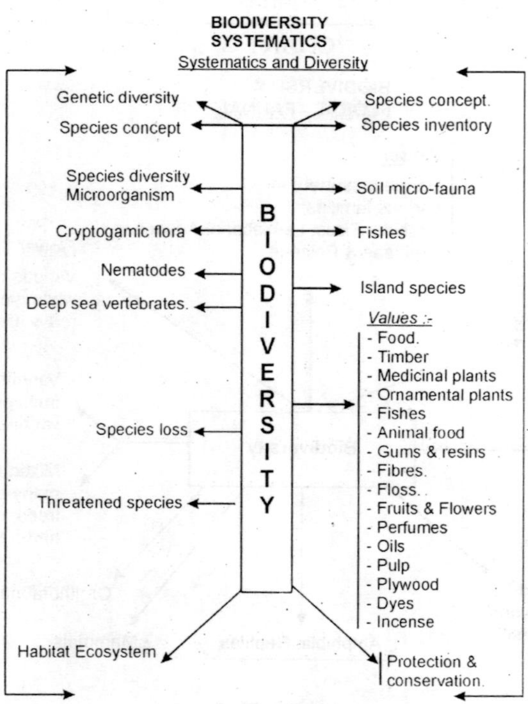

CHART 6

VALUING BIODIVERSITY

Valuing the Forests in Sustainable Development

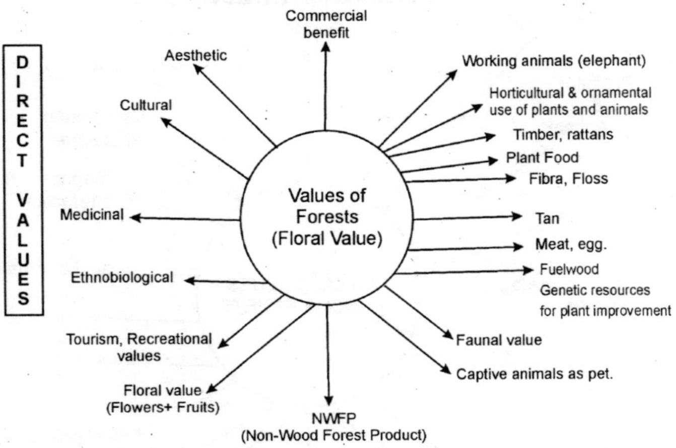

INDIRECT VALUES

(i) Soil Productivity and Protection.
(ii) Protection of Water shed.
(iii) Carbon store.
(iv) Climate stability.

CHART 7

VARIOUS OTHER CONCEPTS OF VALUES OF ENVIRONMENT

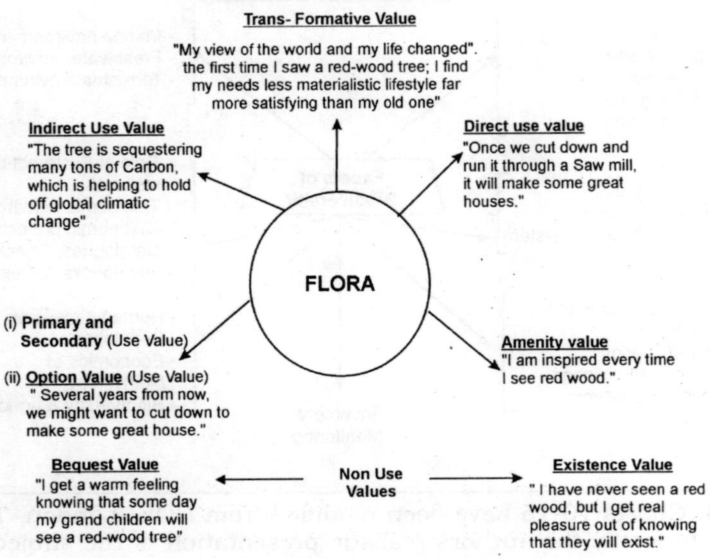

Source : D.L. Perlmun in "Biodiversity" (1980).

CHART 8

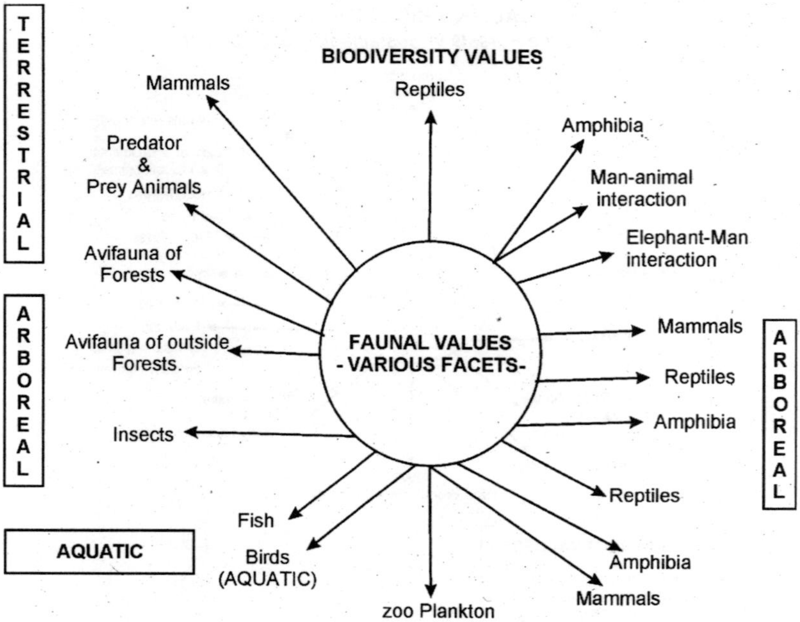

CHART 9

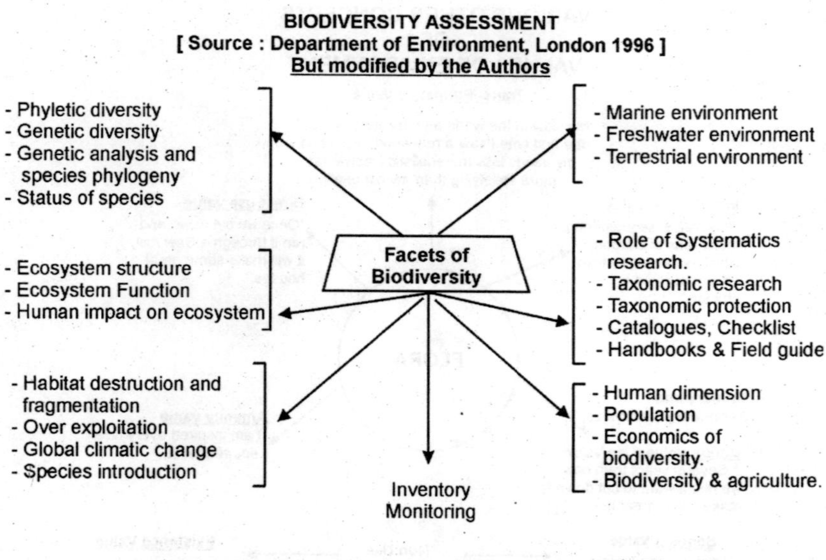

N.B. Charts 10-15 have been modified from DOE, London. This is to testify the authors realistic presentation of the subject.

CHART 10

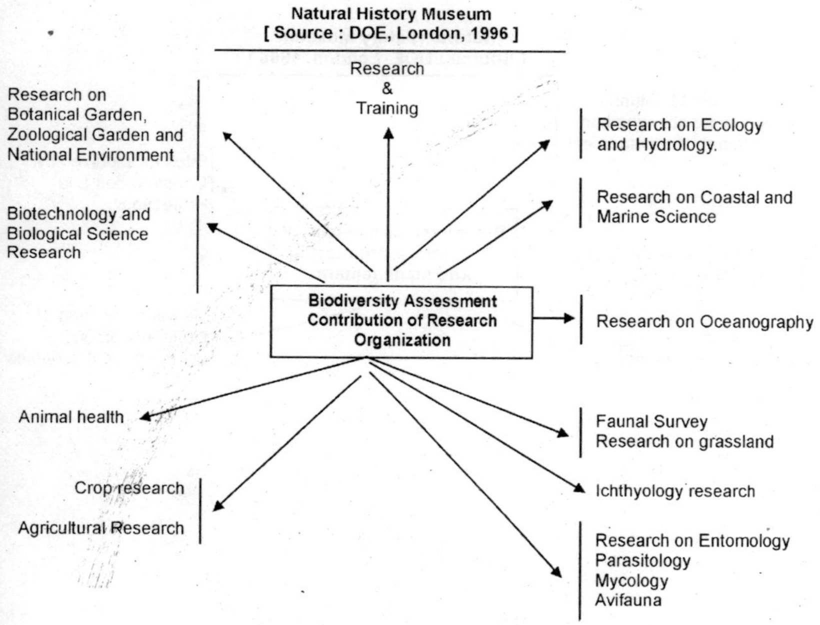

CHART 11

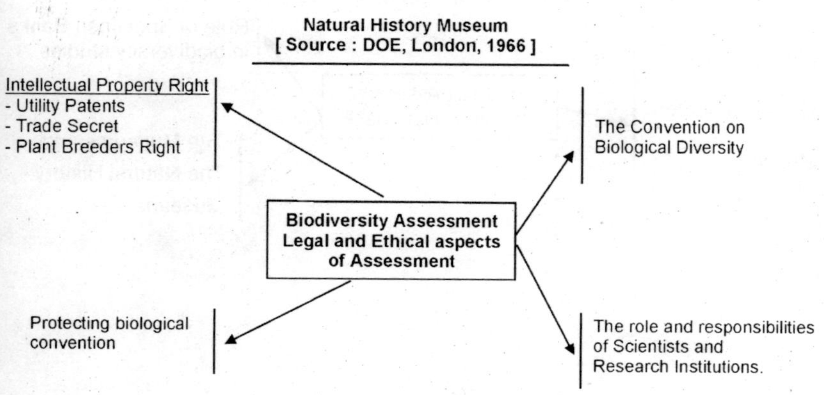

CHART 12

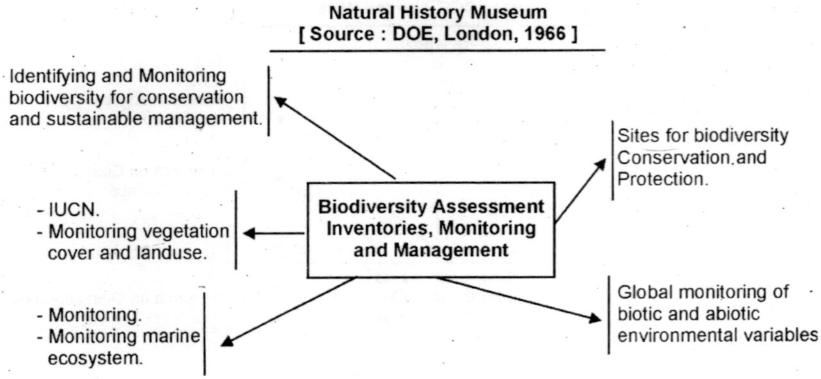

CHART 13

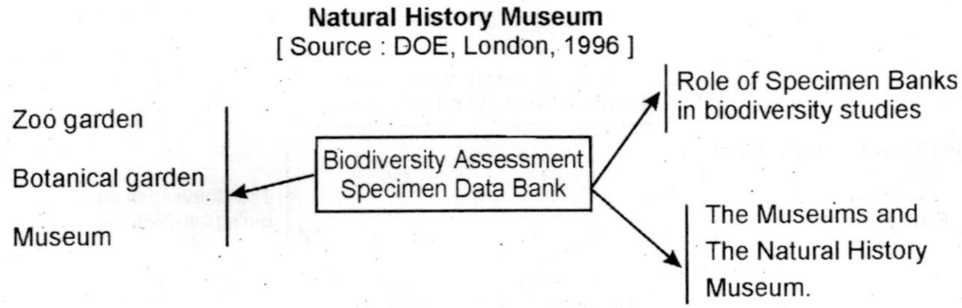

CHART 14

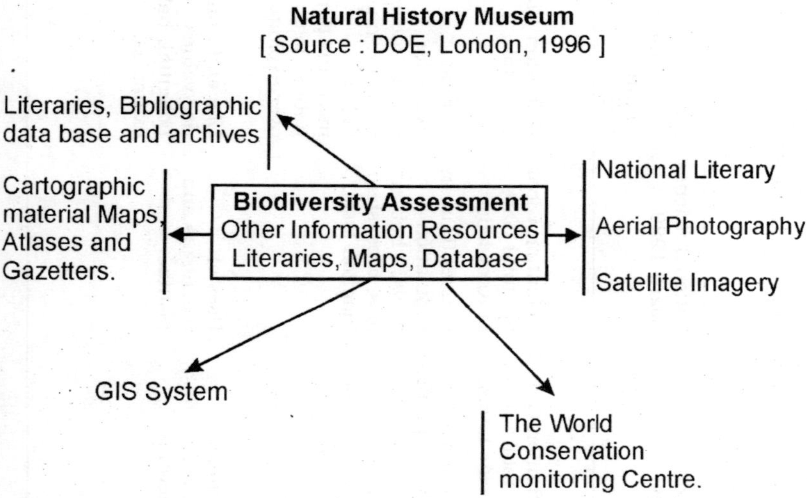

CHART 15

1.4 BIODIVERSITY CONCEPT

Michael. J. Jeffries (1997) has elaborated (in Biodiversity and conservation) this concept by arranging the major themes in a tabular form which read as follows (Table 1).

Jeffries further summarises the importance of biodiversity concepts which touches all areas of human lives and frames the following Table 2.

The concept of biodiversity also covers the following :

Evolution : Ecological processes have evolutionary consequencs. They interact with genetic diversity via adaptation, micro-evolution and specialisation.

Diversification of Genes : Selection and Maintenance of genetic diversity.

Specialisation of Diversity : There are three components to biodiversity : genetic, organizational and ecological. It covers conservation of species, genetic diversity and ecosystem diversity.

Table 1. Major themes within biodiversity. The presence of six main themes in texts from the conceptual origin of biodiversity in 1980 Prior to the 1992 UN Convention on Biodiversity.

Paper or publication	Lovejoy 1980	Norse and McManus 1980	Norse *et al.*, 1986	Biodiversity 1988	National Science Board 1989	Biological Diversity developing countries 1991
Theme						
1. Inventory	Definition Global Sp. total	Definition Global Sp. total	Definition Global Sp. total	Definition Global Sp. total Habitat specific totals.	Definition Global Sp. total Gaps in Knowledge.	Definition Global Sp. total
2. Loss rates	Estimated losses due to forest destruction.	Causes (settlement transport, fragmentation, agriculture, forestry, over exploitation, introduction species)		Global and habitat specific estimates. Geological patterns. Case Studies.	Causes (human population growth, habitat, loss, ecosystem function loss).	Habitat examples. Causes (development, market fail- ure, interven- tions, habitat loss, over exploitation.
3. Value	Potential uses.	Potential uses (food, energy, chemicals, raw materials, medicine).	Products (actual and potential). Ecosystem services, Forestry, Psychological Well-being.	Medicine, industrial, food, potential.	Uses of organisms and plants.	Use and non use values potential, habit examples.

Table 1 contd...

Paper or publication	Lovejoy 1980	Norse and McManus 1980	Norse *et al.*, 1986	Biodiversity 1988	National Science Board 1989	Biological Diversity developing countries 1991
4. Economics	Valuation of ecosystem function.		Timber production	Market failure to price biodiversity new economic approaches, pricing biodiversity	Economics of utilisation (clothes, food, medicine, Shelter).	Economic approaches to value biodiversity.
5. Conservation		Gene banks, botanical gardens, species specific schemes, ecosystem management, legislation	Concepts. Conservation and management, of forests and rare species. Laws.	Priorities. Case studies, technologies, reintroductions.	Recommendations for research on inventories, education training and funcing in developing countries.	Aid Policy procedures for priosites, habitat examples, criteria.
6. Attitude		Philosophica, psychological, affinity to life, right to exist.	Ethics and stewardship	Green movement, society links to ecosystemmorals, religion, philosophy.		Ethical value of biodiversity.

Table 2. Historic, current and foreseeable importance of biodiversity to human

Biodiversity	Historic	Current	Foreseeable
Ecosystem services	Pollination of Crop plants by bees.	Wetlands to cleann pollution, effective as buffers against many pollutants, especially nutrient over- enrichment.	Soft coastal engineering against sea level rises from global warming.
Medicine	Penicillin, antibiotic derived from fungs.	Horse Shoe Crab blood used in bioassays of toxins.	Rainforest plants New Drugs, to combat existing and new diseases.
Biotechnology	Rattahs, Asian forest palms used for fibres for many building and artifacts.	Bacterial genes introduced to crops to confer resistance to insects.	Metal digesting bacteria. Potential use to clean contamination and extract valuable metals.
Environmental monitoring	Losses of birds species as evidence of DDT impact.	European forests damaged by acid precipitation.	Species range and habitat changes in response to global warming.
Food	Irish potato famine due to over-reliance on one vulnerable cultivar.	The weedy tomato, wild gene improved domestic cultivars.	Several high value exotic rainforest fruits.
Recreation	Victorian Fern craze, widespread hobby.	Birdwatching, major recreational activity in developed world.	Eco-tourism. Growth area for global tourism.

Table 2 contd...

Biodiversity	Historic	Current	Foreseeable
Pets and domestic animals	Exotic zoo animals as gifts and status symbols.	Value of current pet trade.	Potential from use of historic breeds as we rediscover useful attributes.
Political and social	Symbolic function in Celtic religion; animals as companion to St Cuthbert.	Eco activists, e.g. greenpace and Earth First; highly visible political impact.	Local ownership and control of resources in developing world.

In a simple language K.J. Gaston et al., have summerised biodiversity concepts as follows :

— Biodiversity is the variety of life, in all its mainfestations.

— Key elements of this variety can be reorganized, comprising three nested hierarchies of genetic, organizational and ecological diversity.

— There is no single overall measure of biodiversity; rather there are multiple measures of different facets.

— The measure of biodiversity chosen may reveal something about the values of the investigator.

— As many desired measures of biodiveristy are difficult or impractical to obtain, surrogate measures (correlates the actual measures desired) may have to be used instead. Some surrogates take into account more of the patterns of numbers and difference in biodiversity in which we are interested than do others.

— Despite its significant limitations species richness is one of the most commonly used surrogates as it 'integrates' many different facets (genetic, organisamal, ecological) or diversity. It is relatively easy to measure and, despite significant limitations, has to some extent become the common currency of the study of biodiversity.

1.5 VALUING BIODIVERSITY

Opening lines of the United Nations Convention on Biological Diversity state — "Conscious of the intrinsic value of biological diversity and of the ecological, genetic, social, economic, scientific, educational, cultural, recreational and aesthetic values of biological diversity and its components....." (Charts 7-9) indicate some values and the necessity to conserve and sustainably use biological diversity for the benefit of present and future generations. It conceives enumerating values of biota an analysis of which is as under —

Sample statements of different types of worth that might be attributed to a large Sal wood (*Shorea robusta*) tree

Following are the value attributed to a large sal tree (i.e., Plant kingdom as a whole) —

— *Direct Use Value* : Once we cut it down and run it through a saw mill, it will make some great houses.

— *Indirect Use Value* : The tree is sequestering many tons of carbon, which is helping to hold off global climate change.

— *Option value* : Several years from now, we might want to cut it down to make some great houses.

— *Amenity value* : People get inspired everytime, when they see a Sal tree.

— *Existence value :* One, who has never seen a Sal tree but gets real pleasure out of knowing that they still exist.

— *Bequest value* : One gets a warm feeling knowing that someday his grand children would see a Sal tree.

— *Transformative value* : One's view of the world and his life would change as soon as he sees a Sal tree. He find his new, less materialistic life style far more satisfying than his old one.

Values attributed to animal kingdom

Following are the values attributed to the animal kingdom including human beings (Keller's classification)

— *Utilisation* : Practical and material exploitation of nature.

— *Naturalistic* : Satisfaction of direct experience of and contact with nature.

— *Ecologistic Scientific* : Systematic study of structure, function and relationship in nature.

— *Aesthetic* : Physical appeal and beauty of Nature.

— *Symbolic* : Use of nature for metaphorical expression, language expression and thought.

— *Humanistic* : Strong affection, emotional attachment and love for nature.

— *Moralistic* : Strong affinity for spiritual reverence and ethical concern for nature.

1.6 SUSTAINABLE DEVELOPMENT : NEED FOR VALUING RESOURCES

Sustainable development means improving and maintaining the well-being of people and ecosystem. It entails integrating economic, social and environmental objectives. The present concept centres round the facts that 'Biodiversity' can be conserved through an alliance of local fringe people; but the context varies since some threats can be solved by communities themselves, while others are too potent for local people to meet without help and they can help conserve the ecosystem on which they depend, but to do so they must have enough basic education, concept and confidence to protect the country's and their own interest and a data base of varied resources.

When conservationists began thinking of the biodiversity protection and sustainable development, they found the necessity to learn more about nature and about the people living in the vicinity, about how communities work and how decisions are made. The science of biological education, anthropology, politics and economics have to be known which led to the creation of the new, synthetic, discipline of conservatin.

All protected areas* should have a management plan to determine priorities, guide the allocation of resources, control all development activities and implement management actions. Such plan takes into account a wide range of biological, physical and social factors and the changing nature of ecosystem.

So long, protected area management was primarily a biological challenge, while the present concept is that the management of wildlife and habitat is only a small part of the picture. It is agreed that challenging management task is finding the means to ensure that human behave in ways that are consistent with conservation objectives; and that, social science plays a much more prominent role in training of protected area managers and that insights offered by economics, psychology, history, anthropology, political science and sociology be harnessed for the benefit of conservation.

Therefore, modern protection area manager needs to be a sort of renaissance person with a solid grounding in both biological and social sciences as well as resource management. Environmental management is a multilayered process associated with the interaction of state and non-state environmental managers with the environment and with each other.

Conservation management of a site involves knowing what species and communities are present and also the understanding the ecology of the site, identifying management

* (i.e. management of wildlife Sanctuaries, National Parks and Garden, Conservation areas etc.)

prescription and work programm. The management plan is a convenient mechanism for bringing these elements together.

The concept of sustainability emphasizes several basic principles when applied to rural communities that are to be met

- that basic needs must be met.
- that local communities must have a decisive voice in planning, and that they should represent themselves through their own institution.

> "Sustainable development refers to the means by which 'development' is made to meet the needs of the present without compromising the ability of future generations to meet their ends". *[World Commission of Environment and Development WCED, 1987 of United Nations]"*

The modern concept of biodiversity conservation - the wise maintenance and utilisation of the earth's resources is no more than the combination of these two principles : the need to plan resource management on the basis of accurate inventory : and the need to take protective measures to ensure that resources do not become exhausted.

The World Conservation Strategy (WCS), prepared by four of the World's leading conservation agencies — The international Union For Conservation of Nature and Natural Resources (JUCN), World Wildlife Fund (WWF), Food and Agriculture Organization of the United Nations (FAO), and the United Nation's Environment Programme (UNEP), launched in 1981 demonstrate how the conservation of living resources is essential for sustaining development by : (i) Maintaining the essential ecological processes and life-support system on which human survival and development depend. (ii) Preserving genetic diversity on which depend the breeding programme necessary for the protection and improvement of cultivated plants and domesticated animals. (iii) Ensuring that man's utilization of species and ecosystem, which support millions of rural people as well as major industries is sustainable.

Protection of natural areas are essential for the conservation of a nation's living resources thus ensuring that :

- Representative samples of important natural regions are retained in perpetuity;
- Biological and physical diversity is maintained.

– Wild genetic materials are conserved.

Protected areas also contribute to the conservation of living resources and to sustainable development by :

– Maintaining the environmental stability of the surrouding region and thereby reducing the intensity of floods and droughts, protecting the soil from erosion and limiting the extremes of local climates;

– Maintaining the productive capacity of ecosystem, thus ensuring the continuing availability of water and plant and animal products;

– Providing opportunities for research and monitoring of wild species and ecosystems and relationship to human development

– Providing opportunities for conservation education for the general public and for policy makers;

– Providing opportunities for complementary rural development and the rational use of marginal lands;

– Providing a base for recreation and tourism;

– Management activities should conserve or restore natural ecosystem disturbance pattern. (Merrill Kaufmann *et al.*,)

– Natural processes will be relied on to control populations of native species to the greatest extent possible. (National Park Service Management Policies) [Source : Boyce M.S. *et al.*, in "Ecosystem Management" (1997)].

In order to outline the complexity of the subject, the expert authorities should consider —

– That Conservation Biology is a multi-disciplinary subject

– That wildlife and floral management has several components

– That wildlife and floral management require application abundance and diversity of information.

– That Environmental management has various facets.

– That wildlife management has several well defined sectors viz. Scientific, Socio economic, Management etc.

– That Conservation has symbolic relationship.

All these requirements have to be met with a solid baseline qualitative and quantitative data on economy, socio-economics and various facets of ecology.

1.7 BIODIVERSITY AND ECOSYSTEM PROCESSES

The maintenance and enhancement of species diversity almost defines ecological conservation. It is therefore, imperative to establish what factors influence species diversity at local level. One such factor is habitat size. Other factors involve physical and biotic processes occurring in the community. Important variables include the type and intensity of both physical disturbance and grazing and nutrient availability.

Recent experiments have shown that there may be a positive relationship between productivity and diversity in communities. In fact, excessive nutrient enrichment tends to reduce species diversity, both on land and water. It appears that a minority of species are able to take advantage of the elevated nutrient levels and proceed to out-compete the other species.

Minor physical disturbance may encourage species diversity by reducing the uniformity of a habitat, while an absence of disturbance may have the reverse effect.

Grazing and grazing intensity also have a major influence on species diversity. Experiments have suggested a unimodal relationship between plant diversity and grazing intensity, although this relationship is influenced by variables such as the type of grazing animal, and the period of the year when grazing occurs.

Absence of grazing may encourage the growth of a few coarse grasses, and eventually may not result invasion by woody plants. The presence of grazing at a low level may contain the growth of more competitive grasses allowing more species to coexist. Above a certain threshold, however, diversity tends once again to decline because a more restricted groups of species is adapted to the very disturbed and unstable conditions arising from the loss of vegetation. A similar effect can be observed with large herbivores on forests (Charts 16-20).

The authority has analysed the relationship between predator and prey animals, vis-a-vis grazing pressure, relating to species diversity. This aspect needs meticulous experimentation.

Possible relationships between species *Diversity* and *Biotic productivity* have also attracted attention. Tree species richness and evapotranspiration have been shown to be closely related. This is significant because evapotranspiration, which integrates both temperature and moisture, is corrleated to primary productivity.

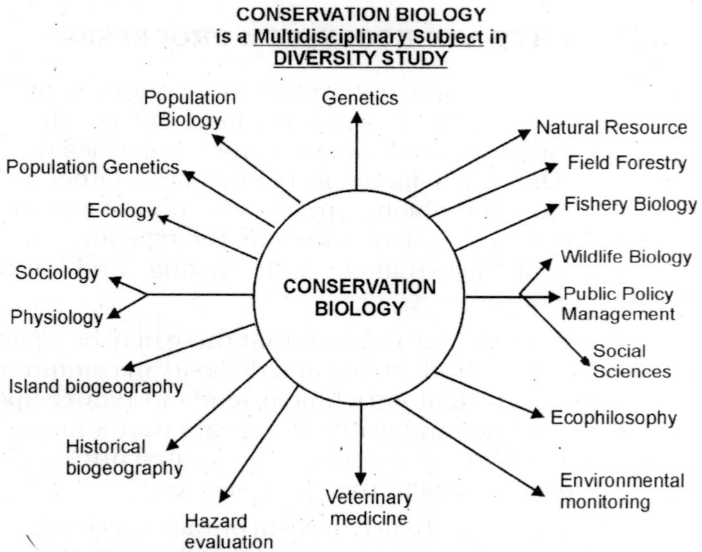

Source : "Principal and Practice of Plant Conservation" by D.R.Given, 1994.

CHART 16

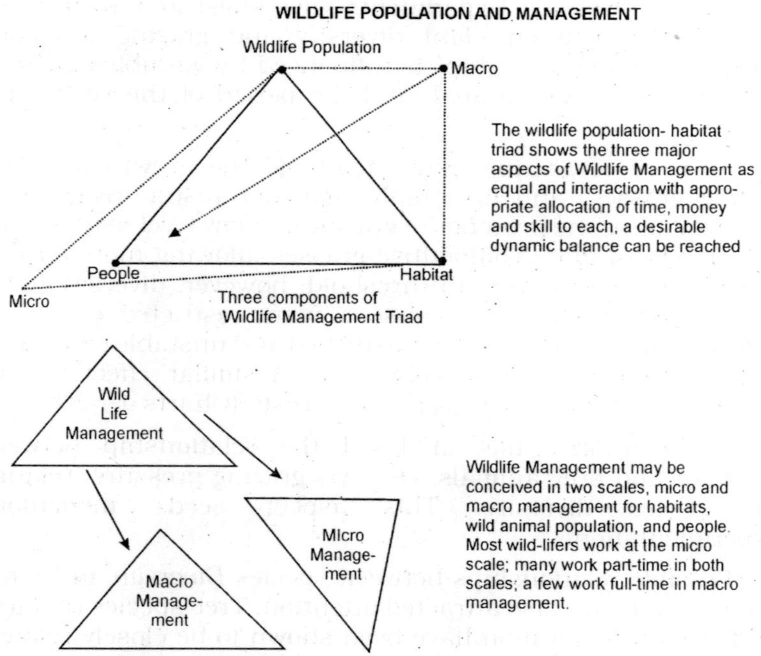

Source : "Wildlife Management" by Robert H. Giles, 1978.

CHART 17

WILDLIFE MANAGEMENT : ABUNDANCE AND DIVERSITY INFORMATION

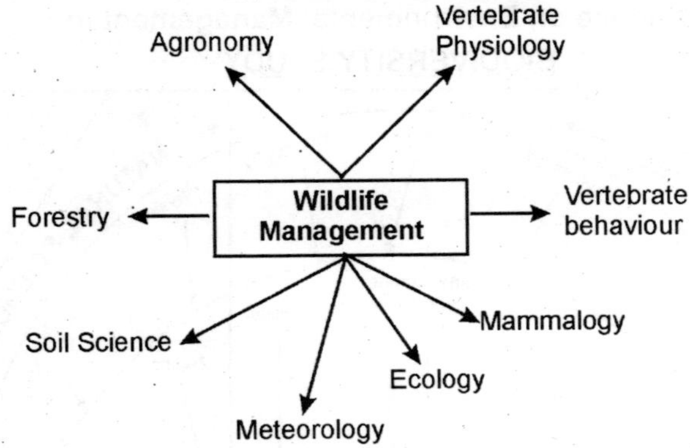

The principles of wildlife management, portrayed by the Central square include many concepts shared with several sciences and other profession only a few of which are shown. Wildlife management requires application of an abundance and diversity of information and is one of the most complex of occupations.

Source : "Principles of Wildlife Management" by J.A. Bailey, 1984
CHART 18

As such interpretation of various issues in any ecosystem is a very complex job as very many factors are involved in the ecosystem process. [Modified from "The Biosphere" by Ian. K. Bradbury, 1998].

1.8 MEASURING BIODIVERSITY

Monitoring and measurement of biological diversity is a complex job; the state variables (composition and structure) are easier to measure and understand than the function variables (ecological processes) although the latter are critical to the existence to the former.

Gaseous NH_3, NO_2 and HNO_3, and as NO_3 and NH_4^+ in precipitation, has been found to influence species composition. The mechanism of exchange of the most nitrogen sources (NH_3 and NO_2) between vegetation and the atmosphere. Leaf surfaces are the sites on which important chemical reactions between atmospheric trace gases occur. Forest canopies capture wind driven fog and cloud droplets.

To monitor atmospheric effect on biological diversity it is necessary to measure both atmoshperic and biotic components of ecosystem.

Way of Understanding Disciplinary Influence on Environmental Management in BIODIVERSITY STUDY

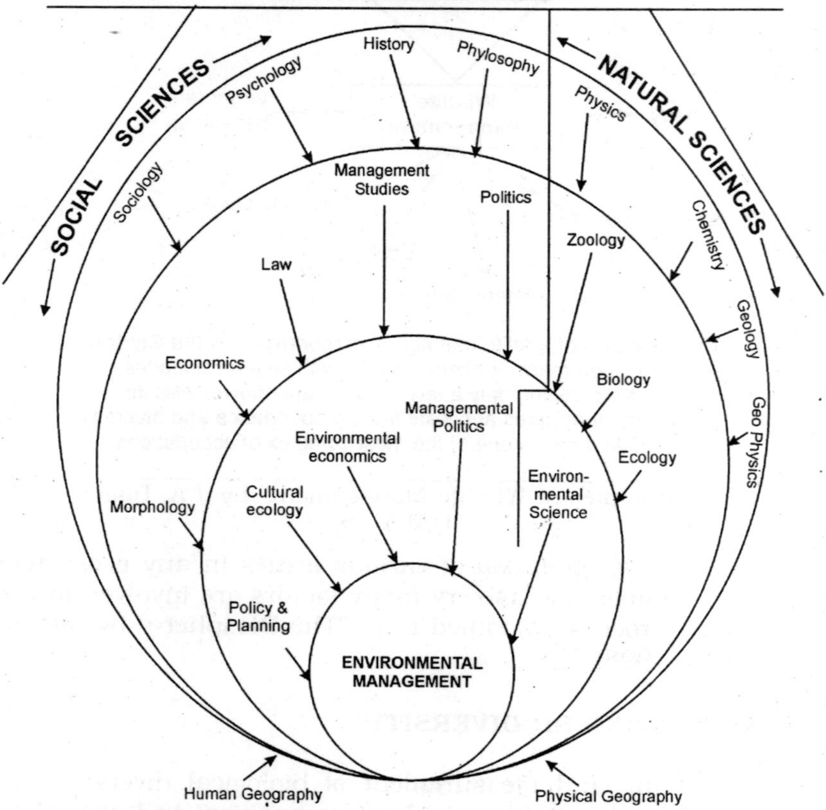

Source : `Environmental Management" by Goeff A. Wilson, R. Bryant,1987.

CHART 19

Air pollution influence the genetic structures of sensitive plant species : ecotypic differentiation is likely to be more common than has been documented.

From various definitions mentioned earlier in this chapter it stands that biodiversity is the variety and variability among living organisms and the ecological complexes in which they occur. Diversity has been defined as number of different items and their relative frequency and these items are organized at many levels, ranging from complete ecosystems to the chemical

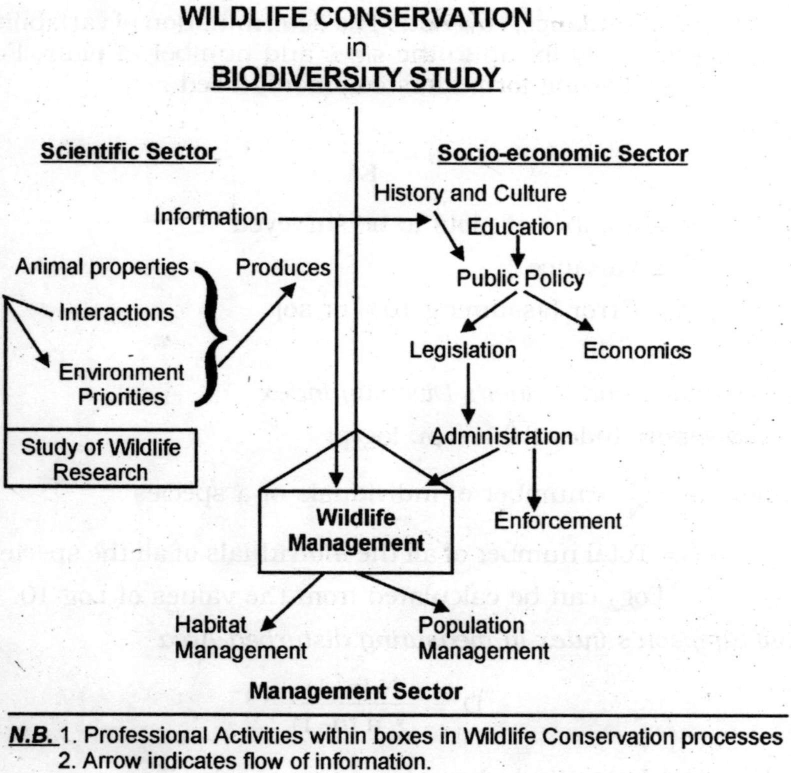

Source : J.A. Bailey, 1984. In 'Principles of Wildlife Management'

CHART 20

structures that are the molecular basis of heredity. Thus, the term encompasses different ecosystems, species, genes, and their relative abundance (Source : The US congessional Office of Technology Assessment). [OTA, 1987].

This definition also suggests the necessity of a mathematical approach in determining relative abundance and evaluating each community on the basis of Diversity Index. The term also suggests cytological investigation in determining species status, besides some other parameters.

Precursor to the assessment of Biodiversity

(i) Quantitative study is an essential pre-requisite to any assessment of the status of species and genera at a site. Such study involves assessment of density, frequency, abundance, relative abundance, cover etc. The determination of variability is necessary to fix up to the sizes and number of plots. For this the following formula is commonly used :

$$n = \frac{t^2 S^2}{E^2}$$

where n = number of plots to be surveyed

S^2 = Variance

E^2 = Error (assuming 10% or so)

t = 1.96

(ii) Shannon and Weiner's Diversity Index

Diversity Index ($\overline{H}$) = – pi. $\log_2$pi

where pi = $\frac{ni}{N}$ = number of individuals of a species

N = Total number of all the individuals of all the species.

Log_2 can be calculated from the values of Log 10.

(iii) Simpson's index in measuring disturbed Area

$$D = \frac{N\,(n-1)}{\sum n\,(n-1)}$$

where D = Diversity Index

N = Total number of individuals of all species.

n = number of individuals per species

$\sum$ = Sum.

While comparing species at two sites the present value. If falls short of past value, will indicate disturbed environment.

(iv) Index of Species richness or variety (say V)

There are three formulae as under :

(i) $V = \frac{S-1}{\text{Log}_N}$, (ii) $V = \frac{S}{N}$, (iii) V = S per 1000 individuals

where S = Number of species.

N = Number of individuals.

(v) Index of Evenness (E)

$$E = \frac{H}{\text{Log}^S}$$

where H = is the Shanon Index.

S = Number of species.

(vi) Index of Dominance (D)

The index (D) represents the percentage of abundance contributed by two most abundant species :

$$\text{Index (D)} = \frac{100 \times Y_1 + Y_2}{Y}$$

where,

Y_1 = most abundant species.

Y_2 = Second most abundant species.

Y = total abundance for all species.

(vii) Simpson's Species Richness Similarity index —

It is same as Sympson's species richness in which one lists the species present at each of the sites being studied, then compares the different lists to see how much overlap exists between sites.

By this it is possible to observe the impact that the variation in underlying environmental conditions (e.g. Soil Composition and Slope) has over the species composition.

D.L. Parlman *et al.*, in their book of Biodiversity (1997), stress on Alpha, Beta and Gama Diversity for assessment of impacts of various factors on biodiversity. They name it "Ecologists' characterizations of diversity across the geographic landscape."

Characterization of diversity

Alpha diversity (in Homogenous site)	Measured in terms of either species richness or combined with equitability i.e. distribution of numbers in individuals among the species.
Beta diversity (in Heterogeneous region) or Community ecology	To what extent do species turnover (replace one another) between different sites or along an enrichment gradient.

Low Beta diversity value indicates no change in environment.

High Beta value diversity indicates variation in environment.

Both Alpha-diversity and Beta-diversity contribute the Gama diversity of a landscape. A region contains many sites with high Alpha-diversity, but if there is little change in the species composition between communities (i.e. if there is low Beta- diversity), then the Gama-Diversity of the region will not be especially high. In comparison, if the sites of a region, each have low Alpha-diversity but there is high turnover of species between communities (i.e. if Beta-diversity is high), then the Gama-diversity of the region may be quite high. Regions with both high Alpha and Beta diversities will have the highest Gama diversity.

1.9 CONSERVATION : APPROACHES AND RESEARCHES

UNEP consider the following :

- Traditionally conservation has been polarized into two separate approaches e.g. *ex-situ* and *in-situ*. The ecologists and conservationists prefer *in-situ* conservation while agriculturists and species oriented biologists prefer the former.
- The recognition and characterization of biodiversity depends critically on the works of three scientific disciplines such as:
    - Toxonomy, which provides the reference system and depicts the pattern of diversity.
    - Genetics, which gives a direct knowledge of the gene variations found within and between species.
    - Ecology, which provides knowledge of the varied ecological systems in which taxonomic and genetic diversity are located and of which it provides the functional components.
- Biodiversity within an area can be characterized by measures of species richness, species diversity, taxonomic diversity and functional diversity each highlighting different perspective.
- A species is a group of organisms that recognize each other for the purpose of mating and fertilization. It is a group of interbreeding natural population that are unable to successfully mate or reproduce with other such groups.
- Genetic diversity is a critical component to issues of biodiversity. It is the genetic diversity within species that allows a species the opportunity to evolve under changed environmental conditions.

Genetic diversity is a component of Biodiversities. The variability that we observe among individuals (phenotype) results partly from the interaction of genetic difference (genotype).

- Pattern of differences under domestication : such as giganticism, suppression of dispersal mechanism, suppression of sexual reproduction; changed growth form; changed life form; changed breeding system; loss of seed dormancy; Biochemical changes; physiological changes.
- Types of diversity Alpha (α), Beta (β) and gama (γ); alpha i.e, at local level, gama i.e. at regional level and beta i.e. interaction of species at local level (degree of species change along a given habitat or physiographic gradient).

UNEP also consider the ratio of biodiversity, closely associated with that of its conservation, preservation and sustainable use [refer - World Conservation Strategy (IUCN 1980), World Commission on Environment and Development (WCED, 1987), A Caring for the Earth : A Strategy for Sustainable Living (IUCN/UNEP/WWF 1991) and the Global Biodiversity strategy (WRI/IUCN/UNEP, 1992)].

The term conservation has not been defined in the Convention of Biological Diversity. The term preservation or maintenance of some or all of the components of Biological diversity which includes also sustainable use of components. Sustainable use means the use of components of biological diversity in a way and at a rate that does not lead to the long-term decline of biological diversity.

The successes and failures of conservation efforts over the last decade have demonstrated that the erosion of biological diversity can be stopped only through novel and complex mixtures of economic, social and political action based on sound knowledge of ecosystems. It is an accepted fact that-

- Biodiversity is greatest in the developing nations of the Tropics.
- Industrialised countries and developmental organizations are devoting substantial resources to these issues.
- Biodiversity in developing countries can not be conserved over the long run unless local peoples and national economics simultaneously derive social and economic benefits.
- Loss of biological diversity if irreversible.
- Species that are lost are lost for ever.

The conservation of biodiversity is a global responsibility.

The challenge of biodiversity research entails not only the gathering of information, but its management, application and communication for an acceptable standard of knowledge about the diversity of the World's biota which calls for the following actions :

- **National Biological Inventories** : This should be organized in a befitting manner, adequately funded and strengthened according to the demand of the situation as and when neccessary.
- **Global biological survey** : For this purpose a strategy for ganging the magnitude and patterns of distribution of biological diversity on earth should be coordinated and implemented. The National Academy of Sciences Study on Research Priorities in Tropical Biology recommended that a comprehensive multidisciplinary Worldwide Survey of well-known groups of tropical organisms be undertaken.
- **Screening of Organisms** : The screening of plants, animals, fungi and micro-organisms for features of potential human benefit should be systematized and accelerated through strengthened programmes.
- **Monitoring** : To detect major changes in the status of biological diversity, appropriate monitoring methods, employing specific indicators of the biodiversity attributes should be implemented.

Site specific research

Site-specific Research is necessary to advance the understanding of ecosystem composition, structure and function, sustainable land use and development and conservation of biological diversity.

- Sustainable use of Biological Resources for which research should be conducted.
- Restoring and Utilising Degraded lands for which in depth research should be conducted.
- Computer Data Bases and Inventories for which resources should be developed to the development of computer data bases, inventories, and information networks for collection of information.
- Remote Sensing and Geographic information system, for which additional research and technical development are

needed to advance the utility of remotely sensed data for ecosystem monitoring in developing countries.

- Strengthening Scientific Network, for which financial and institutional resources are to be established.
- Developing Taxonomic Expertise.
- Expanding Cooperative Research.

Socioeconomic factors are responsible for erosion of biological diversity. The economic aspects of biodiversity conservation demand sophisticated analysis involving economists, ecologists and other researchers. It is also necessary to have country level socioeconomic research. Stress has to be given to International Macro-Economic Research.

Every sphere of research on biodiversity and its conservation-biological, economic and cultural-will require activities across ecological zones. Each developing country should build their own competent researchers and develop solid institutions (Museum, Library, Fund, Equipment etc.). There should be short-term and long-term researches.

It has been appreciated that the conservation of biological diversity has come to be understood as an essential tool of sustainable development World Wide. Biodiversity is a basic determinant of the structure and function of all ecosystems and provide the foundation on which the future well-being of human society rests.

1.10 BIODIVERSITY IN VARIOUS LIVING ORGANISMS

The man made depletion of vast tropical forests has been widely discussed and documented by the Scientists. The loss of the forest cover has been continuously monitored during last two decades. It has been accepted that such depletion may change regional climate (especially rain) patterns, changes in biological productivity, accelerated rates of soil erosion, disruption of watershed stability and increasing emissions of green house gases.

The proximate causes of biodiversity loss are biological, but the root causes of the problem include sociological and economic processes that operate on a global scale.

Evolution and extinction are natural processes, new species has been created in evolution through geologic time.

Loss of biodiversity may be due to direct destruction, conversion, degradation of ecosystem, over-exploitation, habitat disturbance, pollution etc. As genetic variability is lost however the species as a whole becomes more vulnerable to other factors, susceptible to problems of inbreeding, and less adaptable to environmental change.

To date, 1.4 million kinds of organisms have been assigned scientific names, but only a fraction of these species have been studied. Current estimates of the Earth's total species diversity range from 10 million to 100 million.

There are about 2,50,000 species of vascular plants in the World. About two thirds of these are found in the tropics. At least one sixths of earth's diversity of plantlife- 45,000 species are found in Latin America (in Equador, Peru and Columbia). There may be twice as many species in Costa Rica.

Numbers of Described Species of Living Organisms

Kingdom and Major Subdivision	Common Name	No. of Described Species
Virus	Viruses	1,000 (order of magnitude only)
Monera		
Bacteria	Bacteria	3,000
Myxoplasma	Bacteria	60
Cyanophycota	Blue-green algae	1,700
Fungi		
Zygomycota	Zygomycete fungi	665
Ascomycota (including 18,000 lichen and fungi)	cup fungi	28,650
Basidiomycota	Basidomycete fungi	16,000
Ocmycota	Water moulds	580
Chytridiomycota	Chytrids	575
Acrasiomycota	Cellular slime molds	13
Myxomycota	Plasmodial slime molds	500
Algae		
Chlorophyta	Green algae	
Phaeophyta	Brown algae	
Rhodophyta	Red algae	
Chrysophyta	Chrysophyta algae	
Pyrrophyta	Dinoflagellates	
Euglenophyta	Euglenoids	
Plantae		
Byrophyta	Mosses, liverworts hornworts	16,600

Psilophyta	Psilopsids	9
Lycopodiophyta	Lycophytes	1,275
Equisetophyta	Hosetails	15
Filicophyta	Ferns	10,000
Gymnosperma	Gymnosperms	529
Dicotolydonae	Dicots	170,000
Monocotolydonae	Monocots	50,000
Protozoa		
	Protozans: Sarcomastigophorans, ciliates, and smaller groups	30,800
Animalia		
Porifera	Sponges Jellyfish,	5,000
Cnidaria, Ctenophora	corals, comb jellies	9,000
Platyhelminthes	Flatworms	12,200
Nematoda	Nematodes (roundworms)	12,000
Annelida	Annelids (earthworms and relatives)	12,000
Mollusca	Mollusks	50,000
Echinodermata	Echinoderms (starfish and relatives)	6100
Anthropoda insecta other anthropods	Anthropods, insects	123,161
Minor invertebrate phyla		9,300
Chordate		
Tunicata	Tunicates	1,250
Cephalochordata	Acorn worms	23
Vertebrate	Vertebrates	
Agnatha	Lampreys and other jawless fishes	63
Chrondrichthyes	Sharks and other cartilaginous fishes	843
Osteichthyes	Bony fishes	18,150
Amphibia	Amphibians	4,184
Reptilia	Reptiles	6,300
Avea	Birds	9,040
Mammalia	Mammals	4,000
Total all organisms		

Source : Wilson, 1988

It will be very relevant to reproduce a table on described species of living organisms prepared by a panel of the Board on Science and Technology for International Development of U.S. National Research Council.

The role of plants, along with algae and photosynthetic bacteria is well known. Very many plants are sources of food, medicines, fibres, waxes, oils, timber, ornamentals etc. Yet new uses for plants are being discovered regularly (bark of *Taxus brevifolia* has been found to have anti-cancer agents).

Fungi : Fungi have been recognised as an agent for human food, medicine, antibiotics, maintenance of soil fertility, decomposition of biomass etc., Mycorrhizal fungi form symbiotic relationship with plant roots and enhance mineral nutrient uptake and are critical link between the soil and plant components of most terrestrial effort. All these have been found to have significant impacts on sustainable crop and forest management and environmental restoration. Man has derived several benefits from microorganisms. Actinomycetes alone have been the source of 3,000 antibiotics.

This situation offers opportunity for further indepth research.

Microorganisms : Microorganisms contribute plant nutrient movement and constitute biological bridges between trophic levels, between abiotic and biotic factors and between the biogeosphere and the atmosphere. They contribute to the maintenance of soil fertility. They have ability to catabolize organic matter, produce organic compounds and control disease outbreaks. Future research on biotechnology promises to increase the use of microorganisms in solving medical, agricultural and environmental problems.

Invertebrates : Human knowledge on this subject is still very poor. It is said, only 10 percent of invertebrates have been described. About two-thirds of the 1.4 million described species are invertebrates (vast majority of these are insects). On a single tree in the Tambopata Reserve of Peru Wilson (1987) collected 43 species of ants, belonging to 26 genera. Ants alone probably comprise between 5 and 15 percent of the biomass of the entire fauna of most terrestrial ecosystem.

The role of the various groups of invertebrates in ecosystem function including pollination, decomposition, disease transmission is not fully known. The interactions of soil mesofauna (e.g. nematodes, collembolans and mites) and soil microorganisms are vital in maintaining the plant soil system. Nematodes both feed on and act as dispersal agents for soil bacteria.

Marine invertebrates play a major role in ecosystem function. Protozoans, crustacean (e.g. Copepodes, euphausids, isopods, amphipods and larvae of other species) link marine primary producers (phytoplankton) with higher levels of the marine food *web* (fish, mammals). Food habits of Squid, Octopus, Corals, mollusc, echinoderms etc. reflected several aspects of ecosystem functions.

Vertebrates : 41,000 species have been described, but many have yet to be discovered. Fish form almost half of the known vertebrates. Many species have not been classified and life history of many unknown. There are few undescribed species of reptiles, birds and mammals. Nonetheless, new species continue to be discovered, fairly regularly (The black-faced lion tamarin, a previously unknown primate was discovered in 1990 by two Brazilian biologists).

PROPORTIONS OF DIFFERENT TAXA FROM DESCRIBED SPECIES.
The diagram includes taxa likely to contain 1,00,000 + species + vertebrates.

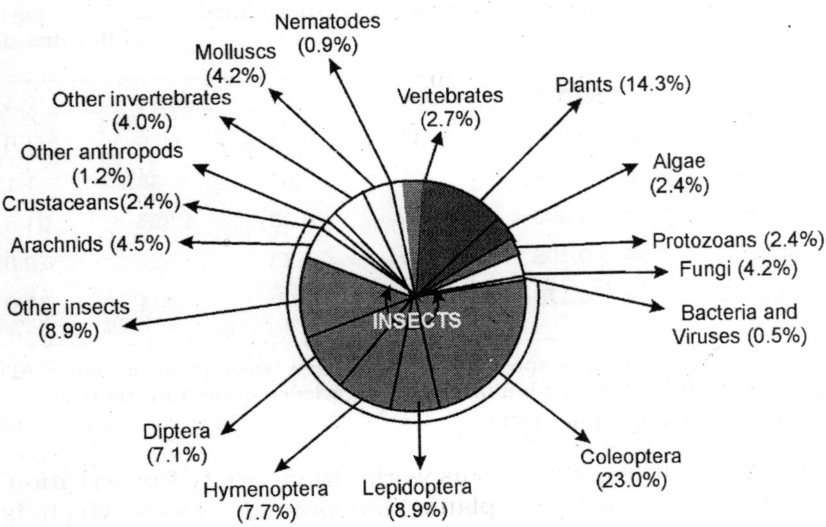

Source : Redrawn from Groombridge, 1992

Numbers of described species and estimates of actual number for selected taxa, in thousands

	Species described	Estimated number of species : high	Estimated number of species : low	Working figure
Viruses	4	1,000	50	400
Bacteria	4	3,000	50	1,000
Fungi	72	2,700	200	1,500
Protozoa	80	1,200	210	600
Nematodes	25	1,000	100	400
Insects	950	100,000	2,000	8,000
Molluscs	70	200	100	200

[Source : Groombridge, 1992; Heywood 1995]

Current status of species of plants and animals.

	Total species	Number extinct	Percentage extinct	Number threatened	Percentage threatened
Plants	250,000	384	0.15	18694	7.4
Fish	19,056	23	0.12	320	1.6
Amphibians	4,184	2	0.05	48	1.1
Reptiles	6,300	21	0.33	1,355	21.5
Birds	9,198	113	1.23	924	10.0
Mammals	4,170	83	1.99	414	10.0

Notes : '*Extinct*' means species known to have become extinct since AD 1600, subject to the limitations in knowledge quoted in the text.

[Source : McNelly *et al.*, 1990]

M.S. Jeffries, 1997 in his work- 'Biodiversity Conservation' has quantified various plants and animal species which is reproduced below, as thought very relevant in the present study.

Species richness and endemism : higher plants

	Flowering plants	Gymnosperms	Ferns	No. of Endemics	% Endemism	Estimates count	Completion	Date
India	15,000	–	1,000	5,000	31.3	e2	2	83-84
Indonesia	20,000	–	2,500	15,000	66.7	e3	4	1991
Iran Islamic Rep.	6,500	33	–	–	[30-35%]	e2	1	89-91
Iraq	2,914	7	16	190	6.5	c	1	66-86
Israel	2,294	8	15	155	6.7	c	1	82-84

Species richness and endemism : higher vertebrates

	MAMMALS		BIRDS		REPTILES		AMPHIBIANS	
	Species knwon	Endemic species	Species knwon	Endemic species	Species knwon	Endemic species	Species knwon	Endemic species
India	317	38	969	69	389	156	206	110
Indonesia	515	165	1519	258	511	150	270	100
Iran, Islamic Rep.	147	4	–	1	164	3	11	5
Iraq	81	1	145	1	81	–	6	0
Israel	–	2	169	0	–	–	–	0

1.11 RESTORING BIODIVERSITY

This is a multifaceted subject which requires the efforts of many talented, knowledgeable and dedicated people, only a few of whom are available for implementation of project findings.

Reed. F. Noss (Oregon State University) observes :

"In these times of crumbling natural ecosystems of rapid political resistance on conservation, it may seem utopian or even downright stupid to dream of restoring ecological diversity at any meaningful scale.... wild land now buried under asfalt and concrete seems impossible to revive. Species that have gone extinct cannot be brought back, at least not with current technologies. Few rare species show signs of recovering, even with the expanditure of millions of conservation dollars on their behalf. How then, can massive ecological restoration and species introduction be technically or politically feasible?"

Charts 15 to 20 fits well for the contents of paragraphs from 1:3 & 1.11 in Biodiversity study with various facets.

1.12 A WORLD BANK GUIDELINES FOR MONITORING AND EVALUATION FOR BIODIVERSITY PROJECTS

Executive Summary Printed in Biodiversity Series is reproduced

Biodiversity projects are designed on the assumption that project interventions will lead to conservation of key biological resources. Monitoring and evaluation are the primary mechanisms to assess whether a project is meeting its targets and objectives. These guidelines are intended primarily to assist World Bank task teams and consultants in the design and implementation of *monitoring and evaluation* (M&E) plans for biodiversity conservation projects or projects with biodiversity components. It is hoped that they may also serve as useful reference materials for government agencies, non-governmental organizations and others involved in the design, implementation or evaluation of biodiversity projects.

M&E plans are mandatory for all Bank projects. *They must be developed as integral elements of projects to provide information on whether project interventions are successful in achieving project objectives and on how social, economic, political and institutional* factors are affecting project performance. Monitoring and evaluation for biodiversity projects involves two kinds of indicators : *Implementation performance indicators* (project inputs and outputs) and *Project impact indicators* (project impact on biodiversity). These guidelines focus primarily on the latter (*impact indicators*).

An M&E plan is a detailed program of work which defines what monitoring activites will take place, when and by whom, and how that information will feed back into management decisions. The plan should include an estimate of costs of implementation, and identify training and capacity building needs among the staff and institutions responsible for this M&E. The plan should also spell out how M&E activities begun under the project will contribute to the estabilishment of a long-term monitoring and evaluation capability in the host country. In the developing country context it is especially *important to develop an M&E plan that is simple, inexpensive, and sustainable in terms of the financial, institutional, and technical resources available.*

A biodiversity M&E, plan should, therefore :

- answer a clearly stated set of questions (i.e. have clear objectives);
- state clearly what indicators will be chosen;
- specify how often monitoring and evaluation will be done, and by whom;
- outline any necessary training or financial inputs that are required;
- State the intended audience for the evaluations;
- Specify how information will feed back into management decisions; and
- State clearly the decision points at which action must be taken to address negative trends.

Monitoring of biodiversity is not the same as measuring biodiversity. Measuring biodiversity provides a snapshot of biodiversity (e.g. number of species present) at the time of measurement. Monitoring is a continuing process which allows managers to identify changes and trends over time so that they can assess whether interventions are achieving biodiversity goals and adapt management accordingly. While preparation of a biodiversity project may require a comprehensive biodiversity survey, future monitoring does not need to update this full set of data. In most cases managers are concerned with monitoring trends rather than measuring absolute values.

Scoping of the M&E plan should be done during project preparation. Experience has shown that there is seldom sufficient information available at this stage to prepare a detailed M&E plan. Nevertheless it is important to develop either the terms of reference for preparation of an M&E plan, or the preliminary outline of the M&E plan. This provides a basis both for the costing of M&E project components and for the development of the full M&E plan as soon as possible after project commencement. Further development and refinement of the M&E plan can be done as necessary during project implementation, based on local experience and field testing.

It is particularly important to define the spatial and temporal scales of monitoring activities. Because biodiversity management deals with ecological processes which are generally long-term (e.g. changes in numbers of a population of a key species) changes resulting from management interventions may be slow to emerge sometimes beyond the project time frame. This suggests the need to establish a monitoring framework that will extend beyond the project term and, if appropriate, feed into a national monitoring system. The appropriate spatial scale for project monitoring will be determined by the specific goals and objectives of the project, and depend on whether it focuses at the landscape, ecosystem or species level.

The most important aspect of any M&E project is the choice of suitable and meaningful indicators. Clearly identifying the assumptions for project interventions will help identify indicators for monitoring both changes in threats and the effectiveness of project interventions in mitigating those threats. Most importantly, indicators must be practical and realistic, and should, whenever possible, be meaningful at both the national and site level, as well as consistent with the main objectives of the project. For example, monitoring changes in forest extent and integrity may provide useful information on park integrity, but will not provide information on the conservation status of tigers or other wildlife if the main threat is poaching.

Most threats to biodiversity result from human activities which, in turn, depend on social and economic factors. Monitoring of socio-economic factors, therefore, is an important part of biodiversity M&E. However, it is necessary to recognize that the relationships between biodiversity health and the socio-economic characteristics of human groups causing impacts are far from clearly established. In fact they are likely to vary from one location to another. This needs to be taken into account in designing M&E plans and particularly in indentifying and interpreting socio economic indicators.

Similarly a range of institutional factors can exert impact on biodiversity health and the effectiveness of biodiversity management and should also be monitored.

Preparation of M&E plans and identification of relevant indicators should, as much as possible, involve those communities and institutions likely to be affected by project interventions. The identification of indicators and appropriate sampling regimes should also take into account existing monitoring programs and data sets at the local and national level, capacity at these levels, and the need to establish agreed sampling and recording protocols at the national level. Consistency of monitoring approaches across local areas and protected area systems should have a high priority.

A meaningful and operationally relevant biodiversity monitoring system will encompass a broad range of subjects to be monitored, including landscape or species dynamics, socioeconomic factors and community involvement and institutional and regulatory factors. No universal set of indicators will apply to all projects but most projects will be measuring a range of indicators. The specific indicators chosen for any individual project will depend on the particular objectives and goals of that project and the activities that are proposed to meet those goals. For any project it will be important to choose a minimum set of a few indicators that are feasible, useful and relevant to that project and can be sustained with local capacity and resources after the end of the project. The guideline provides general guidance on identifying appropriate biodiversity M&E indicators in different thematic areas while the annexures provide examples to assist in the development of indicator sets relevant to particular situations.

Monitoring and evaluation are integral parts of biodiversity management and require adequate resources, including budget and institutional capacity, clear institutional responsibilities and reporting mechanisms. It is important to build incentives and capacity to collect, use and maintain data for monitoring and evaluation. Since M&E will require additional capacity, work and budget beyond the lifetime of the project it is important to develop M&E plans that can be resourced sustainably. The information gathered through M&E activities is useful both for assessing the impacts of the individual project and to provide input into the design and implementation of future biodiversity projects and ongoing biodiversity management programmes.

CHAPTER - 2

QUO VADIS, MEDICINAL PLANTS ?

2.1 FLORAL AND FAUNAL RESOURCES - FACE A CRISIS

Gone are the days when India's floral and faunal resources were considered very rich. In the past sixty years the biotal scenario has undergone a drastic change due to undercontrolled biotic factors; some abiotic factors have also aggravated the situation. The Department of Forests has been in existence since the sixties of the nineteenth century and on its recommendations Wildlife Protection Acts have been enacted for conservation of 'Birds', 'Elephants' and 'Rhinoceros'. More recently the Wildlife (Protection) Act of 1972 for protection of all biotal species whose existence is threatened has been enacted. This Act takes care to protect various categories of flora and fauna in consideration of their present status. In the various management plans drawn up for Forest Divisions efforts were made to preserve the threatened tree species; but little care was taken for the preservation of herb and shrub flora which form the bulk of medicinal plant material.

In this chapter the authors have made an endeavour to bring to light the various adverse factors responsible for depletion of flora and fauna. Several maps and charts have been presented to give a clear picture of the status of forest area, population pressure and the stress on raw forest resources for meeting the increased demand of growing population. Some selected species of medicinal plants have been listed from herbs, shrubs, climbers and trees and their present status has been discussed. Useful information on the impact of biotic and abiotic factors on the regional biota in general has also been presented. Other causes of the depletion of medicinal herbs have also been brought to light.

It has already been said that the purpose of this work is to expose the present vulnerability of many species of medicinal plants whose occurrence was considered adequate in the past. Several paragraphs have also been compiled to give an idea of the versatile qualities of many species of medicinal plants to remedy a wide range of diseases.

However much we boast of our rich floral and faunal wealth especially the vast resources of medicinal plants, the fact remains that such resources have suffered a severe setback during the last 60 years. Many species may not yet have become extinct but surely their number has gone down to an alarming level. India is now limping into a new millennium with a much depleted floral and faunal resources, caused by uncontrolled increase in human and cattle population, rampant grazing, forest fire, acute soil erosion, and ruthless exploitation of biotal resources.

Medicinal plants are mostly herbs, some shrubs and climbers, and a few trees which occur mostly in forest areas, some in scrublands, waste lands and in marginal lands. Forestry practices in upgrading mixed miscellaneous forests and forests comprising less valuable inferior species to plantations of species of high commercial value have depleted ground vegetation, resulting in reduction of biodiversity.

Several statements, diagrams and maps have been presented in this chapter to provide a realistic picture of the types of impact on vegetation. These data bring out the linkage between population growth and the qualitative and quantitative parameters of forest resources in the country and highlight the colossal degradation of forest cover (density per unit area).

India's evergreen, moist deciduous and temperate forests are the primary home of medicinal plants apart from the country side in general. However, the dry-deciduous forests and the dry zones of the country also bear stray herbaceous plants (including medicinal plants) to some extent. Maps showing phyto-geographic zones and forest types with relevant information are provided in order to present a realistic picture of the herb flora of India. India does not hold inexhaustible floral resources, especially of medicinal plants, and a thorough quantitative study will testify to this statement which has not been done so far. The Forest Survey of India has made a very useful assessment of the density, frequency, abundance and volume assessment of important tree species all over the country. Some states also have such figures worked out for their forests. But data on herbs, shrubs and climbers are barely available. The authors of this book have done some limited quantitative assessment of herb/shrub flora. The data are recorded in Chaudhuri's publications listed in the bibliography.

In this context one has to evaluate the effectiveness of the conservation efforts of the Government, whether these steps are adequate to maintain the ecological balance, biodiversity and quality of environment by controlling all the adverse factors. There is general awareness today that increasing population, rampant poverty, and inadequate institutional framework are the underlying causes of forest depletion and degradation.

Complicated Study

Any study on the status and availability of medicinal plant resources requires a thorough qualitative and quantitative ground survey of vegetation cover, especially of herbs and shrubs. Besides, the impending pressure on such resources posed by a rapid growth of human and cattle population in a situation of poverty and faulty land use has to be taken into consideration. It is a multifaceted issue which involves a thorough knowledge of systematic botany, chemistry and bio-chemistry (phyto), pharmacology and of the socioeconomic disciplines.

There is a very large number of species of medicinal plants, each having a wide range of remedial properties for various diseases; there are also plant species which are primarily effective in one or several specific types of diseases and both need special study.

The chemical compositions of various plant parts change from season to season and these parts are to be harvested at the right time to get the efficacy of the exact chemical composition used in the preparation of medicines. But it is a difficult job to synchronize the collection with the time when the plant parts are rich in such chemicals. So the druggist often fail to derive the best efficacy out of a plant to remedy disease.

The tribal people and fringe forest dwellers have been utilising forest plants since time immemorial. Rev. P.O. Bodding's book entitled 'Studies/Medicine Connected Folklore (1986)' is a valuable work on Ethnobotany and Ethnobiology that have anchored deeply among the tribal and rural masses of this country.

Plants are considered to be medicinal if they possess pharmacological activities of possible therapeutic use. In simple language it may be said that plants which are useful to human beings or benefit them are medicinal plants. According to the later definition most plants have some medicinal value. Besides, human knowledge may have failed to discover medicinal properties in some plants which cannot be discarded outright as having no medicinal value. One cannot pinpoint the actual number of plant species in India.

A fruit or flower may be specific for one disease while the leaves, the bark or the root may be specific for other diseases. Ayurveda provides detailed accounts of such diversities in the use of medicinal plants. But at what stage the leaves or bark of plant or roots have to be collected remains an unanswered question.

Medicinal plants have been fairly gainfully utilized in the Allopathic System itself. A large number of medicines and drugs of wide applications prepared from medicinal plant sources are in the circulation in the Allopathic system as well. However, besides the Ayurvedic System, the largest number of medicinal plants are utilised in the preparation of Homeopathic medicines. There is also the Unani System of medicine which is but a modified version of the ancient Ayurvedic System developed in Arabia, Iran, etc. Leaf, bark, stem, root bark, root, flower and fruits have been separately used for medicines, but in some cases the whole plant with roots, leaves, flowers and fruits is used to prepare medicine. Indeed the prevalent practices in the use of medicinal plants and their parts are wide ranging.

Diversity

The diversity of floral species in India is due to the occurrence of 16 different agroclimatic zones, about 10 broad phytogeographical divisions, more than 25 biotic provisions, 450 biomes (habitats of specific species) and 200 forest types. Besides, the region of Eastern Nepal, India (Sikkim) and Bhutan, being the confluence of biogeographic elements, viz. Malayan, Japanese, Indochinese, Ethiopean, Mediterranean, Oriental and palearctic, contributes to diverse biota.

R.P. Chaudhuri in his work 'Herbal Drug Industry' mentions 130 chemical substances extracted from 100 species of higher plants now being used in medicines. A UNDP report (1994) mentions that medicinal plants obtained from developing countries may be valued at more than Rs. 1,00,000 crores. Innumerable plant species in India have yet to be brought under proper research to assess their medicinal value.

Several Indian research centers have started to seek patentable genes, using the latest genetic engineering techniques. More than 400 useful plants have been identified so far. (Tropical Botanical Garden and Research Institute - TBGRI).

Readers may refer to the chapter on Biodiversity in this regard. The Himalayan regions - Western, Central and Eastern- have tridimensional facets (altitudinal, latitudinal and Spatial) which cover various climatic zones (tropical, subtropical, subtemeprate, temperature and alpine). There is a rich rain forest belt in East Himalayas from Arunachal/Assam to Tripura across Manipur, Mizoram, Meghalaya and Nagaland that have more than 45 percent of the plants of India. The Indogangetic and Brahmaputra plains have a poor assemblage of plant species due to intesive cultivation of agricultural crop. The dry zone of Rajasthan, Punjab, Gujrat, Madhya Pradesh, Andhra Pradesh, Maharashtra, Karnataka and Tamilnadu, South-western Karnataka and the Chotanagpur hills in Singhbhum, Mayurbhanj and Sundergarh which have a rich assemblage of flora. But shifting cultivation in all the states of Eastern India and some areas of central and western India has depleted flora and fauna over at least half the forested area in the plains and hills all over the country. Grazing in the forest has deplected herb, shrub and regeneration of flora in the ground floor beside disturbing the ground biota.

India's Critical Forest and Landuse Scenario at a glance

Population increase is linked with forest resource shrinkage.

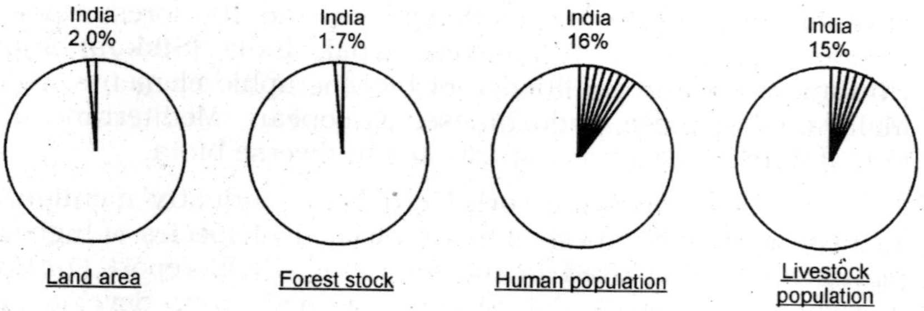

The serious land use situation is reflected in the above diagrams. With a sizeable share of human population (16%) and livestock population (15%) in the world, India has only 1.7% of the total area under forest stock in the world.

The resources of land and forest area are not commensurate with human and livestock population.

[Source : Population & Forest - UNFPA report of 2000]

2.2 FACTORS CAUSING DEPLETION OF MEDICINAL PLANTS

General

Although India has a rich floral and faunal resource having a large number of genera and species, several adverse factors have threatened their existence which are, -

- repeated heavy grazing by a vast population of domestic cattle.
- annual forest fire killing ground herbs and having derogatory effect on soil.
- shifting cultivation (Jhum) in the hills.
- ruthless exploitation of plant parts eradicating the mother plant materials.
- expansion of human population and mushrooming settlements.

- extension of cultivation,
- heavy erosin of land,
- repeated flood and drought,
- excessive deforestation.

Some of these factors have assumed a serious dimension - damaging habitat, reducing effective areas of ground vegetation, density and frequency. City or town outskirts, village locations and peripheries and marginal lands have now been encroached upon or utilised for settlements and cultivation thereby reducing density of herb flora. In forest areas repeated fire and rampant grazing have forced only a few tough and resistant species to survive the onslaught and by this process eradicating the softer species (Lantana, Eupatorium, Mikania, Ipomoea, Parthenium etc. predominate, while Dioscorea, Rauwalfia, Podophyllum, Swertia, Aconitum etc. alarmingly get eradicated). Even though some medicinal plants escape eradication, their occurrence is sporadic and density is extremely low. Of late, various types of pollution also affected regeneration of herb flora.

In support of some of the aforesaid facts a few charts, maps and tables are appended which will show the serious impact now being forced on vegetation indicating a gloomy future.

Population Pressure

India's population has been grown from 361 million in 1951 to 846 million in 1991; and has crossed one billion mark on 11.05.2000 (Table -1)

Table -1. Population : Size, growth and distribution pattern in India 1951-91.

Category	Population millions		Decadal growth rate of population (in percentage)				Ratio of 1991 to 1995 population
	1951	1991	1951-61	1961-71	1971-81	1981-91	
Total	361	845	21.5	24.8	24.7	23.9	234
Rural	299	629	20.5	21.9	19.3	20.0	210
Urban	62	218	26.4	38.2	46.1	36.5	349
Class-I Town	28	139	44.9	53.5	54.4	46.9	504
Million+ Cities	12	71	54.1	53.7	51.3	67.8	601

Pressure on forests, pasture, agricultural land coupled with rural people's poverty and resulting migration to urban and industrial areas may be seen from the Table 2.

So the vegetation is depleted, especially the herbs, which form bulk plant resources for medicine.

Live Stock Population Pressure

Table 2. India's Livestock population : 1951 to 1992 (in million)

Category of Livestock	Decadal popuation of Livestock (in million)				
	1951	1961	1972	1982	1992
Cattle	153.30	175.50	178.30	198.50	204.50
Buffaloes	43.30	51.20	57.40	69.80	83.50
Sheep	38.40	40.00	40.00	48.80	50.80
Goats	47.10	60.90	67.50	95.30	115.30
Horse/Ponies	1.50	1.30	0.90	0.90	0.80
Pigs	4.42	5.20	6.90	10.10	12.80
Camels	0.60	0.90	1.10	1.10	1.10
Others	1.30	1.20	1.10	1.30	1.30
Total	292.92	336.20	353.20	419.80	470.10
Standard Unit	228.00	259.70	270.50	304.50	336.00

Vast increase of sheep and goat is a bad indication in forest conservation. The carrying capacity of India's Forest cover for grazing is estimated at only 31 million live stock, far less the number of livestock actually grazing in forested tract (Simon - 1986).

This situation will deteriorate as the demand for milk, egg, meat will go high with the increase in urban population and there will be further increase in livestock population.

Recession of Forest Cover

Forest cover of India has declined from 72 million hacters in 1951 to around 63 million hecters in 1997. A histographic representation for about 50 years is presented below, which shows decline in per Capita forest area.

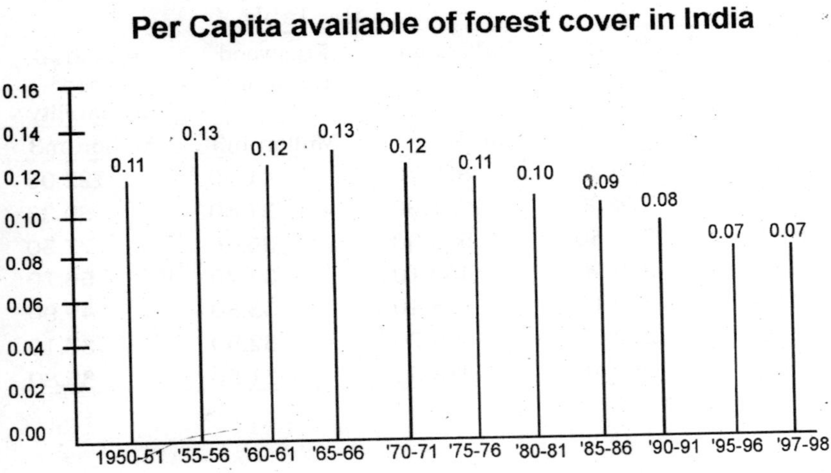

CHART - 21

However, in the Table 6, the situation of forest cover and forest area per capita have been depicted which speak for themselves.

Fuelwood and Industrial wood situation

In India firewood is still the primary source of energy for cooking. At present 78 percent of rural household and 30 percent of urban household (National Survey Organisation, 1997) use firewood as the source of energy for cooking.

According to Tata Energy Research Institute, New Delhi, 2000 estimated annual consumption of firewood in India is around 325 million cubic meters, much above the carrying capacity and have presented a table (Table 3) and a Chart trend as follows :

"Overall production of fuelwood in India has been stagnant over the period 1980-1994, increasing insignificantly from 275 to 281 million cubic meters. Despite this slight increase in total fuelwood production, per capita availability of fuelwood has been consistently declining (Table 4). In 1980 each person had 0.041 Cu.m. of fuelwood available by 1994 this figure declined to 0.031 Cu.m."

Table 3. Availability of Fuelwood and Industrial wood 1980-1994

Year	Fuelwood Production	Industrial wood production	Fuelwood availability	Industrial wood availability
	Million m^3	Million m^3	Million m3	Million m3
1980	275.20	818.70	41.00	122.00
1984	276.30	651.90	37.80	89.30
1986	277.60	590.00	36.40	77.50
1988	279.90	450.10	35.20	56.70
1990	280.10	413.40	33.80	49.90
1992	280.60	320.70	32.50	37.10
1994	281.40	322.30	31.50	35.80

Trends in per capita availability of Fuelwood and Industrial wood (in Cu.m. per 1000 population)

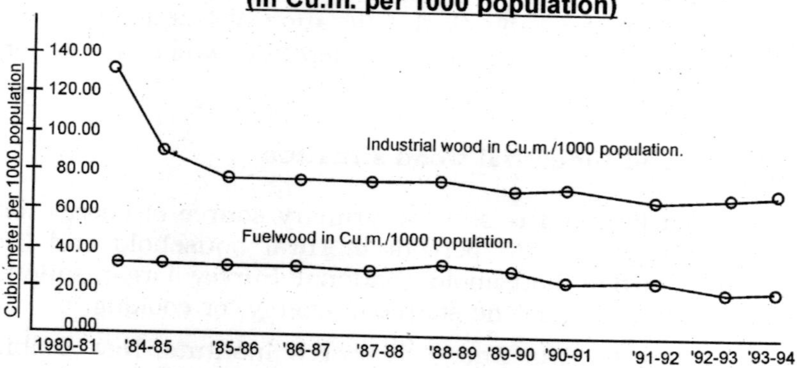

CHART - 22

Likewise, industrial fuel production in India has consistently declined throughout the 1980 - 1994 period. Total production had declined from 819 million cubic meters in 1980 to only 322 million cubic meters in 1994. In this period per capita availability declined from 0.122 cubic meter to 0.036 cubic meter.

Table-4. Over-exploitation of Forests

Fuelwood Situation		Timberwood Situation	
Sustainable Capacity of Forests. (in million m^3)	Current Withdrawal from forests. (in million m^3)	Production capacity (in million m^3)	Annual demand (in million m^3)
48	235	12	28

[Source : UNFPA - 2000]

Poverty and Forest Linkages

Population and Forests : A report on India, Published by UNFPA (2000) : On population and forests, quotes a figure from Economic Survey, 1998-99, which is reproduced below (Table 5).

Table 5. Number and Percentage of Population Below Poverty Line (In India, 1973 to 1993)

Year	Rural India		Urban India		All India	
	No. of population below poverty line (in millions)	Poverty ratio (%)	No. of population below poverty line (in millions)	Poverty ratio (%)	No. of population below poverty line (in millions)	Poverty ratio (%)
1973	261	56.40	60	49.00	321	54.90
1977	264	53.10	65	45.20	329	51.30
1983	252	45.70	71	40.80	323	44.50
1987	232	39.10	75	38.20	307	38.90
1993	244	37.30	76	32.40	320	36.00

"This reveals a downward trend in both rural and urban poverty. India's anti-poverty and employment generation programmes - alongwith economic growth in recent years seem to have helped in reducing the overall incedence of poverty from 55 percent to the total population in 1973 to 36 percent in 1993. However the absolute number of persons below the poverty line has remained more or less the same in the last two decades, i.e., around 320 millions, basically because of the pace of population growth during the period".

Pressure on natural resources like forests would reduce appreciably through generating employment opportunities in agriculture and other fields by poverty alleviation. Vast forest are easily accessible to the poor forest fringe dwellers for collection of fuelwood, fodder, food and minor forest produce.

That the growing population and heavy livestock population are linked to forest depletion and degradation resulting in biodiversity loss, which are clear from the fact presented in various tables.

India's forest resources have been declining for decades, while population has continued to increase. A Herculian effort on both forests population programmes and more holistic forest conservation schemes are needed now more than ever.

Forest Survey of India regularly monitor forest cover situation of the country. The position as in 1997 is presented in a table (Table 6). It shows the country's forest area to be 19 percent of total land-area. Actual situation of ground flora (obviously of medicinal herbs/shrubs) can not be ascertined from this figure. Open forests have hardly a good cover of herbs/shrubs/climbers. Close forests (Dense forests) which are also grazed and fire damaged have in many areas a thin cover of medicinal herbs/shrubs. While inventory of density and freqency of tree crop has been done by Forest Survey of India but no such assessment has been done in respect of herbs/shrubs.

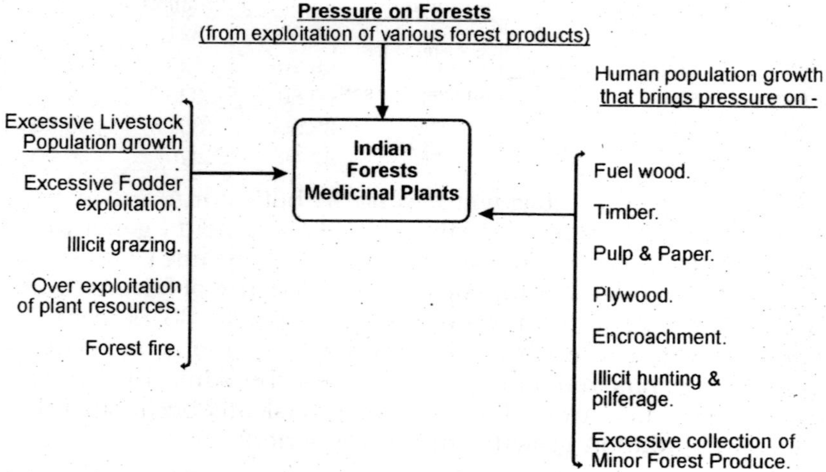

CHART - 23

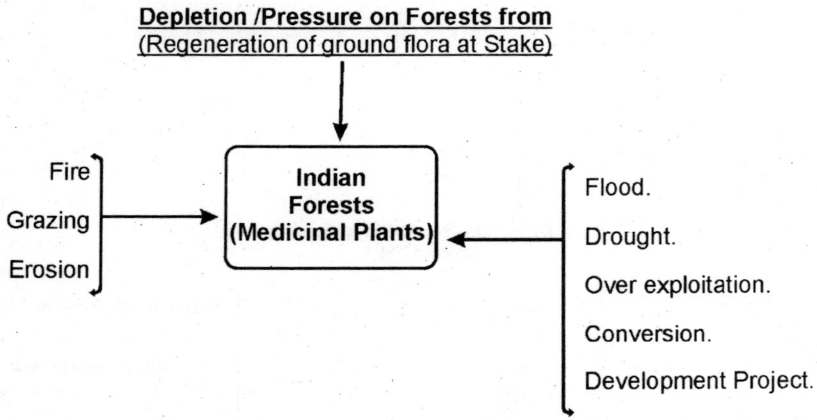

CHART - 24

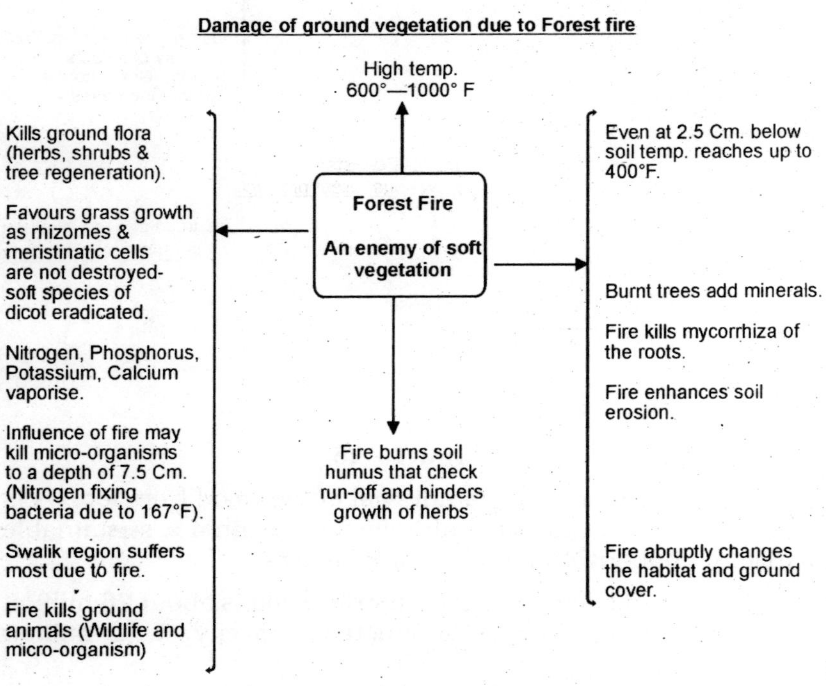

CHART - 25.

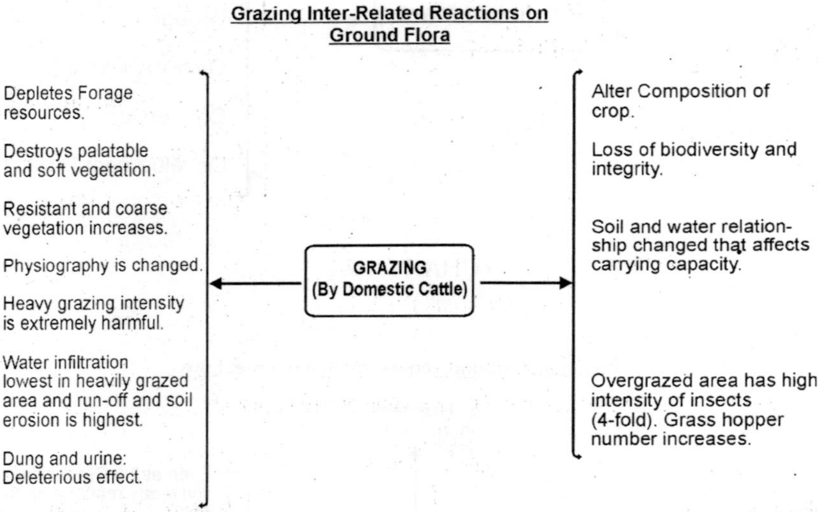

CHART - 26

"The current annual withdrawal of fuelwood from forests is estimated at 235 million Cubic meters, against a sustainable capacity of about 48 million Cubic meters".

"The annual demand for industrial wood is about 28 million Cubic meters against the production capacity of 12 million cubic meters".

"The area affected by forest fire ranges from 33% in West Bengal to 99% in Manipur".

Table 6. Forest Cover Area and Composition in 1997 (Square Kilometers)

State	Geographical Area	Recorded Forest Area	Forest Cover Area	Dense Forest Area	Open Forest Area	Mangroves	% of covered by Forest	Forest Area per Capita (Hectares)
Andhra Pradesh	275,068	63,814	43,290	23,048	19,859	383	16	0.07
Assam	78,438	30,708	23,824	15,548	8,276	0	30	0.11
Bihar	173877	29,226	26,524	13,300	13,224	0	15	0.03
Goa	3,814	1,424	1,255	995	255	5	34	0.11
Gujarat	196024	19,393	12,578	6,337	5,250	991	6	0.03
Haryana	44212	1,673	604	370	234	0	1	0.00
Himachal Pradesh	55,673	35,407	12,521	9,560	2,961	0	23	0.24
J & K	222235	20,182	20,440	11,020	9,420	0	9	0.26
Karnataka	191791	38,724	32,403	24,854	7,546	3	17	0.07
Kerala	38,861	11,221	10,334	8,454	1,880	0	27	0.04
M.P.	443446	154,497	131,195	82,745	48,450	0	30	0.20
Maharashtra	307690	63,842	46,143	23,622	22,397	124	15	0.06
Manipur	22,237	15,154	17,418	4,937	12,481	0	78	0.95
Meghalaya	22,429	9,496	15,657	4,044	11,613	0	70	0.88
Mizoram	21,081	15,935	18,775	4,348	14,427	0	89	2.72
Nagaland	16,579	8,629	14,221	3,487	10,734	0	86	1.18
Orissa	155707	57,184	46,941	26,101	20,629	211	30	0.15
Punjab	50,362	2,901	1,387	511	876	0	3	0.01
Rajasthan	342239	31,700	13,353	3,690	9,663	0	4	0.03
Sikkim	7,096	2,650	3,129	2,423	706	0	44	0.77
Tamil Nadu	130058	22,628	17,064	8,676	8,367	21	13	0.03
Tripura	10,486	6,292	5,546	1,819	3,737	0	53	0.20
U.P.	294411	51,663	33,994	22,958	11,036	0	12	0.02
W.B.	88,752	11,879	8,349	3,557	2,669	2123	9	0.01
Arunachal Pradesh	83,743	51,540	68,602	54,155	14,447	0	82	7.93
India	3,287,263	765,209	633,397	367,260	261,310	4827	19	0.07

2.3 LOSS OF HABITAT

Table 7 gives an arithmetic picture of the habitat loss in Indo-Malayan regions. The situation in India, Nepal, Bhutan and Bangladesh gives a picture of the trend of forest loss in the Indian Sub-continent.

Table 7. Wildlife habitat loss in Indomalayan Nations

	Original wildlife habita (km^2)	Balance remaining (km^2)	Habitat loss (%)
Bangaladesh	1,142,777	68,567	94
Bhutan	34,500	22,770	34
Brunei	5,764	4,381	24
Burma	774,817	225,981	71
China	423,066	164,095	61
Hong Kong	1,066	32	97
India	3,017,009	615,095	80
Indonesia	1,446,433	746,861	49
Japan	320	138	57
Kampuchea	180,879	43,411	76
Laos	236,746	68,656	71
Malaysia	356,254	210,190	41
Nepal	117,075	53,855	54
Pakistan	165,900	39,816	76
Phillipines	308,211	64,724	79
Sri Lanka	64,700	10,999	83
Taiwan	36,961	10,719	71
Thailand	507,267	130,039	74
Vietnam	332,116	66,423	80
TOTAL	8,169,860	2,487,683	68

[Source : Forest Ecosystem by Andrew Goudie : 5th Edition 1995]

2.4 BACKGROUND HISTORY OF AYURVEDA

A search for documents on background history of the origin and evolution of Ayurveda in India revealed that enormous amount of literature exists on the subject; several experts have summarised various facets of the subject from those and from innumerable recent publications spread over last sixty years.

The authors of this book has reproduced such findings sometimes in the language of the experts to give a lucid reading of this vast subject supplemented by their findings juxtaposed in next few pages.

In vedic and post vedic periods since 4500 B.C. some celebrated Indian Physicians e.g. Atreya, Mahabharathi, Nagarjun Sagar, Vagabhata, Sushruta, were legendary figures in Ayurvedic medicines, when the rest of the world was in darkness on the field.

To begin with, the treatises compiled by great sages charaka and Sushruta known as 'Charaka Samhita' and 'Sushruta Samhita' respectively are considered to be the most vital documents on Ayurveda. It is believed that Sushruta, son of Great saint Viswamitra, took lessons from Dhanyantari, the divine physician and after successful completion of the Courses Sushruta compiled a treatise which is known as 'Sushruta Samhita'.

Charaka primarily dealt with medicines and Sushruta on the other hand, was concerned with the advanced state of knowledge on the general principles and details of treatment. Their works are believed to have been compiled not later than 900 BC. But the Encyclopaedia Britanica recorded the period to be 800 BC to 100 AD.

Subsequently in the, 7th Century, A.D. 'Astango Hridaya Samhita' was written* by Vagabhata comprising of (1) Salyatantra (Major Surgery), (2) Shalakyatantra (Minor surgery), (3) Kayachikitsa (Science of medicine), (4) Bhutavidya (Psychiatry), (5) Kaumarabhritya (Paediatrics), (6) Agadatantra (Toxicology), (7) Rasayantantra (Chemistry of drugs), (8) Vajeekaranatantra (the Science of aphrodisiacs).

The period between 800 BC, and 1000 AD could be considered as the golden age of the Indian System of Medicine. Till the remote ancient times after Charaka & Sushruta many other sages and Ayurvedic physicians largely contributed to enrich Ayurveda. But prolonged muslim and British rule resulted in introduction of Arabic and Western medicines, causing not only the stagnation of Ayurvedic System of Treatment but also extinction of some of its systems.

Garcia D'orta, a Portugeese physician studied and practiced the Indian Indigenous system of medicine, who published a treatise on drugs used in Indian medicine on 1563 in Goa.

'Consideraing the efficacy of Indian System of medicine sir William Jones started compilation of available information in this regard in early 19th Century followed by Flemmings

N.B. : As per write up of K.P. Biswas, Rajbhatta instead of Vagbhata.

(1810), Ainslie (1813), Roxburgh (1834). Waring (1868) was first to publish Pharmacopoeia of India, supplemented by Mohidin Sheriff (1869) and later by Dymock, Warden and Hooper (1890-93). George Watt (1889-1904) published the Dictionary of Economic Products of India, which was updated by the Wealth of India series, published by the Scientific and Industrial Research (New Delhi).

During Muslim rule in India Muslin Hekims compiled books on Indigenous medicines in Arabian, Persian and Urdu languages. Amongst them Talip Sheriff and Makhazan-Ul-Adwiya deserve nention.

For enrichment of Ayurveda the contributions of Dr. Wallich, Dr. Wilson, Dr. Wight, Sir O'Shanghnessy, Dr. Birdwood, Dr. Dranoy, Dr. Baden Powel, Roy Bahadur Kanailal Dey, Sir George King, Dr. Voigt, Sir David Prain, Dr. Kirtikar, Major B.D. Basu, Dr. R.N. Chopra, Dr. Nadkarni, Kaviraj Birajcharan Gupta are to be remembered with gratitude.

Charaka divided his works into eight divisions e.g. (1) Sutra or Cloka, (2) Nidana, (3) Vanana (4) Carira, (5) Indriya, (6) Chikitsa, (7) Kalpa, (8) Siddhi.

Mention must also be made of Dravygunasamgraha compiled by Chakkrapanidatta, a renowned physician of the 11th Century, primarily based on the "Senior Triad". Chakrapani is not only considered as the commentator of Charaka, Sushruta and Vegbhata but also a pioneer in evolving the remarkable combined therapy with mineral and herbal products.

On this effort India was found to be self supporting in Agar, Cantharids, Colophony, Datura, Ephedra, Gentian, Glycyrrhiza, Hyocyamus, Lobelia, Oleum amygdalal, Oleum limonis, Oleum Terebinthinae, Psyllum, Rhuburb, Santonin, Scilla, Tragacanth etc.

Chopra mentions that indiscriminate exploration of plants like Rauwolfia, Belledona etc. has depleted natural growth. Some productive areas have gone over to Pakistan after Partition (Glycyrrhiza, Ephedra, Artemisia, Asafoetida).

Chopra has also made a wide range of survey on this subject. He took a realistic stock of the species under drugs and narcotics and found that hemp, tobacco, opium etc. occupy a comparatively small area. He felt that with the exception of cinchona and a few small experimental farms for other drugs in places like Kashmir, Saharanpur, Coimbatore and

Ootacamund very little was done for cultivation of medicinal plants. The cultivation of drug plants such as Digitalis, Belladona, Hyoscyamus Pyrethrum, Seena, Wattle, Derris, Geranium, Peppermint, tung oil has gained some importance. Several exotics like Datura innoxia, Duboisia sp. etc. were also tried successfully. Chopra feels a lot of progress has been made in drug cultivation and research and mentions the following species in particular-Holarrhena antidysenterica, Rauwolfia serpentina, Butea monosperma, Alstonia scholaris, Caesalpinia bondac, Justicia adhatoda, Daemia extensa, Cissampelos pareira, Terminalia arjuna, Psoralea corylifolia, Sida cordifolia, Swertia chirata, Andrographis paniculata, Plantago ovata and several other species.

India is currently at the crossroads between the efficacy of synthetic and natural compounds in the alleviation of human ailments. Despite the rapid strides on the Credit side made in the fields of chemotherapy and antibiotic therapy, there is much on the debit side if we are to combat many of the tropical diseases and their etiology and syndromes while allopathy has taken strong roots in most of the urban areas. The rural population of our country have faith in the efficacy and healing power of the age old systems of Ayurveda (indigenous system of medicine).

Another 'Triad' referred to as Laghu Trayi (Junior three) of Ayurveda consisted of Madhavanidana by Madhavakar (12th Century), Sharngadhara Samhita by Sharngadhara (14th Century) and Bhavaprakasha by Acharya Bhava Mishra (*Ca* 1550 AD) highly referred to even by the modern Ayurvedic practitioners. Among the large number of Nighantu Granthas or Pharmacy lexicons on medicinal herbs Rajanighantu by Narahari Pandit and Madanapalanighantu by Madanapala are the most outstanding ones.

The influence of Hindu medicine permeated far and wide into Egypt, Greece and Rome and moulded the Greek and Roman medicine and through former, Arabic medicine also.

The Romans took a great interest in Indian drugs. There is evidence to show that an extensive trade in Indian drugs existed between India and Rome many centuries ago. The country, with enormous variabilities of climate and with such wonderful ranges of mountains as the Himalayas, was from the earliest times recongised as a rich nursery of the vegetable meteria medica. In the days of Pliny, this drug traffic assumed such enormous proportions that he actually complained of heavy drainage of Roman gold to India in buying costly Indian drugs and spices. The following extract from the writing of an English student of original literature will be of great interest in this connection. In the course of a lecture, Captain Johnson Saint, mentioned the extra-ordinary advance made both in Surgery and medicine in India when Europe was groping for light in her cradle in Greece. Says he "If then this is what we found in surgery, what may we not find in medicine from India - that vast and fertile country which is a veritable encyclopaedia of the vegetable world. The materia medica of the ancient Hindus is a marvel from which both the Greeks and the Romans freely borrowed".

About 130 pure chemical substances, extracted from some 100 species of plants are used in medicines throughout the World. According to a 1994 U.N.D.P. report, the annual value of medicinal plants derived from developing countries is about $ 32 billion (about Rs. 1,00,000 Crores). Theoretically, there is the possibility of discovering 328 new modern drugs lying hidden in nearly 3,25,000 species in tropical rains forests.

Since Independence in 1947 India has made tremendous progress in agro-technology, process technology, standardisation, quality control, research and development etc. Leading Indian research centres have started to seek patentable genes using the latest genetic engineering techniques. About 400 useful plants have already been identified and further screening is on to identify other plants. The Thiruvananthapuram based Tropical Botanical Garden and Research Institute (TBGRI) has started the World's second gene prospecting programme to identify patentable genes from useful plants. Australia is the first country to start such a programme after the signing of the Bio-diversity Convention in 1992.

Indian Scientists of CSIR recognized laboratories and some other R&D centres were engaged in updating various technologies particularly agro and processing technologies with the result that the production of medicinal herbs and their derivatives like phytochemical, essential oils, oleoresins etc. became economically viable. Particular mention must be made of CIMPA in Lucknow; RRLS, Jammu, Jorehat, Bhubaneswar and Trivandrum; NCL in Pune and Private enterprises like CIPLA - Bombay, Amsar - Indore; Chemiloides - Vijayawada and Mehta Pharmaceuticals - Amritsar.

After the periods of Tantras and Sidhas the glories of the Hindu medicines waned and declined. During the invasion of

India by Greeks, Scythians and Mahamedans successively, no original works were written and the Hindu medicine gradually began to decay.

Dr. Ashima Chatterjee summarises this history in the following ways :

- Being the creation of nature, man's instinctive urge, intuition and accumulated knowledge guided him to discover remedies for common ailments from natural sources even in the remote past.
- The traditional system of medicine in India, including its therapeutic use in the form of plant extractives dates back to the early age of Rig Veda (4500-1600 B.C.). This led to evolution (2500-600 B.C.) of Ayurveda which literally means "Science of Life". According to the sages Bhaba Mishra and Charaka Ayurveda had been described as study of science, enabling man to ensure longivity, maintain good health and to provide relief from vaious ailments. Ayurveda was also considered to be the fifth Veda in addition to Rig, Yajus, Sama and Atharva.
- Charaka Samhita, the first recorded treatise of Ayurveda was followed by Sushruta Samhita, compiled a century apart, believed to be not later than 900 BC. But the Encyclopaedia Britanica recorded the period between 800 BC to 100 AD.
- Charaka dealt primarily with medicine, Sushruta, on the other hand, was concerned with the advanced state of knowledge on the general principles and details of treatment. Charaka Samhita is the edited version of the old Scientific treatise (Tantra) by Agnivesha, who wrote the first treatise on Ayurveda medicine, based on the teaching from his preceptor, Atreya, the great sage.
- Subsequently in the 7th Century, A.D. Astango Hridaya Samhita was written by Vagbhata and according to this Ayurveda comprised eight branches, e.g. (1) Shalyatantra (Major Surgery), (2) Shalakya - tantra (Minor Surgery), (3) Kayachikitsa (Science of medicine), (4) Bhutavidya (Pschiatry), (5) Kaumarabhritya (Paediatrics) (6) Agadatantra (Toxicology), (7) Rasayantantra (Chemistry of drugs), (8) Vajeekarantantra (The Science of aphrodisiacs).
- The period between 800 B.C. & 1000 A.D. could be considered as the golden age of the Indian system of medicine, particularly because of these three treatises, collectively known as Vriddha Trayi (Senior Triad) of Ayurveda though

the place of Kashyapa Samhita in Ayurveda cannot be ignored since it contained invaluable information about Kaumarabhritya Tantra i.e. Maternity and Child care.

- Dravyagunasamgraha compiled by Chakrapanidatta, a renowned physician of the 11th Century, primarily based on the "Senior Triad", who not only is considered as the commentator of Charaka, Sushruta and Vagbhata but also a pioneer in evolving the remarkable combined therapy with mineral and herbal products.
- Another "Triad" referred to as Laghu Trayi (Junior three) of Ayurveda consisted of Madhavanidana (dealing with diagnosis) by Madhavakar (12th Century), Sharnagadhara Samhita (materia medica) by Sharangadhara (14th Century) and Bhavaprakasha by Acharya Bhava Mishra (1550 AD) are highly referred to even by the modern Ayurvedic practitioners. Among the large number of Nighantu Granthas or Pharmacy lexicons on medicinal herbs, Rajanighantu by Madanapala is the most outstanding one. Incidentally, most of the slokas describing the medicinal properties of the plant-drugs discussed in all the volumes in the present series have been taken either from Bhavaprakasha or Rajanighantu.
- The details of the definite properties of drugs, prepared from indigenous plants (over 1250 in number) and their uses, the extensive materia medica which was further enriched by addition of a large number of newer drugs during the Buddhist period (that also witnessed the systematic cultivation and investigation of medicinal plants), and the methods of administration of drugs (including injections) akin to the present day practice clearly showed the extent of advancement of the Indian indigenous system of medicine as well as the depth of knowledge of the then practitioners regarding drug therapy and toxicology.
- Captain Johnson Saint, an English scholar described India as a "veritable encyclopaedia of the vegetable World and the materia medica of the ancient Hindus is a manual from which both Greeks & Romans freely borrowed".
- She presented a table indicating that in the remote ancient times the Indian materia medica could classify drugs based on their physiological actions etc.
- Garcia D'orta, a Portugese physician, studied and practiced the Indian Indigenous system of medicine, who published a treatise on drugs used in Indian medicine (1563) in Goa.

Some of the plants used in the 16th Century had been listed.

- Due to prolonged Muslim and British rule resulting in introduction of Arabic and Western medicines the Ayurvedic system of medicine not only remained static for long time but also some of it were lost.
- Considering the efficacy of Indian system of medicine Sir William Jones started compilation of available information in this regard in early 19th Century, followed by Flemmings (1810), Ainslie (1813) & Roxburgh-Waring (1868) were first to publish Pharmacopoeia of India, supplemented by Mohidin Sheriff (1864) and later by Dymock, Warden and Hooper (1890-93). George watt (1889-1904) published The Dictionary of Economic Products of India, which was updated by the Wealth of India Series, published by the Scientific and Industrial Research (New Delhi).
- Systematic and concerted scientific research was pioneered in India on indigenous plants by Dr. R.N. Chopra and his colleagues in the School of Topical Medicine, Calcutta.
- "Ayurvedic medicine", according to the Encyclopaedia Britanica (Macro, 23, 906, 1988) "is an example of a well-organised system of traditional health care, both preventive and curative. It is still a form of health care in large part of Eastern World, specially in India, where a large percentage of population use this system exclusively or combined with modern medicines".

Dr. K.P. Biswas & A. Ghosh in their work on "Bharatiya Bonousodhi" recorded the following :

- In the remote ancient time the Vedic sages were well conversant about the merits and demerits of medicines and "Atharva" veda is the brightest example of the same. Ayurveda was developed from the compilation by "Dhanyantari" from the Atharva veda. Subsequently great sages like Atreya, Varadwaj and Agnivesh as Ayurvedic teachers compiled some treatises on treatments. From all those compilations and treatises great sage Charaka compiled an invaluable treatise named as "Charaka Samhita" for the welfare of ailing common people.
- The Divine Physician Dhanyantari, it is said, was born in the Palace of Kashiraj and the Great Saint Viswamitra sent his son Sushruta to him to learn Ayurveda and after successful completion of the Course "Sushruta Samhita" was compiled by him.

- Both "Charaka Samhita" & "Sushruta Samhita" dealt with Surgery, physiology & anatomy, selection and application of medicines elaborately.
- Amongst ancient treatises- "Astanga Hriday Samhita" by Rajbhatta, treatise by Chakradatta, treatise by Sharangadhar, Vadhaprakash by Vabha Mishra, Rajanighantu by Madanapala, Compilation by Madhabankara & some other books on medicine have vividly described medicinal property and system of treatment.
- During Muslin rule in India Muslin Hekims compiled books on indigenous medicines in Arabian, Persian and Urdu languages. Amongst them Taleep Sheriff and Makhazan-Ul-Adwiya deserve special mention.
- In 16th & 17th Century A.D. Portugese and Dutch Physicians (botanists) compiled chronological history on merits & demerits of Indian medicines. In South India "Hortus Indicus Malabaricus" by Van Rheede had been published in 12 (twelve) volumes. Moreover great personalities like Thoms Rivers, O. Kerbosa, L. De Costa, Dr. Wight, Beddome and others compiled book on Indian plants of different areas at different times.
- During 18th & 19th Century A.D. eminent Botanist Dr. Roxburgh paved the path of modern Research after studying and publishing Indian system of medicine. He can very well be described as the father of Taxonomy in India. In his book Flora Indica he has discussed various Indian plants, their descriptions (Morphology), habitat and properties in great details.
- Subsequently in 1810 A.D. Dr. John Flemming elaborately published the Hindusthani & Sanskrit names of medicines in the magazine named "Asiatic Research". Dr. N. Wallich and Dr. Wilson, both botanist, have served the people by discovering many facts about the properties of indigenous medicines. Materia medica written by Dr. Ainslie, Mohideen Sheriff, Dr. Wight and Dr. Dymock are very important publications. "Pharmacopoia, Bengal" by Sir O'Shanghnessy & Dr. Wallich is a very rich compilation. Book on medicines of Maharastra (erstwhile Bombay Province), that of Timalnadu (earstwhile Madras), Punjab Products, and Indigenous Drugs written by Dr. Birdwood, Dr. Drury, Dr. Badan Powell and Roy Bahadur Kanailal Dey are very excellent medical books.

- Hortus Suberbans Calcuttensis, by Dr. Voigt, Flora of British India, by Sir J.D. Hooker and Dictionary of Economic Products by Sir George Watt, compiled during 1845, 1897 and 1904 A.D. respectively brought about a new era in regard to identification and use of indigenous medicinal and valuable plants.
- Books on numerous indigenous medicinal plants, and cultivation of Cinchona by Sir George King and Glossary of Hindu materia medica made the difficult identification of medicinal plants very easy. Bengal Plants, Trees of Hooghly, Trees of Howrah, Trees of 24-Parganas and that of Sunderbans, written by Sir David Prain contained indications and identifications of many medicinal plants.
- All the books and treatises compiled since Vedic age were written in Sanskrit, English and Latin and are beyond the reach of common man due to its exhorbitant cost. Moreover these are not suitable for the people not knowing the languages. The "Vanausadhi- Darpan" has been written in Bengali, containing less number of medicines and since it does not contain the photographic reprersentations of the plants it is not suitable for learners.

Due to great demand from all corners another book could be compiled, being initiated by Sir David Prain [Ex-Superintendent of Royal Botanic Garden, Calcutta & Kew] and shri Shyamaprasad Mukhopadhyay. This book was compiled in Bengali language, as desired by shri Shyamaprasad Mukhopadhyay for making it suitable for common people by Biswas and ghosh, This book provides a lot of information such as, scientific name of plants, English reference books, the places of availability, parts of plants utilised for preparation of medicines, medicinal properties and application for specific diseases etc. have been depicted as far as practicable with a sincere effort for better understanding of common people. Photographic representations of different plants have been furnished. This book contains descriptions of about 700 plants.

The Ayurvedic system of medicine in the concept of Nagendra Nath Sengupta.

- It would be no exaggeration to say that, all nations of the earth, the Hindus first turned their attention to the study of disease and the means of its alleviation. The Vedas are undoubtedly the most ancient of written records in the World. The 'Ayurveda' or the Science of Life is believed to have formed a part of the Vedas, viz., those that go by the name of the 'Atharvans'. No complete text is obtainable of the Atharva Veda. Hence it is impossible to ascertain how the Science of Life was treated in it. The first great professor of Medicine coming after the age of Atharva Veda with whom we are acquainted is Punarvasn of the race of Atreya, who was one of the will-begotten sons of Brahmin, the Granssire of the universe. It is not known by how many degrees was Punarvasn removed from Atri. It is impossible to fix his age, even roughly. It may even be doubted whether any such person existed. He is represented as having lectured on the Science of Life unto several disciples amongst whom six are mentioned by name. They are Agnivesa, Bhela, Jatukarna, Parasara, Harita and Kharapani, who became professors in time and compiled a treatise on Medicine each. Agnivesa is generally allowed to have been superior in point of merit, whose compilations superseded the compilations of the others. At present, fragments only exist of those compilations.

Dr. Sibakali Bhattacharyya in his book "Chiranjib Banousadhi" which is a masterpiece compilation on information on medicinal plants of India, has discussed the remedial properties of plant species and their parts for more than 800 diseases. He has also listed more than 3000 local names of medicinal plants of various states of the country. It is one of the most informative book on medicinal plants.

K.M. Nadkarni in his bood 'Indian Materia Medica' describes remedial properties of about 550 plant species. Of these, 90 are tree species, 28 are climber species, 35 shrub species and the rest 397 herb species.

2.5 STATUS REPORT : MEDICINAL HERBS

Of all lifeforms of plants in India, the number of herb species outnember others. Roughly, the number of shrub species is about double the number of Tree species, while the nember of herbs is about four times the shrubs.

Herb has been defined as "Plants with no woody stems above the ground. "Others define it as "A plant the stem of which dies every year" or as, "A plant with no persistent parts above the ground, as distinct from shrubs and trees". It has been defined in many other books ranking it with weeds which are not valued for beauty, that grow wild and hinder the growth of superior species, it has been found to grow where it is not desired and where it grows luxuriantly and plentifully. But little about their values as medicinal plants have been discussed.

No plants are weeds. Each species, has a part to play, either singly or as a member of the community. Each maintains a mutual harmony in it's niche and each niche plays a positive role in moulding the environement. Some herbs may be troublesome, most aggressive, harmful or annoying to man or to his agriculture and are termed as weeds. The herbaceous weeds grow luxuriantly; they are resistant, have a high reproductive capacity, gregarious, have a high level of seed dormancy, are quick growing, often poisonous, deep rooted, annual or perennial and abundantly seed bearing. Several herbs are parasitic, some are hosts of fungal and baterial diseases, some are hosts of nematodes and some are associated with insects. Many are delicious food for animals and man alike; many are useful as medicinal plants. Some weeds are guardians of soil. They provide protective cover against erosive action of rain drops and run off. They are also used as mulch in cultivated land.

The herbs come in profusion in swamps, marshes and on other sites which accumulate water in rains. In permanent ponds, weeds also predominate. Aquatic life burst into activities and complete their life cycle within a short span. Succession of plants and animals goes on simultaneously on vertical and horizontal strata. Herbs (submerged, floating, emergent) growing in profusion have diversified animal periphyton of insects, amphipods, mites, snails, etc. The floral and faunal complex attract aquatic birds. The ponds, marshes and swamps become roosting sites and feeding ground for many species of birds. Heron feed on shallow water, grebe, cormorants and terns feed on fish in open water, the egrets, bitterns eat fish in shallow water, cranes and coots are omnivorous, rails, ibises, stilts and snipes probe around in mud, but various ducks eat seeds, roots and soft parts of weeds. In India the aquatic flora consist of 171 to 200 species of plants belonging to 90 genera and 39 families.

But of late the areas of water bodies have been found dwindling rapidly. Marshes and lowlying lands are being brought under human settlements.

Besides the aquatic sites, herbs also grow on road sides, rice field, dry cultivated land and in various types of forests. In spite of their wide range of occurrence relentless biotic pressure, flood and erosion have reduced the number of species and many have been threatened to the verge of extinction. Over-exploitation of medicinal, decorative and other commercial species has been responsible for shrinkage of many species.

Shifting cultivation has also led to the complete eradication of many species. Many herbs have also been over-exploited for fodder purpose.

The environmental role of trees, shrubs and herbs should be studied in conjuection with grasses, sedges, palms, epiphytes, parasites, etc. as all these lifeforms constitute inseparable components of the plant communities to mould the environment and enrich the medicinal plant rèsources.

Herbs form eighty percent of the medicianl plant species. They need meticulous protection. There are many diffused and inconspicuous herbs which have showy flowers. Many herbs grow in grasslands, many in tall reeds and many in alpine grasslands. Innumerable herbs growing in temperate hills have also showy flowers. Many species of canes, yams, lily, asparagus, etc. which form a number of medicinal plants are being over-exploited. There are many species of aroids besides wild banana,, ginger, turmeric, cardamon which grow over extensive areas and form conspicuous physiognomy of the forest floor. Such resources are being damaged by fire and grazing also.

The enumeration of the essential qualities of herb layer remains incomplete unless some mention are made of climbers, ferns and epiphytes.

Large woody climbers, another lifeform and a biotic association are remarkable because of their fantastic forms. They add complexities to the vertical structure of forests. There are, however, delicate climbers, shrubby climbers, woody climbers and stranglers apart from root climbers like some climbing palms. There are about six hundred climbers the half of which are shurbby climbers.

Ferns which have more than 500 species in India are another type of lifeform which influences the environment. Some are fodder, some provide material for interior decoration, some protect soil and some are essential ingredients for landscape planning. They grow in tropical and temperate forests both in horizontal and in vertical strata. Some limited species are useful as medicinal plants.

Epiphytes, however, are another lifeform that derive support from host plant. Some have remarkable adaptation to hold water. They consume water in a restricted way and many conserve water in their tubers. Base of overlapping leaves of some ferns forms a special niche where soil, water and humus are held and nitrogen accumulates. Arboreal ants colonise, lay their eggs and drag seeds that grow in the niche. Insectivorous birds gather to eat ants and eggs; birds also gather and gradually a faunal association grows. Epiphytes are chief components of the vertical structure of our forests.

The trees, shrubs and herbs, the Trinity of environment yield substantial quality of meidcinal drugs besides protecting the environment. No amount of administrative and legislative and protective measures to save the country from environmental disaster would be effective and lasting without creation of a dense cover of vegetation all over the country.

Medicinal Herbs and their conservation

Lists of medicinal (herbs, shrubs, climbers and trees) have been prepared from the view point of their conservation status. The lists must not be taken as checklists of plants of India for only a few and common species have been selected. The readers are requested to refer to Chaudhuri's work on herbs, shrubs, trees and climbers of India mentioned under reference which gives a present day view of the plants of India in general and Eastern India in particular.

Conservation related status are given categoriwise (Numbers indicate status as detailed below):

- Widespread and presently not threatened (Spordie) (S)
- Occur at selected spots and cultivated (C)
- Occur sporadically and require protection (presently safe) (P.S).
- Depleted (D)

The study of vegetation reveals that more herbs occur in waste places and in marginal lands and along the edges of various forest types. Evergreen, semi-evergreen wet type of forests have comparatively few dicot species than monocot in terms of numbers of individuals.

Situation of herb in India

A study undertaken by the authors reveals the following :

(i) Few herbs have a wide range of distribution.

(ii) Most of the herbs are confined to forested areas and/or are sporadic in distribution.

(iii) Gregariously occurring herbs are few.

(iv) Moist localities have more number of species than drier and arid localities.

(v) Temperate locations have more of gregarious herbs.

Statewise Common and Conspicuous Herb Flora of India

(* Sign preceeding botanical name by species indicates very useful medicinal plants)

Andhra Pradesh

Alysicarpus sp.
Crotalaria willdenowiana
Vernonia cinera
Ipomoea pes-caprae
Indigofera anneaphylla
* *Crotalaria hirta*
Xanthium indicum
Tephrosia tenuis
Indigofera cordifolia
Acanthospermum hispidium
Atylosia sp.
Tephrosia purpurea
Alysicarpus monilifer
Indigofera aspalathoides
* *Boerhavia* sp.
* *Achyranthes aspera*
Cassia occidentalis
Crotalaria prostata
* *Boerhavia diffusa*
Celosia argentea

Assam

Ageratum conyzoides
Inula cappa
Mikania micrantha
Alpinia nutans
* *Centella asiatica*
Melastoma malabathricum
Ipomoea sp.

Andamans

Acanthus ilicifolius
Blumea virens
Desmodium triflorum
Jatropha curcas
Cucurbita sp.
Cassia tora
Centella asiatica

Bihar

* *Cassia occidentalis*
* *Calotropis procera*
* *Cassia tora*
* *Abutilon indicum*
Triadax procumbens
* *Desmodium triflorum*
Croton bonplandianum
* *Amaranthus spinosus*
* *Abrus precatorius*
Euphorbia hirta
Asclepias pseudosansa
* *Atylosia scaraboeoides*
Vernonia cinerea
* *Leucas aspara*
* *Alysicarpus* sp.
* *Tephrosia purpurea*
Barleria cristata
* *Atylosia* sp.
Argemone maxicana
Borreria articulata
Corchorus trilocularis
* *Alysicarpus monilifer*
Indigofera linifolia
Mucuna sp.

Delhi

* *Achyranthes aspera*
* *Casia occidentalis*
* *Crotalaria medicaginea*
* *Indigogera cordifolia*
Rhynchosia capitata
Ipomoea hispida
* *Tephrosia tenuis*
Ageratum conyzoides
* *C.tora*
Digera muricata
* *Tephrosia purpurea*
Eclipta prostata
Borreria articulata
* *Justica diffusa*

Gujarat

* *Alysicarpus longifolius*
Heylandia latebrosa
Clemome viscosa
Indigofera cordifolia
Vicoa indica
* *Tridax procumbens*
Crotalaria medicaginea
Trichodesma indicus
Enicostema verticillatum
* *Alysicarpus* sp.
Digera muricata
Rhynchosia capitata
* *Eclipta procumbens*
Borreria articulata
* *Barleria cristata*
Ipomoea sp.

Ipomea pes-caprae
Euphorbia thymifolia
* *Zornia diphylla*
* *Crotalaria burhia*
Euphorbia hypericifolia
Rhynchosia sp.
Commelina benghalensis
Vernonia cinerea
Boerhavia diffusa
Cardiospermum halicacabum
Tephrosia purpurea
Cressa cretica
Phynchosia halicacabum
* *Phaseolus aconitifolius*
Evolvulus alsionoides
* *Clitoria ternatea*
Aerva sp.

Himachal Pradesh

* *Atylosia scarabaeoides*
* *Cassia tora*
* *Crotalaria* sp.
Euphorbia hirta

Jammu & Kashmir

* *Cassia tora*
Euphorbia jorta
* *Tephrosia purpurea*

Kerala

* *Mimosa pudica*
Sphaeranthes aspera
* *Crotalaria alata*
* *Desmodium floribundum*
* *Cassia tora*
* *Phaseolus* sp.
* *Desmodium triflorum*
* *Achyranthes aspera*
* *Atylosia scarabaeoides*
* *Moghonia chappar*
* *Crotalaria* sp.
Sesbania aculeata

Karnataka

* *Alysicarpus vaginalis*
* *Indigorera linifolia*
Euphorbia hirta
* *Indigofera enneaphylla*
* *Clitoria ternatea*
* *Desmodium diffusum*
* *Tephrosia purpurea*
* *Plectranthus* sp.
* *Tephrosia tinctoria*
* *Pueraria phaseoloides*

Madhya Pradesh

Trimufetta bartramia
* *Alysicarpus vaginalis*
* *Xanthium indicum*
Caesulia axillaris
* *Sesbania bispinosa*
* *Moghania* sp.
* *Zornia diphylla*
Euphorbia hirta
* *Alysicarpus rugosus*
* *Corchorus trilocularis*
* *Urena lobata*
* *Desmodium triflorum*
* *Centella siatica*
* *Phaseolus radiatus*
* *Alysicarpus tetragonolobus*
* *Cassia tora*
* *Atylosia* sp.
* *Indigofera cordifolia*
* *Borreria articulata*
* *Indigofera linifolia*

Maharashtra

* *Crotalaria retusa*
* *Alysicarpus vaginalis*
* *Indigofera cordifolia*
Phaseolus trilobus
* *Tephrosia purpurea*
* *Amaranthus* sp.
Xanthium indicum
* *Desmodium triflorum*
* *Crotalaria prostata*
Heylandia latebrosa
* *Indigofera glandilosa*
* *Atylosia* sp.
* *C. burhia*
* *Alysicarpus longifolius*
* *Crotalaria juncea*
* *Argemone mexicana*
* *Indigofera hebepetala*
* *Aerva lanata*
* *Cassia tora*
* *I. linifolia*
Polygala erioptera
Tephrosia tenuis
* *Desmodium diffusum*
* *Crotalaria prostata*
Heylandia latebrosa
* *Desmodium triflorum*
* *rotalaria sericea*
* *Achyranthea aspera*
* *Crotalaria alata*
* *Alysicarpus rugosua*
* *Sesbania bispinosa*
* *Atylosia scarabaeoides*
* *Crotlaria vestita*
* *Alysicarpus tetragonolobus*

Manipur

* * *Crotalaris* sp.
* * *Desmodium* sp.
* * *C. medicaginea*
* * *Indigofera enneaphylla*

Orissa

* * *Cassia occidentalis*
* * *Indigofera enneaphylla*
* * *Cassia tora*
* * *Crotalaria* sp.
* *Ageratum conyzoides*
* * *Acyranthes aspera*
* *Croton bonplandianum*
* * *Desmodium triflorum*
* * *Tridax procumbens*
* *Triumfetta bartramia*
* *Eranthemum purpurascens*
* * *Alysicarpus vaginalis*

Punjab

* * *Achyranthes aspera*
* *Rhynchosia capitata*
* * *Vicoa indica*
* * *Boerhavia repens*
* *Xanthium indicum*
* * *Atylosia scarabaeoides*
* *Chenopodium album*
* *Trigonella polycerata*
* * *Indigofera cordifolia*
* * *Alysicarpus* sp.
* * *Indigofera linifolia*
* * *Casia tora*
* * *Crotalaria medicaginea*
* * *Tephrosia purpurea*
* *Vernonia cinerea*
* *Euphorbia hirta*
* * *Artemisia* sp.
* * *I. linifolia*

Rajasthan

* * *Achyranthes aspera*
* * *Cassia tora*
* * *Indigofera enneaphylla*
* * *I. cordifolia*
* * *T. tenuis*
* * *Amaranthus blitum*
* * *C. albida*
* * *Alysicarpus vaginalis*
* * *Alysicarpus longifolia*
* * *Crotalaria medicaginea*
* * *I. linifolia*
* * *Tephrosia purpurea*
* * *Xanthium indicum*
* * *Crotalaria burhia*
* * *Eclipta alba*
* *Argemone mexicana*

* *Boerhavia repens*
* *Tridax procumbens*
* *Aerva tomentosa*
Heylandia latebrosa
* *Boerhavia diffusa*
* *Mimosa pudica*
Tribulus terrestris
* *Alternanthera echinata*
* *Crotalaria filipes*
Justicia diffusa
* *Crotalaria umbellata*
* *Phaseolus radiatus*
Dicoma tomentosa
* *Zorina diphylla*
* *Desmodium rotundifolium*

Tamil Nadu

Rhynochosia Psuedo-cajan
Xanthium indicum
* *I. linifolia*
* *Atylosia scarbeoides*
* *Tephrosia purpurea*
* *Alysicarpus vaginalis*
Euphorbia thymifolia
* *Indigofera trifoliata*
* *Atylosia* sp.
* *Indigofera viscosa*
* *Mimosa pudica*
Euphorbia trigona
* *Indigofera trigonelloides*
Stachytrapheta indica
Ageratum conzyzoides
Rhynchosia rufescens
* *Tephrosia* sp.
* *Indigofera enneaphylla*
Stylosanthes sp.
* *Desmodium floribundum*
Stylosanthus fruticosa
Croton banpladianum
* *Leucas hyssopifolia*
* *Achyranthes aspera*
* *Crotalaria prostrata*
* *Pogostemon mollis*
Polygala crotalarioides
* *Alysicarpus monilifer*
Mollugo sp.
* *Jatropha curcas*
* *Sida acutangula*

Tripura

Carex condensata
* *Desmodium trifolium*
* *Atylosia scarabaeoides*
* *Hyptis brevipes*
* *Ocimum sanctum*
* *Tridax procumbens*

Uttar Pradesh

* Alysicarpus sp.
* Euphorbia hirta
* Phyllanthus maderaspatensis
* Vicoa indica

Heylandia latebrosa

* Tephrosia villosa
* Crotalaria medicaginea

Vernonia cinerea

* Zornia diphylla
* Aerva lanata

Boerhavia repens

Cassia occidentalis

Spiraea canescens

Stellaria decumbens

* Polygonum cirrhifolium
* Cannabis sativa
* Leucas aspera
* Atylosia sp.
* Atylosia sp.
* I. linifolia
* Cassia tora
* Desmodium triflorum
* Tridax procumbens
* Xanthium indicum
* Alysicarpus monilifer

Justicia diffusa

* Chenopodium sp.

Stellaria sp.

* Indigofera heterantha

West Bengal

* Centella asiatica
* Leucas aspera

Argemone mexicana

* Cassia tora

Croton bonplandianum

Digera muricata

* Leucas aspera
* Desmodium triflorum

Borreria articulata

* Tephrosia purpurea
* Acanthus ilicifolius
* Cassia occidentalis
* Tephrosia villosa
* Achryanthes aspera
* Polygonum sp.
* Canabis sativa
* Moghania sp.

Vernonia cinerea

* Atylosia scarabaeoides
* Clitoria ternatea
* Datura metel

Euphorbia hirta

Mikania micrantha

* Alysicarpus monilifer
* Vicoa indica
* Tridax procumbens
* Sesuvium portulacastrum

Phyla nodiflora

Ipomoea pes-caprae

Triflorum repens

Inula cappa

* Pueraria sp.

Mapping of Herbs

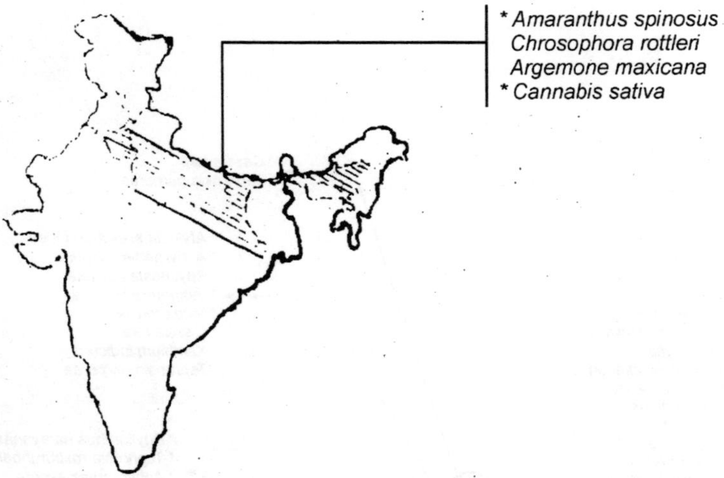

Environment and Ecology of Herb-Shrub Flora

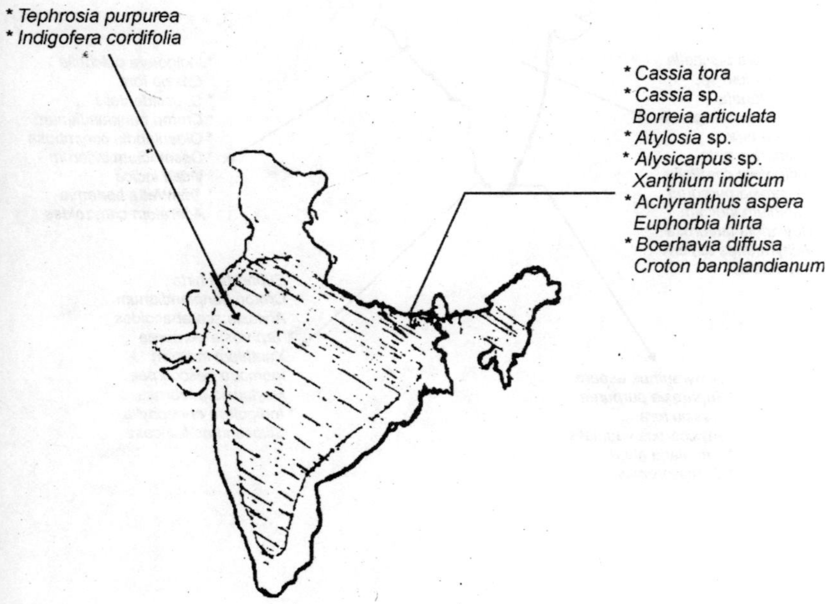

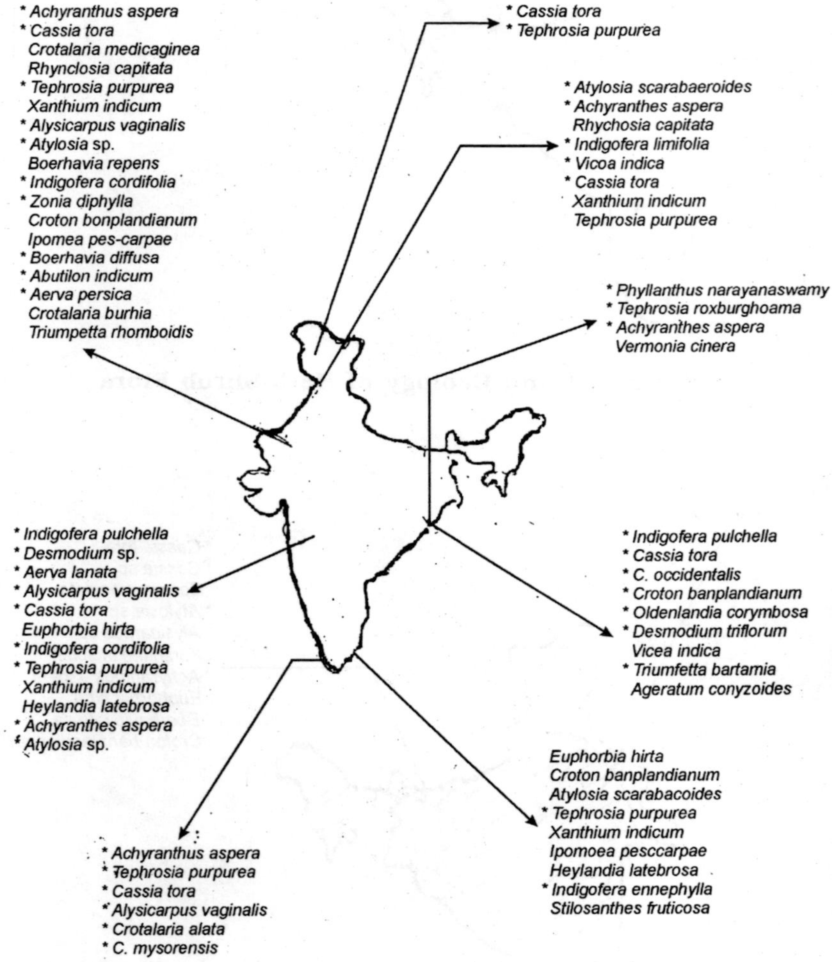
* Achyranthus aspera
* Cassia tora
Crotalaria medicaginea
Rhynclosia capitata
* Tephrosia purpurea
Xanthium indicum
* Alysicarpus vaginalis
* Atylosia sp.
Boerhavia repens
* Indigofera cordifolia
* Zonia diphylla
Croton bonplandianum
Ipomea pes-carpae
* Boerhavia diffusa
* Abutilon indicum
* Aerva persica
Crotalaria burhia
Triumpetta rhomboidis
* Cassia tora
* Tephrosia purpurea
* Atylosia scarabaeroides
* Achyranthes aspera
Rhychosia capitata
* Indigofera limifolia
* Vicoa indica
* Cassia tora
Xanthium indicum
Tephrosia purpurea
* Phyllanthus narayanaswamy
* Tephrosia roxburghoama
* Achyranthes aspera
Vermonia cinera
* Indigofera pulchella
* Desmodium sp.
* Aerva lanata
* Alysicarpus vaginalis
* Cassia tora
Euphorbia hirta
* Indigofera cordifolia
* Tephrosia purpurea
Xanthium indicum
Heylandia latebrosa
* Achyranthes aspera
* Atylosia sp.
* Indigofera pulchella
* Cassia tora
* C. occidentalis
* Croton banplandianum
* Oldenlandia corymbosa
* Desmodium triflorum
Vicea indica
* Triumfetta bartamia
Ageratum conyzoides
Euphorbia hirta
Croton banplandianum
Atylosia scarabacoides
* Tephrosia purpurea
Xanthium indicum
Ipomoea pesccarpae
Heylandia latebrosa
* Indigofera ennephylla
Stilosanthes fruticosa
* Achyranthus aspera
* Tephrosia purpurea
* Cassia tora
* Alysicarpus vaginalis
* Crotalaria alata
* C. mysorensis

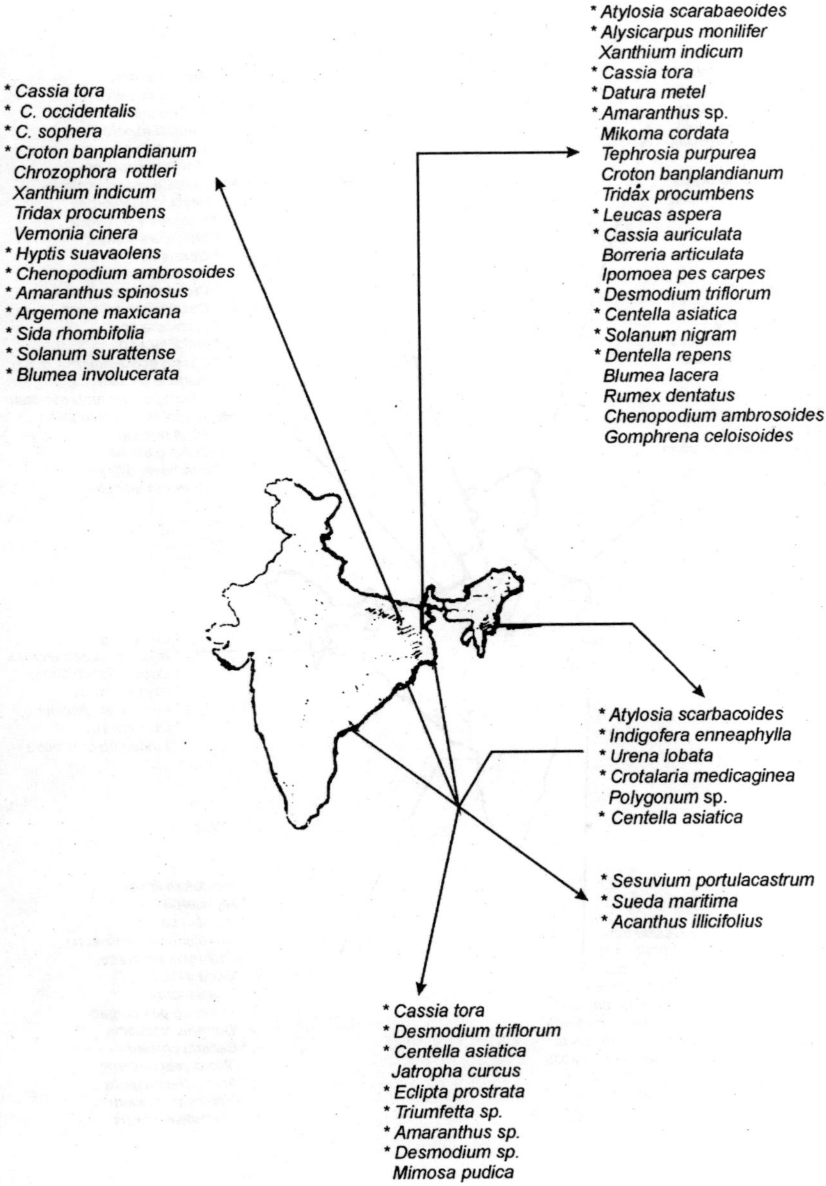
* Atylosia scarabaeoides
* Alysicarpus monilifer
Xanthium indicum
* Cassia tora
* Datura metel
* Amaranthus sp.
Mikoma cordata
Tephrosia purpurea
Croton banplandianum
Tridax procumbens
* Leucas aspera
* Cassia auriculata
Borreria articulata
Ipomoea pes carpes
* Desmodium triflorum
* Centella asiatica
* Solanum nigram
* Dentella repens
Blumea lacera
Rumex dentatus
Chenopodium ambrosoides
Gomphrena celoisoides
* Cassia tora
* C. occidentalis
* C. sophera
* Croton banplandianum
Chrozophora rottleri
Xanthium indicum
Tridax procumbens
Vernonia cinera
* Hyptis suavaolens
* Chenopodium ambrosoides
* Amaranthus spinosus
* Argemone maxicana
* Sida rhombifolia
* Solanum surattense
* Blumea involucerata
* Atylosia scarbacoides
* Indigofera enneaphylla
* Urena lobata
* Crotalaria medicaginea
Polygonum sp.
* Centella asiatica
* Sesuvium portulacastrum
* Sueda maritima
* Acanthus illicifolius
* Cassia tora
* Desmodium triflorum
* Centella asiatica
Jatropha curcus
* Eclipta prostrata
* Triumfetta sp.
* Amaranthus sp.
* Desmodium sp.
Mimosa pudica

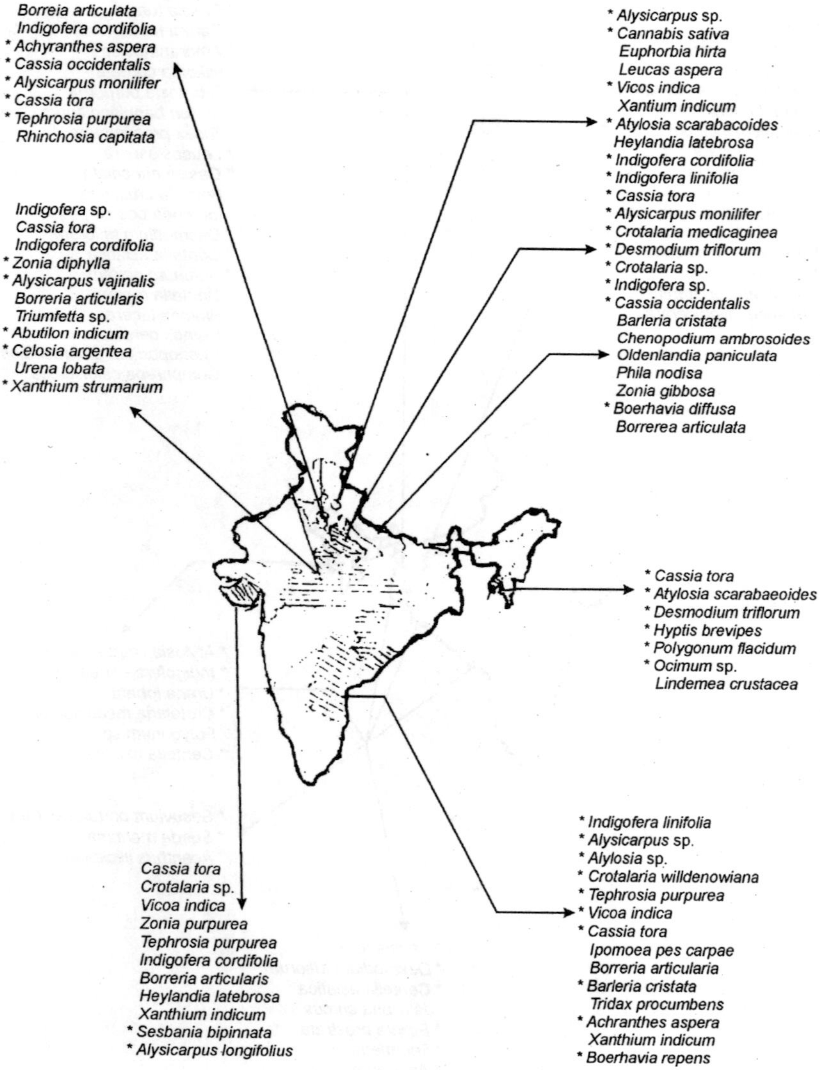
Borreia articulata
Indigofera cordifolia
* Achyranthes aspera
* Cassia occidentalis
* Alysicarpus monilifer
* Cassia tora
* Tephrosia purpurea
Rhinchosia capitata
Indigofera sp.
Cassia tora
Indigofera cordifolia
* Zonia diphylla
* Alysicarpus vajinalis
Borreria articularis
Triumfetta sp.
* Abutilon indicum
* Celosia argentea
Urena lobata
* Xanthium strumarium
* Alysicarpus sp.
* Cannabis sativa
Euphorbia hirta
Leucas aspera
* Vicos indica
Xantium indicum
* Atylosia scarabacoides
Heylandia latebrosa
* Indigofera cordifolia
* Indigofera linifolia
* Cassia tora
* Alysicarpus monilifer
* Crotalaria medicaginea
* Desmodium triflorum
* Crotalaria sp.
* Indigofera sp.
* Cassia occidentalis
Barleria cristata
Chenopodium ambrosoides
Oldenlandia paniculata
Phila nodisa
Zonia gibbosa
* Boerhavia diffusa
Borrerea articulata
* Cassia tora
* Atylosia scarabaeoides
* Desmodium triflorum
* Hyptis brevipes
* Polygonum flacidum
* Ocimum sp.
Lindemea crustacea
Cassia tora
Crotalaria sp.
Vicoa indica
Zonia purpurea
Tephrosia purpurea
Indigofera cordifolia
Borreria articularis
Heylandia latebrosa
Xanthium indicum
* Sesbania bipinnata
* Alysicarpus longifolius
* Indigofera linifolia
* Alysicarpus sp.
* Alylosia sp.
* Crotalaria willdenowiana
* Tephrosia purpurea
* Vicoa indica
* Cassia tora
Ipomoea pes carpae
Borreria articularia
* Barleria cristata
Tridax procumbens
* Achranthes aspera
Xanthium indicum
* Boerhavia repens

Some habitat specific medicinal plants

(a) Specific selected sites in plains

* *Aerva lanata* (Chaya)
* *Chenopodium album* (Bestak)
* *Chenopodium ambrosoides* (Beto sag)
* *Alocasia indica* (Mankachu)
* *Amorphophallus campanulatus* (01)
* *Commelina bengalensis* (Kanachira)
* *Acanthus ilicifolius* (Ḥaraduchkata)
* *Andrographis paniculata* (Kalmegh)
* *Asterocantha longifolia* (Khirok).

(b) Temperate areas

* *Oxalis corniculata*	*Impatiens* sp.
Rubus sp.	*Potentilla* sp.
* *Embelia ribes*	*Maesa indica*
Calamintha sp.	* *Brunella vulgaris*
* *Achyrathes bidentata*	* *Blepharis edulis* (Ulangan)
Clematis nepalensis (Churaha)	*Clematis triloba*
Clematis smilacifolia	*Clematis gouriana*
Thalictrum foliolosum (Mamira)	*Corydalis govaniana* (Bhutkishi)
Nasturtium palustre	*Viola serpens.*
Viola sp.	*Geranium* sp.
* *Plantago ovata* (Isabgul)	* *Plantago major*
Verbascum thapsus	* *Rubia cordifolia* (Monjista)
* *Polygonum molle.*	* *Polygonum chinense*
* *Coptis teeta* (Mismitita)	*Delphinium denudatum* (Nirbishi) (rare)
Climicifuga foetida (Jiunti) (rare)	* *Paeonia emodi* (udsalap) (rare)
Illicium griffithii	* *Digitalis purpurea* (rare)

* *Swertia chirata* (rare)
* *Solanum dalcamara* (rare)
* *Frilillaria* sp. (rare)
Geranium nepalense
Viscum sp.
* *Lilium gigantium* (rare)

(c) Alpine areas (Threatened)

* *Anemone obtusiloba* (Ralanyog)
* *Aconitum palmatum* (Alibisa)
* *Aconitum ferox*
* *Genum* sp.
* *Hyoscyamus niger*
* *Picorhiza kurroa*
* *Caltha palustris*
* *Aconitum heterophyllum*
* *Aconitum* sp. (10 sps.)
* *Gaultheria fragrantissina*
* *Saussuria lappa.*

(d) Tropical areas

Tephrosia purpurea
* *Curcuma zedoaria* (Ekangi)
Curcuma aromatica (Halud)
Polygonum plebejum
Polygonum barbatum
Polygonum orientale
Indigofera pulchella
Indigofera trifolia
Desmodium gangeticum
Desmodium Pulchellum
Crotalaria verrucosum
Crotalaria albida
Alysicarpus longifolius
* *Uraria jogopoides* (Chakulia)
Costus speciosus
Alpinia allughus
Tacca pinnatifida
Pouzolzia indica
Curcuma angustifolia
Tephrosia villosa
* *Curcuma amada*
* *Cannabis sativa*
Polygonum glabrum
Polygonum hydropiper
* *Curcuma longa* (cultivated)
Indigofera linifolia
Indigofera Sp.
Desmodium polycarpum
Desmodium Sp.
Crotalaria juncea
Crotalaria Sp.
* *Uraria picta* (Sankajota)
* *Psoralia corylifolia* (Buchki)
* *Alpinia galanga*
* *Curculigo orchioides*
Tragia involucrata
* *Sanseveria roxburghiana*
* *Zingiber officinale* (cultivated)

Zingiber cassumaner
Pothos scandens
Amomum subulatum (cultivated)
Rauwfolia serpentina (rare)
Commelina obliqua
Scindapsus officinalis (Rajpipal)
Elleteria cardamom (Cultivated)

(e) Grasses and Sedges

Hygroryza aristata
Polyloca barbata
Vetiveria zizanioides
Cynodon dactylon
Killyngia monocephala
Cyperus rotundus
Cyperus esculentus
Scirpus grossus
Coix lachryma Jobi
Saccaharum officinarum
Paspalum scrobiculatum
Imperata cylindrica
Fimbristylis fusciformis
Cyperus esculentus
Cyperus iria
Scirpus articulatus

Herbs : Occur sporadically

Chlorophytum arundinaceum
Commelina obliqua
Cyanotis sp.
Commelina bengalensis
Floscopa sp.

(f) Aquatic, Marsh and Semi-aquatic

* *Ipomea aquatica*
Pragmites karka (Nol, Mohanal).
Spinacia oleracea
Monochoria vaginalis (Nanka)
Pistia stratiotes
Trapa natans var. *bispinosa*
Articularia bifida
* *Aeschynomene aspera* (Shola)
Mentha spicata (Pudina)
* *Mentha piperta* (")
* *Enhydra fluctuans*
Vallisnaria spiralis
Ranunculus sceleratus (Jal-Dhania)
Alternanthera Sessilis
Typhonium trilobatum.
Jussiaea suffruticosa
Cardanthera uliginosa
* *Aeschynomene indica* (Shola)
* *Mentha arvensis* (")
Lassia heterophylla

Alocasia indica
Homoinia reparia
Nymphaea rubra
Nymphaea nouchali
Limanthemum cristatium
Nymphaea alba
Nymphaea pubescens
Ammania baccifera.

SPECIAL GROUPS OF PLANTS

Aromatic Plants

Mimusops elengii (Bakul) - Common
Vetiveria zizanoides (Vetiver) - Common .
Jasminum grandiflorum (Jasmine) - Sporadic
Pandanus odoratissima (Screw-pine). - Sporadic
Myristica fragrans (Nulmig) - Rare
Syzygium aromaticum (Clove) - Rare
Cinnamomum camphora (Camphor) - Rare
Aquilaria agallocha (Aloewood) - Rare
Santalum album (Sandle wood) - Very localised (Protected)

Cosmetic Plants (Sporadic, common)

Lawsonia inermis (Henna)
Sesamum indicum (Sesame)
Hibiscus rosa-sinensis (Hibiscus)
Areca catechu (Areca nut)
Sapindus trifoliatus (Soap-nut)
Curcuma longa (Turmeric)
Mallotus philippinensis (Indian Kamella)
Citrus indica (Lime)
Crocus sativus (Saffron)
Piper betel (Supari)
Indigofera tinctoria (True indigo)
Hemidesmus indicus (Sarsaparila).

Culinary Plants (Sporadic common)

Moringa Olifera (Drum stick)

Tamarrindus indica (Tamarind)
Allium sativum (Garlic)
Zingiber officinale (Ginger)
Cinnamomum zeylanicum (Chinamon)
Mangifera indica (Mango)
Cinnamomum tamala (Tejpata)
Borassus flabellifer (Palmyra)
Coriandrum sativum (Corriander)
Bassia latifolia (Mahua)
Nigella sativa (Kalijira)
Elettaria cardamomum (Cardamom)
Cuminum cyminum (Cumin)
Solanum melonginum (Egg plant)
Carum capticum (Joan)
Carum roxburghiana (Randhuni)
Foeniculum capillaceum (Mouri)
Ferula asafoetida (Hing)
Piper nigrum (Black piper).

Sacred Plants (Sporadic, mostly)

Ficus bengalensis (Benyon)
Ocimum sanctum (Holibasil)
Butea monosperma (Palash)
Cocos nucifera (Coconut)
Aegle marmelos (Boel)
Cannabis sativa (Indian hemp)
Ficus religiosa (Sacred fig)
Syzygium cuminii (Jamun)
Azadirachta indica (Margosa)
* *Nelumbium speciosum* (Lotus) * Sporadic to rare.
Elaeocarpus ganitrus (Rudrakshya)
Catheranthus roseus (Nayantara)

Herbs of Waste Places (Sporadic mostly)

Argemone mexicana (Sealkanta)

* *Cleome gynandra*
* *Polygala chinensis* (Muradu)
* *Polycarpea corymbosa*
* *Portulaca quadrifida* (Chota nunia)
* *Polygonum orientale*
* *Oxalis corniculata* (Amrul)
* *Biophytum sensetivum* *(Ban naranga)*

Ageratum conyzoides

* *Eclipta alba*
* *Sida ueronicoefolia* (Joka)
* *Sida cordifolia* (Barlela)
* *Urena sinucata*

Acalypha indica (Muktajhuri)

* *Evolvulus alsinoides*
* *Solanum nigram*
* *Solanum indicum*
* *Solanum torvum*

Physalis peruviana

* *Scoparia dulcis*
* *Andrographis paniculata* (Kalmegh)
* *Heliotropium indicum* (Srihaslini)

Zornia diphylla

Trianthema portulacastrum

* *Capsella bursapastoris*
* *Cleome isosandra*
* *Polygala crotalarioides* (Nilkanti)
* *Portulaca oleracea* (Barmunia)
* *Malvastrum* sp.
* *Sphaeranthus indica* (Mahashrabani)
* *Oxalis acetosella*
* *Martynia diandra* (Bagh nakhi)

Blumea lacera (Kuksim)

Taraxacum officinale

* *Sida rohmboidea* (Pila Barela)
* *Urena lobata* (Ban Okra)
* *Clitoria turnatea*
* *Centella asiatica*
* *Murremia emerginata*
* *Solanum ferox*
* *Solanum xanthocarpum*
* *Solanum surattense*

Withania somnifera

* *Picrorrhiza kuroa*
* *Cenopodium album*
* *Abutilon indicum* (Paleri)

Flemengia sp.

Oldenlendia corymbosa (Khetkapra)

Mussenda frondosa (Negballi)	* *Gangrea maderaspetane* (Namuti)
* *Loranthus* sp.	*Linum usitatissimum*
Astragalas sp.	*Lindenbergia urticaefolia* (Haldebasanta)
Toremia asiatica	*Vandellia* sp.
Justicia gandurussa (Jagat Madan)	*Rungia parviflora* (Piri)
Achyranthus aspera (Apang)	*Pogostemon parviflorus*
Boerhaevia diffusa (Punarnava)	*Amaranthus spinosus* (Kanta note)
Amaranthus gangeticus (note)	*Amaranthus viridis*
Chrosophora cinera (Tut kukusima)	*Xanthium strumerium* (Ban Okra)
Cannabis sativa	*Rumex dentatus*
Rumex maritimum	*Cyanotis* sp.
Phyllanthus reticulatus (Panjuli)	*Phyllanthus simplex*
Phyllanthus urinaria (Hazarmani)	*Phyllanthus nirusi*
Cenopodium ambrosioides	*Wedelia calendula* (Bhimraj)
Euphorbia hirta (Swadaparui)	*Malachra capitata*
Atylosia scarabioides	*Sophora* sp.
Mallugo hirta	*Oldenlendia herbacea*
Elephantopus scaber (Shaydalan)	
Celosia argentea	*Reinwardtia trigyna*
Gloriosa superba (Languli)	

2.6 STATUS REPORT : MEDICINAL SHRUBS

Shrubs are low woody plants or a bush with no trunks, no crown or foliage, often multistemmed and usually of a height less than 3 to 4 metres. They are mostly shade bearer and form an essential component of lifeform of vegetation community.

Various species of shrub with their intimate assemblages and forms are biologically more important. They provide an ideal cover to the terrestrial animals. The herbivores can graze comfortably on the fodder shrub because of their identical height. In a dense tree forest the shrubby layer contains about

50 percent of regeneration of tree species.

The degree and duration of the shade have a positive bearing on the composition and density of shrub crop. Heavy shade results in a very limited number of shrub flora. The intensity of light is concerned with photosynthesis and food production; it also affects flower and fruit production. Specific insect and avifaunal species are concerned with shrubs for food and pollination. At least eight shrub genera form an imperative source of larvae food of the moths and butterflies.

Over-exploitation of tree crops in all the three storeys cause considerable physical damage to shrubs; these shrubs are essential for soil binding and create a micro climate for the survival of young plant regeneration; besides, they help litter formation and conservation of water.

Shrubs and young regeneration of other plant species suffer irreparable losses due ot repeated grazing and fire. These latter two factors are directly responsible for eradication of plant species where only those with underground stem survive. The authors' research reveal that only about 40 to 50 percent of the species survive such depredation; also the density of occurrence of speices is affected. Vast tracts of forests where grazing and fire play have now very coarse grass and a few fire hardy shrub species.

About ninety species of shrubs are conspicuous in the temperate hills of India which have colourful flowers; they also grow in profusion. Special mention may be made of rhododendron and roses. In the plains, however, more than one hundred species occur but they do not have conspicuous flowers. In eastern India there are at least a hundred species of fodder shrub; it is about two thirds in other parts of the country. There are at least forty genera of shrubs yielding edible fruits which attract avifauna, animals and men alike.

A survey made by the authors all over the country has revealed that only fifty species of shrubs are commonly found, Lantana, Eupatorium, Clerodendrum, Calotropis, Cassia, Carissa, Capparis, etc. are found conspicuous and are occurring in profusion all over the country. It should be considered as an ominous indication of systematic disappearance of species primarily due to anthropogenic factors. The number of shrub species in the country is about 4500 (tree species about 1500); but wide occurrence of only 50 species is hardly to be believed which is too meagre.

Status of medicinal shrubs

Vast shrub resources of this country have not been studied from medicinal, aesthetic and environmental point of view. This country has shrub flora occurring on wide ranges of alkaline, saline, estuarine, sandy soils in varied climatic conditions. The time is ripe to select typical shrubs that have conspicuous flowers, wide range of foliage patterns and shapes and plant them in arboreta, parks and gardens. Some shrubs are indispensible ingredients of parks, gardens, residential quarters premises and in environmental planting to arrest air pollution.

An attempt has been made to list some common and conspicuous shrubs occurring all over the country (Statewise) and to mark some common medicinal plants. Although some of these shrubs are obnoxious weeds (Eupatorium, Cleistanthus, Lantana, Calotropis, Euphorbia, Jatropha etc.), quite a good number occur on dry rocky and sandy areas (Euphorbia, Opuntia, Agave, etc.,) which provide shade, protect soil, shelter birds and bear fruits for man and animal besides, providing plants of medicinal values. Vast avifaunal species in the country side derive a lot of food and seek shelter in such shrub flora.

Results of study over the country

The distribution of a few shrubs as a result of analysis shows the following :

(i) Few shrubs have wide distribution.

(ii) Several shrubs are conspicuous all over the country.

(iii) Few shrubs are gregarious in their occurrence.

(iv) Most of the shrubs are confined to forested regions.

(v) Moist localities have more of shrubs than other habitats.

Common and conspicuous Shrub Flora of India

Andhra Pradesh

Carissa spinarum	*Cassia auriculata*
Carissa carandas	*Zizyphus mauritiana*
Acacia leucophloea	* *Strychnos nux-vomica*

Dodonaea viscosa
Jatropha gossypifolia
Morinda tinctoria
Randia domestorum

Assam

Eupatorium sp.
Phoenix humilis
Rubus ellipticus
Randia dumetorum
Lantana camara
Litsaea umbrosa
Osbeckia sp.

Bihar

* *Phoenix sylvestris*
* *Carissa carandas*
* *Butea monosperma*
Carissa opaca

Delhi

* *Calotropis procera*
Cassia mimosoides
Z. nummularia
Capparis decidus
Zizyphus mauritiana
Prosopis species

Gujarat

Calotropis procera
Capparis decidua
* *Butea monosperma*
Mimosa rubicaulis
A. leucophloea
Zizyphus nummularia
Acacia catechu
Suaeda fruticosa
Zizyphus mauritiana
Cassia auriculata
Cassia mimosoides
Acacia arabica
Euphorbia royleana
Prosopis juliflora
Helicteres isora
Premna coriacea

Himachal Pradesh

Acacia caesia
Carissa spinarum
Berberis aristata
Prinsepia utilis
* *Indigofera gerardiana*
Berberis sp.
Flacourtia indica
Desmodium gangeticum
Colebrookea oppositifolia
* *Woodfordia fruticosa*

Cotoneaster microphylla
* *Adhatoda vasica*
Dodonaea viscosa
Rhus continus
Plectranthus rugosus
Berberis chitria
* *Agave americana*
Zanthoxylum alatum

Jammu & Kashmir

* *Calotropis procera*
Acacia modesta
Lantana camara
Arisaema helleborifolium
Deutzia corymbosa
Rubus lasiocarpus
Robinia pseudo-acacia
Rosa moschata
Plectranthus rugosus
Indigofera pulchella
Parrotia sp.
* *Adhatoda vasica*
Clerodendrum viscosum
Viburnum foetens
Desmodium tiliaefolium
Jasminum humile
Berberis lycium
Pistacia integerrima
Rubus ellipticus
Artemisia vulgaris
Daphne sp.

Kerala

Cassia mimosoides
Randia dumetorum
Pandanus sp.
Lantana camara
Euphorbia tirucalli
Cocos nucifera
Phoenix sylvestris

Karnataka

* *Acacia leucophloea*
* *Azadirachta indica*
Lantana camara
* *Calotropis procera*
* *Butea monosperma*
Dichrostachys cinerea
Rhus wallichii
* *Zizyphus* sp.

Madhya Pradesh

* *Holarrhena antidysenterica*
* *Butea monosperma*
Phoenix sylvestris
* *Acacia catechu*
Lantana monosperma
Osbeckia aspericaulis

Helicteres isora
Dichrostachys cinerea
Vitex negundo
Barleria prionitis
* *Zizyphus* sp.

Maharashtra

* *Calotropis procera*
Indigofera gerardiana
Lantana camara
Zizyphus sp.
Plextranthus rugosus
Gymnosporia spinosa
Cassia auriculata
Zizyphus mauritiana
Carissa carandas
Cassia sp.
Cleistanthus collinus

Manipur

Desmodium cephalotes
Priotropis cytisoides
Cycas siamensis
Pueraria hirsuta
Buddleia sp.
Phoenix humilis

Orissa

Barleria prionitis
* *Calotropis procera*
Anona squamosa
Combretum decandrum
Jatropha glandulifera
* *Zizyphus* sp.
* *Adhatoda vasica*
Cassia sp.
Zizyphus mauritiana
* *Vitex negundo*

Punjab

* *Capparis decidua*
* *Zizyphus nummularia*
Salvadora oleoides
* *Butea monosperma*
Murraya koenigii
* *Accacia arabica*
Capparis sepiaria
* *Tamarix dioica*
Prosopis juliflora
* *Calotropis procera*
* *Adhatoda vasica*
Lantana camara
Carissa opaca
Dodonaea viscosa

Rajasthan

Acacia arabica
Zizyphus nummularia
* *Calotropis procera*
Mimosa rubicaulis
* *Acacia catechu*
Lantana camara
Acacia leucophloea
Solanum seaforthianum
Prospis juliflora
Cassia mimosoides
Capparis decidua
* *Butea monosperma*
* *Euphorbia nivulla*
Prosopis spicigera
Phoenix sylvestris

Tamil Nadu

Ixora parviflora
Zizyphus nummularia
* *Carissa carandas*
Gardenia gummifera
Acacia leucophloea
Zizyphus sp.
* *Randia dumetorum*
Lantana camara
Rubus sp.
Suaeda sp.
Cassia auriculata
Dodonaea viscosa
Cassia siamea
* *Calotropis procera*
Opunita sp.
Hypericum mysorensis
Phoenix sylvestris

Uttar Pradesh

* *Adhatoda vasica*
Moghania sp.
* *Butea monsperma*
Lantana camra
Carissa opaca
* *Acacia arabica*
Cotoneaster microphylla
Rosa moschata
Spiraea linoleyana
Myricaria bracteata
Rhododendron campanulatum
Salix flabellaris
* *Zizyphus* sp.
Carissa sp.
Zizyphus nummularia
* *Phoenix decidua*
* *Calotropis procera*
* *Berberis umbellata*
* *C. acuminata*
Jasminum humile
Viburnum nervosum
Rosa sp.
Syringa emodi

Distribution zones of Conspicuous Shrub Flora in India

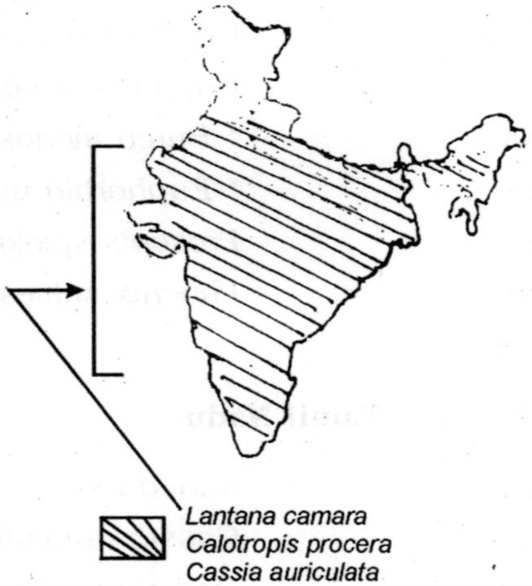

Distribution of Conspicuous Shrub Flora in India

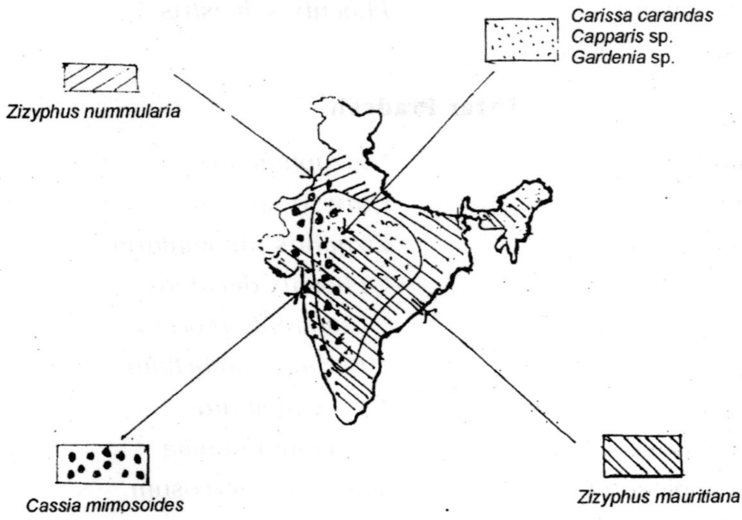

Distribution of Conspicuous Shrub Flora in India

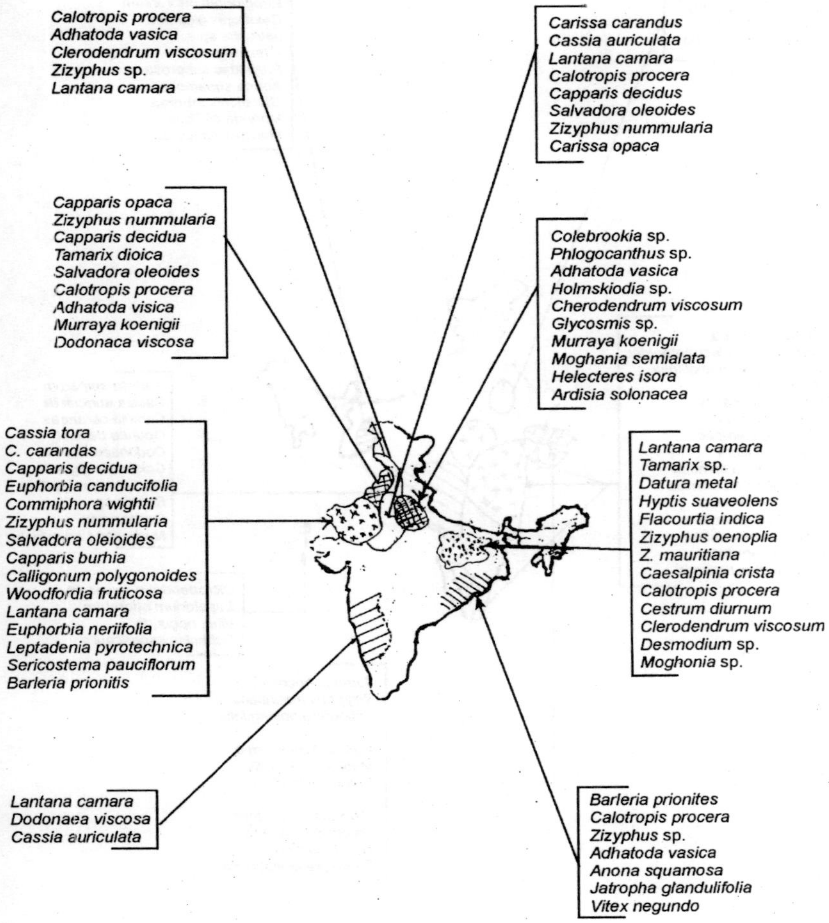

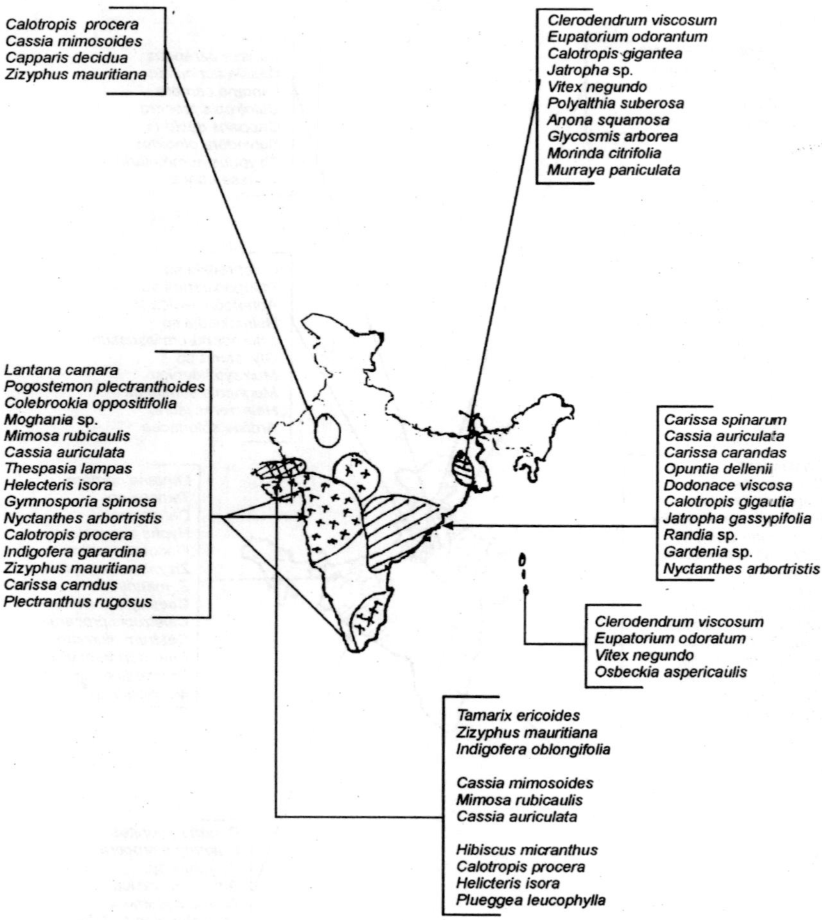
Calotropis procera
Cassia mimosoides
Capparis decidua
Zizyphus mauritiana
Clerodendrum viscosum
Eupatorium odorantum
Calotropis gigantea
Jatropha sp.
Vitex negundo
Polyalthia suberosa
Anona squamosa
Glycosmis arborea
Morinda citrifolia
Murraya paniculata
Lantana camara
Pogostemon plectranthoides
Colebrookia oppositifolia
Moghania sp.
Mimosa rubicaulis
Cassia auriculata
Thespasia lampas
Helecteris isora
Gymnosporia spinosa
Nyctanthes arbortristis
Calotropis procera
Indigofera garardina
Zizyphus mauritiana
Carissa camdus
Plectranthus rugosus
Carissa spinarum
Cassia auriculata
Carissa carandas
Opuntia dellenii
Dodonace viscosa
Calotropis gigautia
Jatropha gassypifolia
Randia sp.
Gardenia sp.
Nyctanthes arbortristis
Clerodendrum viscosum
Eupatorium odoratum
Vitex negundo
Osbeckia aspericaulis
Tamarix ericoides
Zizyphus mauritiana
Indigofera oblongifolia
Cassia mimosoides
Mimosa rubicaulis
Cassia auriculata
Hibiscus micranthus
Calotropis procera
Helicteris isora
Plueggea leucophylla

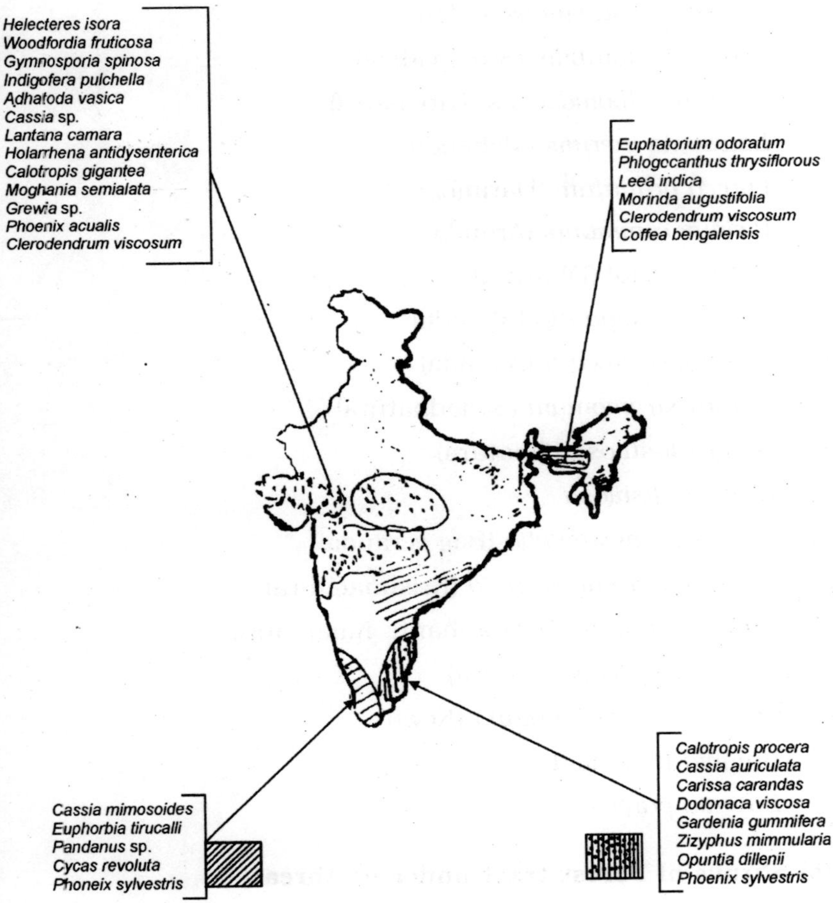
Helecteres isora
Woodfordia fruticosa
Gymnosporia spinosa
Indigofera pulchella
Adhatoda vasica
Cassia sp.
Lantana camara
Holarrhena antidysenterica
Calotropis gigantea
Moghania semialata
Grewia sp.
Phoenix acualis
Clerodendrum viscosum
Euphatorium odoratum
Phlogccanthus thrysiflorous
Leea indica
Morinda augustifolia
Clerodendrum viscosum
Coffea bengalensis
Cassia mimosoides
Euphorbia tirucalli
Pandanus sp.
Cycas revoluta
Phoneix sylvestris
Calotropis procera
Cassia auriculata
Carissa carandas
Dodonaca viscosa
Gardenia gummifera
Zizyphus mimmularia
Opuntia dillenii
Phoenix sylvestris

(a) A broad view of shrub and undertree
(Garden and other places under no threat)

Ervatamia coronania (Tagar)
Nerium odorum (Rakta karabi)
Thevetia nerifolia (Halde karabi)
Hibiscus rosa-sinensis (Jaba)
Hibiscus mutabilis (Sthalpadma)
Hibiscus albomoschus (Latkasturi)
Lawsonia inermis (Mehendi)
Punica granatum (Darimba)
Ricinus communis (Aronda)
Datura metal (Dhatura)
Calotropis gigantia (Akanda)
Calotropis procera (Akanda)
Datura stramonium (Sadadhatura)
Datura fastuosa (Dhatura)
Ipomoea fistilosa
Jatropha gossypifolia (Bag veranda)
Caesalpinia pulcherrima (Krishnachura)
Barleria cristata (Swata jhanti, Kalajhanti)
Pandanus tectorius (Keya)
Pandanus roxburghiana (Keya)
Breynia rhamnoides
Breynia patens.

(b) Shrubs of Forest tract under no threat

Leea crispa (Banchalida)
Helectris isora (Avikush, Athmum)
Murraya koengii (Curry leaf)
Justicia adhatoda (Vasak)
Morinda tinctoria (Abecheck)

Morinda citrifolia (Anch)

Coffea bengalensis (Bancoffee)

Coffea sp.

Murraya exotica (Kamini).

Ixora passiflora (Gandhale, Rangan)

Ixora coccinia (Rangan)

Laportia crenulata (Chutra)

(c) Shrubs of Lower Temperate Hills under no threat

Melastoma matabathricum

Woofodia fruiticosa (Dhaiphal)

Berberis aristata (Daru haridra)

Memecylon edule

Pogostemon plectranthoides

Pogostemon parviflorus

Colebrookia oppositifolia

Anisomeles indica

Hypericum patulum

Hypericum sp.

Vivurnum foetidum

Lonicera glauca

Cinchona succirubra

Cinchona ledgerina

Osbeckia nepalensis

Osbeckia crinita

(d) Stray shrubs under no threat

Cassia tora (Chakmundi)

Cassia occidentalis (Kal kasundi)

Gossipium herbaceum

Lantana camara

Hiptis suaveolens.

Cassia Suphera

Euphorbia nerifolia (Snowii)

Eupatorium odoratum

Leonurus sibiricus

Boerhaavia diffusa

Boerhaaria repens *Artemisia vulgaris*
Clerodendrum serratum *Glycosmis arborea*
Martinia diandra (Baghnakhi).

2.7 STATUS REPORT : CLIMBING MEDICINAL PLANTS

India has about 265 species of woody (Linas) and herbaceous climbers; of these about 125 are woody climbers and the rest are herbaceous.

Of these, a hundred species are used as medicinal plants. Climbers which are trailors and the herbaceous ones which depend fully on the host plants for support have no bright future as their existence depend normally on host plants.

Some woody climbers such as Tinospora, Pureria, Entada, Vitis, Mucuna etc. require strong host plants for their support and survival and therefore, removal of trees bring their obvious depletion.

Quite a large number of herbaceous climbers grow in forest margins, waste places and on marginal lands; they very frequently face biotic hazards and therefore, their existence have now become increasingly threatened; the biotic pressure is in the form of heavy grazing by animals, growth of human population and their dependence on forests and demand of land for settlements.

A list of commonly available climbers are presented.

(a) Over-exploited — May face depletion

Paederia foetida (Gandhal) - very sporadic.

Tinospora cordifolia (Gulancha) : Over-exploited by forest dwellers, it being a good fodder species (enhance milk formation in cow) and also for plant collection, for medicinal purposes. It is typically a forest species. Other plants are -

Cuculus villosa (Jaljamni)
Stephania harnandifolia (Nimutha)
Cissampelos pareira
Cyclea burmani
Vallaris solane (Haparmati)
Tylophora indica (Antomal)

These plants grow in forest margin; as such they are directly or indirectly destroyed due to lopping and pilferage of host trees.

Gymnema sylvestre (Chlagalbeti) *Jasminum multifloram* (Kanda) *Aganosma dichotoma* (Madhabilata) *Tinospora malabarica* *Calamus viminalis* (Bet).	Occur sporadically

Asparagus sp. : Occur scattered mostly in dry-deciduous forest area and less commonly in moist deciduous forests.

Dioscorea sp. : Many species, over exploited by tribal people all over the forests.

Tribulus terrestris (Gokhur) : Occurs scattered in marginal forests and over exploited.

Gnetum sp. : This living fossil is a *rare plant* in some deciduous and semi evergreen forests.

Aristolochia indica : Grows in forest margin and over exploited due to lopping or pilferage of host trees.

Smilax macrophylla : Sporadic occurrence.

(b) Presently safe species, not threatened

Mimosa pudica (Lajjabati)

Cassytha filiformis (Silinga)

Capparis spinosa

Croton tiglium

Mimosa rubicaulis (Dhuna)

Jasminum grandiflorum (Chameli)

Argyreia speciosa (Rishya gandhri)

Ipomoea reptans (Kalmi Sag)

Capparis decidua

Abrus precatorius (Kunch)

Hiptage bengelensis (Madhabilata)

Cucurbetaceous climbing shrubs : e.g. Cucumis, Momordica, Luffa, Lageneria, Trichosanthes and Beniscasa, Phaseolus, Dolichos. Others are :

Marsdenia volubilis (Nach chikni)

Hemidesmus indicus (Anantamul)

Croton caudatus

Scindapsus officinalis (Gajapepul)

Pothos scandens

Cassytha filiformis

(c) Occur-sporadically : (density & frequency poor)
(Future not secured)

Phlogocanthus thyrsiflorus (Tamrapuspa)
Toddalia asiatica (Kunj)
Piper longum (Pippali)
Cardiospermum helicacabum (Lata phutki)
Entada scandens (Gila)
Sarcostoma brevistigma (Somlata)
Mucuna prurita (Almagupta)
Vitis sp.
Millettia auriculata
Millettia pachycarpa
Caesalpinia crista
Melothria maderaspetana
Muchia scabrela
Zehneria umbellata
Cryptolepis buchanani
Cryptostagia grandiflora
Oxystelma esculentum
Daemia sp.
Holostemma annulare
Marsdenia voluvilis
Tylophora asthmatica
Dregea voluvilis
Smilax macrophylla (Kumarica)
Piper cubeba (Kababchini)
Holostemma rheedii
Jasminum sambac

(d) Rare

Gnetum scandans

Gnetum ula

2.8 STATUS REPORT : MEDICINAL TREE FLORA

2.8.1 Some basic information

Indian rains forests are dominated by Dipterocarps. There are only nine species of Dipterocarps besides many other associate tree species. In Malaysian region, there may be 60-250 species of the members of Dipterocarpaceae. The rains forests of Assam is dominated by three species such as Dipterocarpus macrocarpus (Hollong), Shorea assamica (Makai) and Terminalia myriocarpa (Hollock). In south-western region (at Makut, Nilumbur, Coorg, North Canara, Palghat, Wynaad, Tiruneveli and several other places), Dipterocarpus has Palaquium, Mesua, Cullenia, Poeciloneuron and Calophyllum as principal associates besides vatica, Terminalia, Pterocarpus, Artocarpus, Hopea etc. Only 40 speceis, out of total 200 tree species, are being commercially exploited, Tropical forests .of India are primarily providing plywood, match wood and packing case wood besides yielding some quantity of constructional and furniture timber. Production per hactere is about 300 m^3 which is highest in any forest type in India. Large areas of tropical forests in Nagaland, Cachar hills, Assam, Mizoram, Manipur and in Arunachal Pradesh have been felled in the form of shifting cultivation. Forestry operation in achieving natural regeneration by felling of non-commercial species has also changed the nature of forests. Mixed forests have been converted into pure forests of a few selected species. Intense biotic activities have completely changed the nature of forests in both north-eastern and south-western regions. At present, only selected areas remain as preservation plots and biosphere reserves to represent typical tropical rains forests. Wildlife sanctuaries and national parks are also indirectly protecting some isolated spots.

Tropical rains forests of Andaman and Nicobar Islands are the pride of the Country. It extends over about 6,500 sq. km. and still have substantial areas of untouched forests. High degree of endemic flora and fauna is an asset of the tract. Besides, it has several species of animals not found anywhere in the world. Mention may be made of Narcondom hornbill, Nicobar megapode, Andaman teal, Crab-eating macaque and Robber crab. It is said that about 35% of the flora is endemic and, therefore, the forests need complete protection. There are 250 species of avifaunal species, besides many species of invertebrates with substantial number of endemic species.

Destructive forces have also affected these forests. Refugee rehabilitation (25,000 East Bengal refugees) has been responsible for bulk destruction and clear felling of about 1,0000 ha. has augmented substantial depletion of the forests. selection felling done to ensure natural regeneration has opened up extensive areas of forests which have lost the character of tropical forests. Many areas have been exposed to erosion. Most significant change and depredation has been caused by severe cyclone which is a regular feature of Andaman forests. Hundreds of square kilometres of forests have been uprooted and damaged (shallow root system and heavy crown of trees). Ground flora has also been disrupted.

Most disastrous human activities have been the conversion of virgin tropical forests of Little Andaman into oil-palm plantation. This species (Elaeis guinnensis, an exotic) has been planted over 2,000 ha. by felling of forests. Areas planted up with oil palm have shown surface erosion and no plants have come up under the shade of this crop. The human population of the islands now exceeds 130,000 and has perhaps reached the saturation point.

There is no parallel to the richness of faunal elements in tropical forests of India. Mention may be made of tiger, leopard, elephant, various species of deer, sambar, gaur, besides many other species in the forests. There are innumerable species of insects. Most interesting character of the forests is the occurrence of several species of primates such as the hoolock, the bonnet macaque, the rhesus monkey, the Assamese macaque, the lion-tailed macaque, the common langur, the capped langur, the Nilgiri langur, the slow lorris and slender loris.

If the forests of India survives the severe biotic impact medicinal plants will survive, for the trees that constitute forests are part of our amazing resources that have contributed to the prosperity of our nation. They are also sources of history and beauty, gifts of God and friends of man. John Elvin, the English patriot and philosopher, as early as in 1664 warned and advised, "Men seldom plant trees till they begin to be wise, that is, till they grow old and find by experience, the prudence and necessity of it". When trees grow collectively- a forest is born; an environment is created and plant and animal succession follows.

Trees interrupt reflected light, screen out undesirable sights, border the cities, lakes and streams. Trees with the rough, thick and sticky leaf surface collect soot and oil and reduce environmental pollution. Trees adorn our homesteads and our cities, shelter our wildlife; they act as soothing balm on the wearied and harried city dweller. Trees have played an important role in the evolution of humanity; they are greatest benefactors, friends, teachers and preceptors. They have the sense of beauty and instincts of love, service, sacrifice and harmony.

Forests are for human survival, "An age builds up cities, an hour destroys them; in a moment the ashes are made but forest is a long time growing". Forests are protective, productive and aeshtetic. The vital impacts of forests on climate, soil erosion and floods are acceptable to all. They protect us against winds, storms and rains. Where forests abound, the rains are plentiful. Forest soil with its microbes and vegetative cover acts as a natural filter.

Forests are perennial source of raw materials necessary for a variety of industries - pulp and paper, panel products, match wood, sports goods, plywood, fibre board, saw mills, constructional wood and medicinal plants.

Tropical forests are dark and impenetrable and abound with insects, reptiles, amphibians and mammals. These forests throw a baffling and intriguing challenge in their multistoreyed structure which is diverse and complex. Under the cover of trees, grow climbers, shrubs and herbs.

India has more than 1500 tree species of which only few are regenerating. Over-exploitation, selective exploitation, grazing and fire are threatening their survival. Trees which occupy the topmost layer of a forest may not serve a foolproof function against various atmospheric hazards; it also needs a well knit shrub and herb layer. Of the three vertical layers, top middle and bottom, the middle layer has the maximum number of species; next is the bottom layer and it is minimum in the top layer. It is needless to mention that there is hardly such forest left in the country where the bottom and middle storey trees still stand. Even the top layers have been stripeed off of many tree species. All these give a dismal picture of the state of vegetation in this Country.

Extensive areas supporting natural tree growth have come under reckless exploitation and eventual intense environmental degradation. Deforestation of 5 million hectares of land added by ravine affected 7 million and another 48 million hectares lost in river valley projects offer a gloomy picture of tree cover of the country. The much bloated claim that India has 22.8 percent of it's geographical area under forest is far from true. India has hardly 6.8 percent of it's geographical area under somewhat good vegetation cover; more than 13.4 percent of the geographical area is unver poor cover. Even then the situation has turned worse owing to uneven distribution of the forests. Vast erodable areas in the hilly tracts of western ghat, Himalayas and central India have been denuded beyond immediate recuperation. All these leave one to think that the tree vegetation in India are now in a bad shape to offer any viable safeguards against the environmental hazards.

2.8.2 Tree Conservation

Trees form only a small proportion of medicinal plants in India. Of 1200 trees only a few have enough medicinal properties to work economically. A small list has been prepared and the species have been put under as may be seen from the list.

Statistical data on density and percentage of occurrence of a few economically important species are presented in this paragraph which will indicate that such data are needed for all the medicinal plants. The data have been adopted from the work of Forest Survey of India. The figures related to the occurrence of number of trees per hectare in the forested areas of some States of the Country. There is necessity for similar studies on herbs, shrubs and its' evaluation.

The forest selected for presentation of data on trees are moist deciduous and dry deciduous forests. Only the frequency and density of crops have been mentioned. About 500 tree species come under this study. But the situation is fast changing due to the practice of shifting cultivation, forest fire, illicit felling, forestry felling and for various developmental processes. These onslaughts have affected the ground vegetation (herbs and shrubs) much more than the trees. A study of shrub vegetation of the country reveals that a few resistant species only are now widespread in the country while most other delicate ones have sporadic occurrence. Similar is the situation of herb flora which are being depleted all over the sites where they now hold the ground. The converstion of mixed forests into a few species of commercial importance (monoculture) has also depleted very many herb and shrub species occurring in the forests.

Status of Medicinal Plants (Trees)

Trees (Not under threat at present – A few selected species)

Acacia arabica (Babul)
Aegle marmelos (Boel)
Alstonia scholaris (Chatin)
Albizzia lebbek (Sirish)
Anona squamosa (Ata)
Areca catechu (Superi)
Azadirachta indica (Neem)
Anthocephalus chinensis (Kadam)
Artocarpus heterophylla (Kanthal)
Anona reticulata (Nona)
Bauhinia racemosa, (Seta Kanchan)
Bauhinia variegata (Rakta kanchan)
Bauhinia purpurea (Deba kanchan)
Bombax ceiba (Semul)
Butea monosperma (Palash)
Carica papaya (Papaya)
Cassia fistula (Amaltus)
Dalbergia sissoo (Sisam)
Dipterocarpus sp. (Garjan)
Delonix regia (Krishnachura)
Dillenia pentagyna (Tantari)
Eucalyptus tereticornis (Blue gum)
Erythrina indica (Madar)
Emblica officinalis (Amloni)
Fixus bengalensis (Bat)

Fixus religiosa (Aswathwa)

Fixus glomerata (Jagna dumur)

Fixus hispida (Kak dumur)

Ficus cunea (Jaya dumur)

Grewia sp. (Phalsa)

Holarrhena antidysenterica (Kurchi)

Moringa olifera (Sajina)

Nyctanthes arbor- tistris (Siwli)

Nerium odorum (Rakta Karabi)

Stereospermum personatum (Parari)

Streblus asper (Seora)

Vitex negundo (Nishinda)

Ziziphus sp. (Kul)

Trema orientalis (Kuail)

Thespesia populnea (Pipli)

Tamarix sp. (Barjha)

Toona ciliata (Toon)

Tectona grandis (Teak)

Tamarindus indica (Tentul)

Termarindus indica (Arjuna)

Syzygium cuminii (Kalo jam)

Prosopis spicigera

Prosopis cinararia

Putranjiva roxburghii (Putranjiva)

Diospyros melanoxylon (Kend)

Trees occurring sporadically which may be vulnerable in several decades (a few selected species)

Acacia suma (Somi)

Acacia catechu (Khair)

Ailanthus excelsa (Mahanimba)

Anacardium occidentale (Kaju)

Anogeissus latifolia (Dhaw)

Bassia latifolia (Mahua)
Boswellia serrata (Salai)
Capparis decidua
Citrus sp.
Careya arborea (Kumbi)
Cochlospermum gossypium (Galgali)
Gmelina arborea (Gamar)
Feronia elephantum
Flacourtia sepiaria
Hopea odorata
Hydnocarpus sp. (Chalmugra)
Melia azadirach (Goraneem)
Michelia montana (Champ)
Michelia niligerica (Champ)
Mahonia nepalensis
Oroxylum indicum (Sona)
Ougenia oojeinensis (Panjan)
Premna sp. (Ginari)
Phoenix sp. (Khejur)
Pongamia pinnata (Karanj)
Pterospermum acerifolium (Kanak champa)
Pterygota alata (Narikeli)
Terminalia chebula (Haritaki)
Spondius mangifera
Salix tetrasperma
Knema augustifolia
Kigelia pinnata
Mallotus philippinensis (Rohini)
Morus alba (Tut)
Polylthia semiarum (Labsi)
Bischofia javanica (Kanjal)
Aglaia roxburghii (Priangn)

Adansonia digitata (Kalpataru)
Mangium salvaefolium
Artocarpus lakoocha (Dewa)
Barringtonia acutaugula (Hizal)
Crataeva nurvula (Barin)
Lannea coromemdelica (Jeol)
Gardenia gummifera (Blurur, Neriheriger)
Sympolcos racemosa (Lodhra)
Trewia nudiflora (Pitali)
Gynocordia odorata (Bandre, Chalmugra)
Flacourtia catafracta (Baichi)
Ochrocarpus longifolius
Garuga pinnata (Giga)
Aphanamyxis polystachya (Tikiaraj)
Disoxylum hamiltonii (Gab)
Cordia dichotoma (Bahanari)
Elaeocarphus tuberculatus (Rudraksha)
Evodia lunuankenda
Evodia fraxinifolia
Diospyros embriopteris (Gab)
Mesua ferrea (Nageswar)
Michelia champaca (Champ)
Antiarix toxicaria
Elaeocarpus ganitrus (Rudrakshya)
Juniperus sp.
Pterocarpus marsupium (Peasal)
Strichnos nux-vomica (Kuchila)
Strichnos potatorum (Nirmala).

Trees (Need protection – A few selected species)

Balsamodendron Mukul
Bixa orillana
Camphora officinarum

Callophyllum inophyllym
Caryota urens
Ceiba pentandra
Juglans regia
Cupressus semprevirens
Betula utilis
Aquillaria agallocha (Aguru)
Taxus baccata
Juniperus communis
Schrebera switeniodes (Ghanta parul).

2.8.3 Quantitative assessment of medicinal plant species

(i) West Bengal

Moist Deciduous Sal forests in Dooars and Tarai

- Dooars Forests (Stem/acre)

Site-1 Dooars Forests (Stem/area) sal-61, other sp. were Meliosma sumplicifolia 14, Prema benglensis 8, Cinnamomum obtusifolium 5, Cryptocarya amygdalina 3, Turpinia pomifera 2, Toona ciliata 2, Cinnamomum cecidodaphne 2, Macaranga sp. 2. These are a few findings.

Site -2 Sal-11; other species were Elaeocarpus varunua 19, Machilus villosa 12, Aphanamixis polystachya 11, Schima wallichii - 11; Dillenia pentagyna - 8, Bauhinia purpurea 4, Actinodaphe obovata 3, Termi- nalia belerica 2, Lagerstroemia parviflora 2, Meliosma simplicifolia - 2, Eurya sp. 2. These are a few findings.

Site-3 Sal-32; other species are Meliosma simplicifolia 29, Machilus gamblei 12, Prema bengalensis 7, Turpinia pomifera 5, Dysoxylum binectariferum - 4, Cryptocarya amygdalina 5, Syzygium cumini 2, Toona ciliata-2, Cinnamomum obtusifolium. These are a few findings.

Site-4 Sal - 43.1; Other species are Aphanamixis polystachya 11.8, Terminalia belerica 9.1, Sterculia villosa 5.3, Schima wallichii 6.0, Lagerstroemia parviflora 3.6,

Syzyzium sp. 3.6, Garuga pinnata - 3.4, Amoora wallichii - 1.7 Albizzia sp. 1.4, Machilus sp.

Site-5 Sal-29.5, Aphanamixis polystachya 5.6, sterculia villosa 5.6, Castanopsis sp. 6.6, Syzygium sp. 5.2, Machilus sp. 4.7, Terminalia belerica 3.5, Biscofia javania - 3.5, Lagerstroemia parviflora 3.5, Albizzia sp. 3.1, Schima wallichii, Garuga pinnata 1.2, Zanthoxylum budrunga 1.6.

Site-6 Sal 43-1; other species are Aphanamixis polystachya 11.9, Terminalia belerica 9.01, Sterculia villosa 5.3, Schima wallichii 5.9, Lagerstraemia parviflora 3.5, Syzygium sp. 3.42, Garuga pinnata 3.43, Albizzia sp. 1.39, Machilus sp. 0.98.

Site-7 Sal-2.5 stem/ac.; Other species are machilus gamblei 23 Meliosma simplicifolia 22.9, Syzygium formosum 10, Premna bengalensis 7, Casearis sp. 5, Vatica lancaefolia 3.5, Macaranga denticulata 2, Cryptocarya amygdalina 3, Syzygium sp. 2. These are a few findings.

Site-8 Meliosma simplicifolia 13, Aphanamixis polystachya 5.9, Tetrameles nudiflora 5, Dysoxylum binectariferum 4.8, Crataeve nurvula 4.7, Dillenia pentagyna 4.4, Castanopsis tribuloides 4.4, Albizzia lucida 4.4, Bauhinia purpurea 4.2, Meliosma pinnata 5.3. These are a few findings.

Site-9 Sal 1.8 per ac.; other species are aphanamixis polystachia 30.5, Prema integrifolia 11.4, Wrightia tomentosa 8.2, Mallotus philippinensis 7.7, Terminalia belerica 5.5, Syzygium sp. 4, Litsaea sp. 4, Stereospermum chelonoides 2.3, Sterculia villosa 1.8. These are a few findings.

Site-10 Litsaea polyantha 15, Prema integrifolia 11, Mallotus philippinensis 8.9, Terminalia belerica 5.5, Gmelina arborea 4.5, Litsaea sebifera 4.5, Wrightia tomentosa 4, Toona ciliata 3.5, Aphanamixis polystachya 3, Sterculia villosa 3.5. These are a few findings.

Darjeeling Hills of West Bengal

A token survey was carried out in Sinhalila area and following species were listed.

Rhododendron sp. 67.40 stems per ha.

Lithocarpus pachyphylla.	24.00 Stems per ha.
Tsuga dumosa-	20.50 Stems per ha.
Castanopsis sp.-	17.70 Stems per ha.
Abies densa-	14.50 Stems per ha.
Lithocarpus lineata -	12.50 Stems per ha.
Acer campbellii -	5.00 Stems per ha.

Other species are, - *Alnus nepalensis, Simingtonia populnea, Cinnamomum* sp. etc.

Kalimpong Hills of West Bengal

According to a token survey carried out over small area the following species were listed.

Machilus gammieana -	22 Stems per hectre
Lithocarpus lineata -	15 Stems per hectre
Alchimundra cathcartii -	15 Stems per hectre
Lithocarpus pachyphylla -	10 Stems per hectre
Nyssa javonica -	9 Stems per hectre
Castanopsis sp. -	8 Stems per hectre
Acer campbellii -	8 Stems per hectre
Cinnamomum sp. -	8 Stems per hectre
Lithocarpus pamelosa -	7 Stems per hectre.

Overall density per hectare is 275 trees.

Purulia District of West Bengal

(22°-43' to 23°-42' N longitude & 85°-49' to 86°- 54' E latitude)

The forests of purulia district grow over barren red lateritic soil. 80% of the crop are of 5 cm. to 9 cm. dia.

Shorea robusta -	232 Stems per ha.
Terminalia tomentosa -	23.750 Stems per ha.
Semecarpus anacardium -	17.000 Stems per ha.
Lagerstroemia parviflora-	15.750 Stems per ha.
Terminalia tomentosa -	12.000 stems per ha.
Mallotus philippinensis -	36.58
Diospyros sp.	

Holarrhena sp.

Pterocarpus marsupium.

Miscellaneous stratum

Holarrhena antidysentrica	-	41.059
Anogeissus latifolia	-	13.16
Bauhinia purpurea	-	9.73
Cochlospermum gossypium	-	10.00
Launea coromendelica	-	17.89
Mallotus philippinensis	-	36.58
Terminalia tomentosa	-	51.06
Shorea robusta	-	46.32
Emblica officinalis	-	13.16
Pterocarpus marsupium	-	10.26
Gmelina arborea	-	8.94
Saccopetalum tomentosa	-	7.30
Croton oblongifolius	-	6.84
Buchanania lanzan	-	5.79

(ii) Little Andamans

According to F.S.I. Survey following species were listed.

Dipterocarpus sp.	-	9 per ha.
Sideroxylon longipetiolantum	-	7 " "
Plerocymbium tinctorium	-	5 " "
Myristica sp.	-	11 " "
Artocarpus chaplasha	-	5 " "
Pometia pinnata	-	9 " "
Xanthophyllum andamincum	-	3 " "
Terminalia bialata	-	1 " "
Atrocarpus lakoocha	-	1 " "
Parishia insignis	-	3.1 " "
North Andamans	-	275 Stems per ha.
Middle Andamans	-	327 Stems per ha.
South Andamans	-	371 Stems per ha.

(iii) Sikkim

Figures of South Western Sikkim are only available.

Prime species are : *Symplocos theifolia*, (Frequency occurrence is 90, and Stems per ha. are 69) has the maximum density, this species is of immense medicinal value of importance (Frequncy given in the brackets).

Species are Castanopsis sp. (Fr. Ocr. 90, and 13 Stems/ha), *Machilus odorattissima* (F-66 and 8/ha). *Eurya Japonica* (F-80, 8/ha) *Macaranga* sp. (F-28, 7/ha.), *Lithocarpus lamellosa* (Quercus), *Acer Campbellii* (F-50, 8/ha.), *Litsaea* sp. (F-90, 3/ha.), *Machilus gammiana* (F-70)

(iv) Bihar Pradesh (Sal bearing tract)

Survey area 53,000 sq. km. in Chhotanagpur hills.

Forest Type 5 BC_1(C) - Northern Tropical Dry Peninsular Sal Forests.

Stems per hectare

Sal (poor quality) - 450 Stems/ha. (In Sal Forests)

Boswellia serrata - 22/ha. (In Sal Forests)

" " - 111/ha. (In Boswellia forests)

Acacia catechu - 132/ha. (In Acacia forests)

Miscellaneous species : 63/ha.

Species are Terminalia tomentosa, Anogeissus latifolia, Lannea coromendelica, Buchanania lanzan, Syzygium cumini, Ficus species, Diospyros melanoxylon, Ougenia oojeinensis, Bombax ceiba, Madhuca latifolia, Lagerstroemia parviflora, Adina cordifolia, Butea monosperma, Schleichera oleosa and Pterocarpus marsupium.

(v) Melghat Forests of Maharashtra State

Melghat forests of Amravati distict of Maharashtra State lie between the parallels of latitude 21°-11' and 21°-46' N and longitudes of 76°-38' and 77°-34'E. These forests are close to the forsts of Akola, Buldhana of Nimar and Belul districts of Madhya Pradesh.

The forests fall under southern Tropical Dry Deciduous Teak Type (5A/C1b) and Southern Tropical Dry Mixed

Deciduous Forest Type ($5A/C_3$) as per Champion and Seth's Forest Types of India. Teak forests occur extensively with Dendrocalamus as understorey, where Dendrocalamus is absent Lantana and Cymbopogon martinii and other tall grasses from the ground vegetation.

MELGHAT TIGER RESERVE

Trees

Quantitative Survey carried out by Forest Survey of India and Tiger Project Authorities, shows the occurrence of 277 trees per ha. dominant sp. as follows :

Forest Types :

5A		Southern Tropical Dry Deciduous Forests.
5A/C1	–	Dry Teak Forests
5A/C1a	–	Very Dry Teak Forests
5A/C1b	–	Dry Teak Forests
$5A/C_3$	–	Southern Dry Mixed Deciduous Forests.
5E-2	–	Boswellia forests.
5E-5	–	Butea forests.
5/DS1	–	Dry Deciduous Scrub forests.
5/IS1	–	Dry Tropical Riverin forests.

Melaghat Tiger Reserve

No. of species	: 648
No. of Genera	: 398
No. of Family	: 97
No. of Legume Sp.	: 98
No. of Graminae	: 92
No. of Tree sp.	: 86

High Frequency (%)

Tectona	-	10
Anogeissus	-	70
Lagerstoemia	-	70

Wrightia	-	6
Emblica	-	60
Dalbergia	-	10
Flacourtia	-	60
Boswellia	-	60
Acacia catechu	-	70
Bamboo	-	50

Southern Tropical Dry Deciduous Forests : (Average ht. - 20.27 m.)

Species are (Trees)

Tectona grandis
Terminalia tomentosa
Wrightia tinctoria
Bauhinia racemosa
Mitragyna parviflora
Casaeria elliptica
Garuga pinnata
Flacourtia indica
Emblica officinalis
Boswellia serrata
Miliusa tomentosa
Madhuca latifolia
Albizzia procera
Grewia lilaefolia
Bridelia retusa
Lannea coromendelica
Qugenia oojeinensis
Cassia fistula
Terminalia belerica
Schleichera oleosa
Zizyphus xylopyra
Bombax ceiba
Lagerstroemia parviflora
Anogeissus latifolia
Acacia sundra
Terminalia Arjuna
Adina cordifolia
Chloroxylon swietenia
Diospyros melanoxylon
Sherbera swietenoides

Woody climbers

Butea superba
Celastrus paniculatus
Capparis zeylanica
C. trifolia
Hemidesmus indicus
Bauhinia vahlii
Millettia auriculata
Zizyphus rugosa
Cayratia auriculata
Dregea vollubilis
Mucuna prurita.
Pueraria tuberosa

Combretum ovalifolium
Cissampelos pariera
Cissus repanda
Crytolepis buchanani
Acacia pinnata
Ampelocissus latifolia
Canavalia gladiata
Argyreia sericea

Herbs

Cleome viscosa
T. rhomboidea
Cardiopermum helicacabum
Hybanthus aspaspermus
Spoubia delphinifolia
Rungia elegans
Galinsoga parviflora
Faracarym coelestinum
Ipomoea quomoclit
I. pestigris
Euphorbia hirta
Crinum sp.
Urena lobata
I. laxiflorum
Pogostemon bengalensis
S. alba
Leea edgworthii
P. monoica
S. stricta
Lepidagathis cristatta
Conyza stricta
Hemidesmus indicus
Evolvulus alsinoides
Achyranthes aspera
Phyllanthus debilis
Habenaria platyphylla
D. hispida
D. pentaphylla
Triumfetta pentrandra
Biophytum sensetivum
Polygala elongata
Bidens biternata
Hemigraphis latebrosa
Lavendula burmanii
Cryptolepis buchanani
Argyreia sericea
J. obscura
Euphorbia prunifolia
E. hypericifolia
Chlorophyum tuberosum
Desmodium gangeticum
Orthosiphon rubicundus
Sida acuta
S. orientalis
Pinpinela lateriflora
Spermacoca hispida
Dipteracanthus patulus
L. trinervis
Tridex procumbens
Trichodesma indicum
Leucas biflora
Celocia argentea
P. virgatus
Dioscorea belophylla
D. bulbifera
Uraria refescens

Plectranthus rugosus *Anisomeles indica.*

Herbs, growing in Agricultural lands

Argemone mexicana *Sida acuta*
Cassia fumila *Spermacoce stricta*
Ageratum conyzoides *I. annea procumbens*
Vernonia sp. *Echinostemna hyssopifolium*
Solanum migrum *Justicia simplex*
Leucas cephalotes *Celocia* sp.
Phyllanthus sp. *Merdania nudiflora.*
Hibiscus lobatus *Psoraba cerylifolia*
Ammania baccifera *Oldenlandia corymbosa*
Caesulia axillaris *Tridax* sp.
Mollugo pentaphylla *Merremia gangetica*
Striga angustifolia *Lepidagathis cristata*
Achyranthes sp. *Euphorbia* sp.
Cyanotis fasciculata

Shrubs

Helicteres isora *Nyctanthes arbortristis*
Colebrookia oppositifolia *Vitex negundo*

Grasses

Echinochloa frumentacea *Paspalidium flavidum*
P. scrobiculatum *Brachiaria enciformis*
Coix aquatica *Eleusine aegyptica*
Eragrostis ciliaris *E. tenalla*
Setaria intermedia *S. pallidaefusca*
Apluda mutica *Themeda quadrivalvis*
Chloris dolichostachys. *Panicum miliare*
Eleusine coracana *Setaria italica*
Chloris mirgata *Cynodon dactylon*
Echinochloa colona *Iseilema laxum*
Sporobolus diander *Heteropogon contortus*
Sehima nervosum *Capillipidium assimele*

(vi) Western Nepal

Survey was carried out over an area 20,000 sq.km. (80°-30' to 83°-45' E longitude and 27°-45' to 30°-30' N latitude). The table shows the situation of tree stems per hectare.

(vii) Mewasi, Maharastra

Total area - 1,201 Km^2

Forested area - 712.56 Km^2

The forests of Mewasi fall under Southern Dry Deciduous Forests, accouding to Champion & Seth's Classification of Forest Types; Where Tectona grandis predominates. But there are wide areas, where miscellaneous crop predominates with a small proportion of Teak.

Stems per ha. of different species, in the 3 (there) categories of forests are latulated below :

Species	No. of stems per hectare		
	Teak & Misc. Forests	Misc. Forests	Teak Forests
Tectona grandis	24.6	5.1	36.8
Terminalia crenulata	9.0	10.5	6.3
Pterocarpus marsupium	2.3	0.5	–
Dalbergia latifolia	4.3	1.4	3.6
Qugenia oodienensis	2.6	3.1	2.1
Diospyros melanoxylon	3.3	0.5	3.6
Bridelia refusa	5.3	2.8	1.0
Anogeissus latifolia	11.6	3.1	8.5
Lagerstroemia parviflora	5.9	2.5	3.6
Boswellia serrata	5.3	3.7	4.2
Lannea coromendelica	12.6	4.2	6.3
Garuga pinnata	5.0	3.7	0.8
Schleichera oleosa	4.3	0.5	0.4
Terminalia belerica	1.9	3.1	0.4
Adina cordifolia	2.6	0.5	1.0
Mitragyna parmiflora	1.3	2.5	0.4
Albizzia procera	0.6	0.8	–
Butea monosperma	15.0	16.5	6.3
Acacia Catechu	2.3	0.8	0.6
Dalbergia paniculata	3.3	1.4	1.0
Grewia tileaefolia	2.3	1.7	1.0
Morinda tinctoria	7.6	4.0	3.8
Saccopetalum tomentosum	2.6	1.1	0.6
Wrightia tinctoria	4.6	3.1	2.3
Holarrhena antidysenterica	3.6	2.8	–
Other species	13.9	12.0	–

Table (West Nepal)
Stems/ha. of Important Species in different Forest Types

Species	Forest types					
	Chir	Kail-Deodar	Fir-spruce-Tsuga	Conifers mixed with H.W.	High Level Broad leaved	Low level Broad leaved
Chir (Pinus roxburghii)	109.136	—	—	22.553	1.485	8.862
Kail (Pinus wallichiana)	0.021	214.228	15.015	31.672	1.975	—
Deodar (Cedrus deodara)	—	2.532	—	—	—	—
Fir (Abies pindrow)	—	4.630	132.862	33.183	3.094	—
Spruce (Picea smiliana)	—	6.622	18.223	13.246	0.399	—
Tsuga (Tsuga dumosa)	—	4.630	6.506	8.450	0.211	—
Thuner (Taxus baccata)	—	—	19.165	5.325	1.824	—
Cyprus (Surai) (Cupressus toralosa)	—	5.658	34.001	—	—	—
All oaks (Quercus sp.)	6.040	35.254	24.807	43.335	125.997	0.693
Bhojpatar (Betula sp.)	—	9.461	30.381	22.808	8.970	—
Kanjal (Bischofia javanica)	—	—	3.638	6.584	3.232	—
Sal (Shorea robusta)	5.298	—	—	23.644	0.080	178.580
Sain (Terminalia sp.)	—	—	—	2.060	—	21.631
Dhawra	—	—	—	0.501	—	25.638
Up land hardwoods	17.882	16.992	56.248	45.651	237.572	10.718
Low land hordwoods	11.623	—	—	4.786	3.130	166.784
Total conifers	109.157	238.300	225.772	114.429	8.988	8.862
Total Broad leaved	40.843	61.707	115.074	149.369	379.001	404.044
Grand Total	150.000	300.007	340.846	263.798	387.989	412.906

Forest Types

Southern Moist Deciduous Teak Forests (3B\C1b)

Dry Bamboo Brakes - (5E\9)

DANG, SURAT, BULSAR & BHARUCH DISTRICT FORESTS OF GUJRAT.

Catchment Area : 23,736 Km^2

Forest Area : 5,816 Km^2

Situation : 20°-75' to 22°-15', North Latitude.

72°-32.5' to 74°-20' East Longitude.

Forest survey of tree crop shows the occurrence of about 200 stems per ha. Tectona grandis predominates, forming one eighth of the total trees per ha. It has the tree species in Sl. Nos. 2,4,8,16 & 17 as principal associate in the top storey. In the second storey wrightia tomentosa forms most conspicuous and widely occurring species.

Trees per hectare

Tectona grandis	25.20
Terminalia crenulata	7.80
Pterocarpus marsupium	0.90
Dalbergia latifolia	3.00
Ougenia oogeenensis	1.50
Diosphyros melanoxylon	1.30
Bridelia retusa	1.40
Anogeissus latifolia	3.40
Lagerstroemia parviflora	2.00
Boswellia serrata	0.60
Dalbergia paniculata	0.50
Ficus species	0.80
Grewia tilliaefolia	2.60
Morinda tinctoria	2.00
Saccopetalum tomentosum	1.50
Lannea coromendelica	3.40

Garuga pinnata	4.60
Schliechera oleosa	0.60
Terminalia belerica	1.60
Adina cordifolia	1.00
Sterculia urens	0.10
Albizzia sp.	0.50
Butea monosperma	2.00
Acacia catechu	1.80
Wrightia tinctoria	13.30
Holarrhena antidysenterica	2.00
Bombax ceiba *Madhuca latifolia* *Emblica officinalis* *Syzygium cumini* *Bauhinia racemosa* *Cassia fistula* *Zizyplus xylopyra* *Aegle marmelos*	200.098
Other species	11.50
Mallotus philippinensis *Bambosa arudinacea*	

Pine stratum : (Number of stems per ha. is 103.6)

Species are, - *Pinus roxburghii* - 34.60, *Pinus excelsa* - 22.00, *Quercus graffithii* - 11.80, *Quercus* sp. - 11.60, *Rhododendron* sp. - 13.20, *Betula alnoides* - 1.40, *Phoebe* sp. - 1.40.

(viii) Meghalaya

This survey pertains to a very small area in the foot hills of Southern slpes where Shorea robusta occurs. Meghalaya has varied species occurring in Sub-tropical & Sub-temperate climate. Most of the species of medicinal value.

Trees per hectare

Tectona grandis	135.2293
Terminalia tomentosa	5.1211
Ougenia oojeinensis	13.6774
Anogeissus latifolia	12.7263
Lagerstroemia parviflora	11.3199

Boswellia serrata	3.5248
Lannea coromendelica	5.3636
Madhuca latifolia	1.4197
Adina cordifolia	2.9277
Mitragyna parviflora	2.2456
Grewia tiliaefolia	7.9847
Wrightia tinctoria	13.3616
Other species	62.6827

In addition following species have also been recorded

Shorea rovusta	33.90
Bauhinia sp.	12.28
Terminalia sp.	8.22
Callicarpa sp.	7.08
Dillenia indica	5.97
Schima wallichi	6.22
Albizzia sp.	5.83
Careya arborea	5.75
Others	87.75

Shrubs (Melghat) : Lantana camara, Sida - 5 sp., Grewia - 5 sp., Alysicarpu - 8 sp., Bhima - 6 sp. Justicia - 5 sp. Leucas - 6 sp., Commelina - 5 sp., Fimbristylis - 6 sp., Eragrostis - 9 sp. Desmodium - 5 sp., Indigofera - 8 sp., Bauhinia - 5 sp., Cassia - 7 sp., Ipomoea - 10 sp., Euphorbia - 6 sp., Ficus - 7 sp., cyperus - 7 sp., Brachiaria - 5 sp.

(ix) Lohit and Tirap forests of Arunachal Pradesh

Principal species of economic importance are :

Mesua ferrea, Dipterocarpus macrocarpus, Shorea assamica, Michelia champaca, Amoora wallichii, Mansonia dipikae, Canarium resinifolium, Kydia calycina, Pterospermum acerifolium, Bischofia javanica, Schima wallichii, Ailanthus grandis, Gmelina arborea, Terminalia citrina.

Stems per hectare

Dipterocarpus Sp.	- Shorea forests	: 226 Stems\ha.
	- In miexed miscellaneous forests	: 150 Stems\ha.
	- In Lohit forest	: 215 Stems\ha.
	- In Changlang	: 185 Stems\ha.
	- Khonsa	: 294 Stems\ha.

So far volume per hectare are concerned, Terminalia myriocarpa, Castanopsis sp., Dillenia indica, Toona ciliata, Cinnamomum cecedodaphne and Chukrasia tabularis show prominence, besides that of most other principal economic species.

Principal Forest Types are :-

- Assam Valley Tropical Wet Evergreen Forests (Diptero-caupus)
- Northern Indian Moist Deciduous Forests (Terminalia myriocarpa).

Kameng District, Arunachal

Situation : Latitude : 26°-54' & 28°-01' N.

Longitude : 91°-30' & 93°-22' E.

The forests have been thoroughly surveyed in the eastern and western districts. The surveryshows a rich assemblage of species of modferate density. The area, however, has been suffering depletion due to shifting cultivation and large scale Army settlements.

A critical review of both the districts as regards tree vegetation is as follows :

East Kameng district

Broad-leaved stratum : 168 econimically impotant species have been indentified - (229 trees per hectare). Occcurrence of species per hectre are furnished below :

Castanopsis indica\Castanopsis sp. - 14.108, Altingia excelsa - 7.233; Ficus dp. - 6.566; Quercus sp. - 6.464; Sygygium sp. - 6.361; Alnus nepalensis - 4.668; Canarium sp.- 4.66; Macaranga indica - 4.412; Dysoxylum binecteiferum -

4.104; Amoora sp. - 2.719; Canarium resiniferum - 2.616; Magnolia sp. - 2.462; Trema orientalis - 2.360; Engelhardtia spicata - 2.155; Michelia sp. - 4.925.

Bamboo stratum : 9 species are porominent e.g. - castanopsis, Altigia, Bischofia, Macaranga, Sterculia, Quercus Aostonia, Michelia, Sehiuna (Per hectare occurrence of each species is between 2 to 3).

Altogether 168 species of trees have been recorded alongwith their occurrence per hectare. Of these species most widely occurring trees have been depicted. The occurrence of Ficus, Gmelina, Glochidion, Artocasrpus, Antidesma, Cordia, Dillenia, Dysoxylum, Erythrina, Eugenia, Terminalia species and species of several other trees indicate a bio-ecological bonanza for bird and animal fauna to survive.

Density (Forest) of both part of the district are listed below:

East Kameng			**West Kameng**
Dense forest	:	343	328
Moderately Dense forests.	:	268	234
Open Forests	:	95	107

West Kameng District

The surveryed area represents many tree species. About 39.438 stems per hectre have been left unidentgified. The area commands a beautiful landscape having about 35.756 stems of Rhododendron per hacteare. Besides, the occurrence of Terminalias (3.779), Eurya (5.394), Acer, Betula, Amoora, Canarium, Aquillaria, Castanopsis, Ehretia, Carya, Eriobotrya, Eugenia, Feronia, Ficus Gmelina, Juglans, Machilus (5.943), Magnolia, Michelia (3.325), Morus, Myristica, Spondius etc. species make the area biologically very interesting.

The survery reveals the presince of may more that 200 tree species. Conspicuous species occuring in large number per hectre are presented below :

Broadcleaved stratum : (Number of stems per hectare is 223.678).

Economically important species are :

Quercus griffithii	- 13.631
Quercus lineata	- 4.490
Quercus sp.	- 16.892

Castanopsis indica	- 7.623
Abies pindrow	- 6.718
Alnus nepalensis	- 8.491
Eurya japonica	- 5.394
Machilus odoratissima	- 3.585
Phoebe sp.	- 3.359
Acer Sp.	- 2.972
Schima wallichii	- 2.972
Sygygium cumini	- 2.907
Dysoxylum binecteriferum	- 2.132
Eugenia Sp.	- 2.100
Terminalia Sp.	- 3.773

There were plenty of Rhododendron species most of which were Rhododendron arboreum (35.756).

(x) Manipur State

The forst area of this state is 7621.44 Km^2 of which 17% is under Wet temperate 20% under pine, 51% udner Wet hill, 6% under Semi- evergreen and 6% under Teak-Garjan forests.

Commercially important tree species are - Pinus keseya Quercus sp., Alnus nepalensis, Castanopsis hystrux Schima wallichii, Gmelina arborea, Dipterocarpus tubinatus, Tectona grandis etc.

Semi-evergreen forests (upto 900 m. altitude)

This forests have preponderences of Leuraceous sp. (Cinnamomum Phoebe, Litsaea, Machilus, Lindera, Actinodaphne etc.); Meliaceous sp. (Aphanamixis, Amoora, Toona, Dysoxylum etc.).

Sub Tropical Wet Hills Forests (Alt. - 900 m. to 1800 m.)

This zone comprises of the following conspicuous genera : Saurauia, Phoebe, Beilschmiedia, Schima, Cinnamomum, Morus, Ostodes, Lithocarpus, Engelhardtia, Castanopsis, Toona, Tetrameles etc.

Wet Temperate Forests (Alt. - 1700 to 2700 m.)

Principal species are Quercus lamellosa, Q. lineata, Q. glanca, Nichelia champaca, Magnolia compbellii, Schima Khasiana, Castanopsis tribuloides, Evodia fraxinifolia, Simingtonia populnea, Michelia manipurensis, Ilicium griffithii and Magnolia insignis.

(xi) Tripura

Survey was carried out over 10,477 Km^2. Several Forest Types occur within the surveyed area e.g. -

- 1/3/3C_1 - Eastern Himalayan Bhabar Sal Forests.
- 1/1B/C_3 - Cachhar Tropical Evergreen Forests.
- 1/3/3C/C_3 - Moist Mixed Deciduous Forests.

Most important families are -

Fabaceae	- 44 genera, 94 species.
Poiceae	- 49 genera, 91 species.
Rubiaceae	- 38 genera, 78 species.
Euphorbiaceae	- 35 genera, 69 species.

Principal tree species (according to percentage) are -

Albizzia procera	- 7.37%
Vitex peduncularis	- 5.21%
Bombax ceiba	- 1.56%
Schima wallichii	- 4.40%
Dillenia pentagyna	- 0.48%
Terminalia belerica	- 1.41%
Gmelina arborea	- 2.91%
Lannea coromendelica	- 4.47%
Shorea robusta	- 4.50%
Artocarpus chaplasa	- 1.53%
Toona ciliata	- 6.00%

Other species are -

Syzygium cumini

Sterculia villosa

Garuga pinnata

Michelia montana

Careya arborea

Mallotus phillippinensis.

(xii) South Bhutan

Forest Types : Broad leaved & conifer. The surveyed area along with area distribution including forest sub-types are tabulated below -

Total surveyed area - 10,318.28 Km^2

Forested area - 8,822.02 Km^2 (85.50%)

Broad Leaved (Stems per ha.) : Betula sp. - 19.084, Acer Sp. - 22.802, Schima wallichii - 1.115, Albizzia sp. - 0.162, Castanopsis sp. - 0.246, Prunus sp. - 3.453, Quercus sp. - 0.312, Engelhardtia spicata - 1.674, Ficus sp. - 0.099, Rhodod- endron sp. - 311.387, Symplocos spicata - 2.804, Juglans regia- 0.083, Shorea robusta - 0.983, Tectona grandis - 0.511, Gmelina arborea - 0.304, Michelia sp. - 5.714, Morus laevigata- 1.265, Phoebe goalparensis - 5.705, Terminalia myriocarpa - 1.729, other Misc. Sp. - 187.207.

Conifer forests : Abies densa - 164.599, Juniperus Sp. - 11.223, Larix griffithii - 6./813, Pinus roxburghii - 14.907, pinus roxburghii - 14.907, Pinus wallichiana - 0.303.

Types	Area (Ha)	Percentage of total forest area
Chir forests (Pinus roxburghi)	1.272	0.20
Fir & Spruce forests	44.495	5.00
Mixed confier	3,883	0.40
Broad leaved mixed with conifer	37.035	4.20
Upland and low land Hardwood	7,45,189	84,50
Alpine pasture	35,028	4.00
Blank area	15,300	1.70

North Central and North Eastern Bhutan

North Central are	- 6,760.72 Km^2
North Central area	- 4,137,19 km^2
Total :	- 10,897.91 km^2

Tree forests	: 8,241.26 Km^2
Alpine pasture	: 844.43 "
Non-forested	: 1,723.63 "
Cloud/Gap	: 88.59 "
Total	: 10,897,91 Km^2

Stems per ha. in different Forest Types

Chir - 101.653

Kail - 321.690 - Mostly below 30 cm. dia.

Fir & Spruce - 171.528

Mixed Conifer forests

Conifer - 255.195

Broad leaved - 377.610

Conifer mixed broad-leaved forests

Conifer - 239.496 - Mostly below 38 cm. dia.

Broad leaved - 417.232

Hardwood forests

Conifer - 22.538

Broad leaved - 749.837

North Western Bhutan

Total surveyed area	: 7,620.21 Km^2
Tree Forest area	: 5,609.29 Km^2
Alpine pasture	: 499.23 Km^2
Blank area	: 79.27 km^2

Forest Type	Area (Ha)	Percentage of total forest
Chir forests	19,820	3.50
Kail forest	56,023	10.00
Fir & Spruce forests	1,28,021	22.80
Mixed Conifer forests - (Tsuga, Larix Junipers etc.)	92,512	16.50

Broad leaved mixed with conifer forests	48,791	8.70
Upland hardwood and Lowland hardwood forests	1,58,112	28.20
Alpine pasture	49,723	8.90
Blank areas	7.927	1.40

Forest Types

- Eastern Sub-montane Semi-evergreen Forests - 2B/ C1b: Species are, - Phoebe hainesiana, Beilschmeidia sp., Syzygium sp., Castanopsis sp., Michelia sp., Elaeocarpus sp., Tetrameles nudiflora, Ailanthus grnaids, Cinnamomum sp., Altingia sp., Quercus sp., Schima Wallichii.
- Eastern Hill Sal Forests - 3C/C1a.
- Eastern Bhabar Sal forests - 3C/C1b.
- Eastern Himalayan Moist mixed Deciduous Forests-3C/C3b
- Eastern Hollock Forests - $3C/C_3/S2$ (a,b)
- Upper Himalayan Chir Pine Forests - 9/C1 (a,b).

  Pinus roburghii, Quercus sp., Rhododendron sp. Symplocos sp.
- East Himalayan Wet Temperate Forests : 11B/C1 (a,b,c).

  Quercus lamelosa, Q, pachyphylla, Michelia sp., Acer sp., Magnolia sp., Carpinus sp., Exbucklandia sp.
- East Himalayan Mixed conifer Forests - 12/C3(a).

  Tsuga dumosa, Abies densa, Quercus sp., Acer sp., Taxus baccata, betula sp., Rhododendron sp., Birch - Rhododendron, Alpine.

  Sub-Alpine Birch Fir Forests.

  Temperate Conifer Forests : Larix, Tsuga, Picea, Pinus sp.

Gedu and Changekha area of Bhutan (Western Bhutan)

Area - 524 Km^2 (Plywood resources)

Situation : 26°-49'-56.7" to 27°-25'-04" N

89°-22'-41.6" to 89°-39'-14" E.

Stems/ha. at Gedu

Acrocarpus fraxinifolius - 0.060, Ailanthus grandis - 0.975, Amoora wallichii - 2.511, Betula sp. - 7.708, Mangifera indica - 1.374, Schima Wallichii - 7.122.

Albizia sp. - 5.250, Cedrela sp. - 1.809, Duabanga grandiflora - 0.150, Michelia sp. - 15.823, Quercus sp. - 63.700, Bombasx ceiba - 0.120, Terminalia myriocarpa - 0.460.

Abies densa - 0.181, Acer sp. - 18.244, Exbucklandia populnea - 0.313, Tsuga burniniana - 0.019.

Atnus sp. - 2.793, Castanopsis sp. - 0.253, Machilus sp. - 28.458.

Aesculus assamica - 7.020, Alangium sp. - 0.539, Aphanamixis polystachia - 0.262, Bauhinia sp. - 0.794, Beilschmiedia sp. - 9.164, Bischofia javanica - 0.122, Bridelia retusa - 0.273, Callicarpa arborea - 1.268.

Cassia fistula-0.306, Cinnamomum cecedodaphne-0.197, C. tamala-1.980, Glochidion assamicum- 0.396, Cordia obliqua-0.538, Croton tiglium-0.163, Drypetes lancifolia-2.317, Echinocarpus sterculeaceous-1.506, Elaeocarpus sikkimensis-2.1, Daphnephyllum himalayana - 4.473, Engelhardtia spicata-2.021, Eruthrina sp.-11.993, Garuga pinnata-1.231, Gynocardia odorata-0.857, Gynocardia sp.-6.715, Jambosa formosa-70.206, Juniperus sp. -0.702, Kydia calycina - 0.408, Litsaea sp. - 0.767 Macaranga sp. - 6.765, Meliosma sp. - 6.741, Morus laevigata - 1.739, Ostodes paniculata - 8.066, Phoebe goalparensis - 0.298, Phoebe hainesiana - 0.296, Prunus sp. - 1.976, Rhododendron sp. - 53.402, Rhus sp.- 6.151, Sapium eugenifolium - 0.065, Saurauria nepalensis- 0.689, Spondius mangifera - 0.070, Sterculia sp. - 1.118, Stereospermum sp. - 0.024, Syzygium sp. - 0.657, Talauma hodgsoni - 0.968, Taxus baccata - 0.046, Vitex heterophylla-0.560, Other species - 233.196.

Less common trees are : Aglaia roxburghiana, Adansonia digitata, Aleurites moluccana, Antiaris toxicana, Bixa orelana, Balanites eagiptica, Camphora officiarum, Cinnamomum camphora, Elaeocarpus ganitrus, Evodia lunuankhanda, Engenia caryophyllata, Juniperus communis, Juniperus macropode, Juglans regia, Gordonia obtusa, Hydnocarpus sp., Feronium elephantum, Garcinia gostina, Melaleuca leucodondron, Myristica fragrans, Pterocarpus marsupium,

Pterocarpus santalinus, Pistacia integerina, Schrebera swietenioides, Saraca asoka, Zanthoxylum sp., Z. alatum, Strichnos nuxvomica, Symplocos racemosa, Terminalia chebula, Taxus baccata, Ternstroemia Japonica.

Bhutan (General) : Stems per hactaere

Chir Pine Forests	Pinus roxburghii	102		Mixed conifer Forests		
	Broad leaved	141	243	Tsuga dumosa		
Kail Pine Forests	Pinus wallichiana	322		Larix griffithii Juniperus & Other conifers.	253	
	Other conifer	18				
	Broad leaved	120	460	Broad leaved	377	630
Fir Spruce forest	Abies densa	140				
	Pinus spinulosa	32				
	Other Conifer	43				
	Broad leaved	562	777			

Conifer and Broad leaved forests : (Stems per ha. - on a few species)

Acer- 50, Betula - 28, Quercus - 170, Broad leaved - 168, Conifer - 239 = 655; other types - 770.

(xiii) Assam

Assam has a large forest area. Forest areas adjacent to Bhutan, Arunachal Pradesh, Nagaland, Manipur, Mizorum and Cachhar bear similar types of forests as listed below :-

Tropical Evergreen and Tropical Semi-evergreen types with many sub-types are in existence.

Broadly there are three formations —

(a) Mesua - Dipterocarpus (Evergreen Forests) - in Cachhar.

(b) Kayea assamica formation (Evergreen forests) in Upper Assam.

(c) Assam Valley (Semi-evergreen forests).

Formation (a) : Species percentage are as follows :

Ammora wallichii	- 2.40
Dipterocarpus macrocarpus	- 29.50
Shorea assamica	- 10.32
Mesua ferrea	- 13.24
Mangolia pterocarpa	- 7.00
Cinnamomum glanduliferum	- 2.17
Vatica lancaefolia	- 4.70

Terminalia myriocarpa	- 1.60
T. chebula	- 4.60
T. belerica	- 1.08

Formation (b) - Kayea Forests : Species percentage are as follows :

Kayea assamica	- 68.08
Dysoxylum sp.	- 6.35
Echinocarpus assamicus	- 3.70
Canarium sp.	- 4.31
Pterocarpus lancaefolium	- 6.25
Mesua ferrea	- 2.69
Terminalia chebula	- 2.57
Amoora wallichii	- 1.50

Formation (c) - Assam valley Evergreen forests :

Species percentage are : Castanopsis indica - 3.90, Amoora wallichii - 5.02, Canarium sp. - 2.80, Magnolia sp. - 4.00, Stereospermum chelonoides - 11.00, Mansonia dipikae - 4.50, Cyclostemon assamicus - 5.5 , C. ellipticus - 4.50.

Other areas of Assam :

North of Brahmputra river : Species percentage are : Amoora wallichii - 4.8, Castanopsis sp. - 3.4, Cinnamomum cecedodaphne - 4.5, Dysoxylum procerum - 5.0, Dillenia indica- 4.0, Albizzia lucida - 1.5, Phoebe goalparensis - 9.1, Tetrameles nudiflora - 3.4, Machilus bombycina - 4.9, Nyssa javonica - 2.5.

Nowgoang area : Stems per hectare are -

Lagerstroemia speciosa - 4.6, Bombax ceiba - 2.1, Albizzia procera - 3.0, Castanopsis sp. - 4.6, Syzygium praecox - 1.6, Vitex peduncularis - 1.6, Terminalia belerica - 2.6, Lannea coromendelica - 2.0, Shorea robusta - 5.2, Cassia fistula - 6.0, Lagerstroemia speciosa - 8.0, L. parviflora - 4.9, Bombax ceiba- 2.1, Albizzia procera - 2.2, mansonia dipikae - 1.6, Pterospermum acerifoloum - 1.9, Toona ciliata - 2.2, Ulmus lancaefolia - 4.0.

(xiv) North Nagaland

The forests are of Tropical Evergreen and Semi-evergreen with many sub-types. Juli forests have patches ovf bamboo and reeds.

Types of Forest	No. of Stems per ha.
Evergreen & Semi-evergreen forests :	205
Pure Bamboo Shrubs :	69
Bamboo and Miscellaneous Shrub :	85
Reed Strata :	44

Principal species with stems per hectare

Dipterocarpus macrocarpus - 10.54, Lagerstroemia speciosa - 0.09, Gmelina arborea - 0.53, Dysoxylum procerum - 4.44, Tetrameles nudiflora - 0.69, Terminalia belerica - 1.75, Pterosperum lancaefolium - 6.14, Canarium resinifolium - 6.24, Castanopsis indica - 6.58, Terminalia myriocarpa - 1.17, Syzygium cumini - 2.00, Vatica lancaefolia - 5.17, Endospermum chinensis - 2.29, Magnolia/Michelia sp. - 2.29, Schima wallichii - 2.44, Phoebe goalparensis - 0.14, Shorea assamica - 6.68, Mesua ferrea - 6.49, Artocarpus chaplasha - 0.68, Bischofia javanica - 2.78, Michelia champaca - 1.70.

Most of the trees have medicinal value.

Plants of North Eastern States of India

(xv) Mizoram State :

Dipterocarpus sp., Terminalia sp., Artocarpus chaplasa, Amoora wallichii, Sterculia alata, Michelia champaca, Schina wallichii, Toona ciliata.

Pinus insularis, Quercus sp., Castanopisis sp., Podocarpus nerifolia, Rhododendron sp.

Diyan/Manabhum Reserve : Dipterocarpus macrocarpus, Gmelina arborea, Altingia excelsa, Mesua Ferrea, Magnolia sp., Michelia sp., Kydia calycina, Vatica lancaefolia, Bombax ceiba, Amoora wallichii, Terminalia citrina.

(xvi) North Cachar (Cachar hills)

Situation : 92° 31' 28" - 93° 28' 30" E.
25° 02' 45" - 25° 49' 17" N.

Geographical Area : 4,890 Sq. km.

Forests : Reserve Forest : 663.16 Sq. Km.
Protected Forest : 44.36 sq. Km.

Unclassed Forests : 4,177.30 Sq. km.

The forests of North Cachar are in the State of Assam but form a part of hilly terain, that joins Nagaland and Manipur hills in east and Meghalaya hills in the west.

The forests fall under the following types of Champion and Sethi's classification :-

1B/C_3 - Cachar Tropical Evergreen Forests.

2B/C_2 - Cachar Tropical Semi-evergreen forests.

Northern Sub-tropical Broad-leaved forests.

Bamboo forests (moist and secondary moist).

The forest Survey of India has listed 146 species of trees besides many other un-identified species. There are 13 species of bamboos in the forests. The area is very rich in tree flora, some species of importance are listed below :

Stems per hectare : (Less common trees)

1.	*Amoora wallichii*	0.138	22	*Schima wallichii*	6.227
2	*Lagerstroemia parviflora*	3.669	23	*Phoebe cooperiana*	0.027
3	*Phoebe goalparensis*	0.528	24	*Mesua ferrea*	0.111
4	*Ailanthus grandis*	0.305	25	*Dillenia indica*	5.976
5	*Tetrameles nudiflora*	1.167	26	*Kydia calycina*	2.974
6	*Chukrassia velutina*	0.555	27	*Garuga pinnata*	2.112
7	*Morus lavaegata*	0.333	28	*Artocarpus sp.*	1.612
8	*Terminalia sp.*	8.228	29	*Bombax ceiba*	0.416
9	*Michelia sp.*	0.361	30	*Artocarpus lakoocha*	0.083
10	*Gmelina arborea*	2.196	31	*Adina oligocephana*	1.918
11	*Cinnamomum sp.*	0.139	32	*Albizzia odoratissima*	0.222
12	*Terminalia myriocarpa*	0.027	33	*Bauhinia sp.*	12.287
13	*Castanopsis indica*	0.945	34	*Carega arborea*	5.754
14	*Pterospermum acerifolium*	0.555	35	*Ficus sp.*	1.584
15	*Albizzia sp.*	5.837	36	*Podocarpus nerrifolia*	0.027
16	*Lannea cormendelica*	3.169	37	*Shorea robusta*	33.915
17	*Syzygium cumini*	2.668	38	*Semicarpus anacardium*	0.583
18	*Toona ciliata*	1.362	39	*Trewai nudiflora*	0.585
19	*Duabanga grandiflora*	1.666	40	*Callicarpa arborea*	7.088
20	*Anthocephalus chinensis*	0.139	41	Others	57.490
21	*Machilus sp.*	0.361			

Amongst the listed species, economically important are Sl. Nos. 8, 15, 22, 25, 33, 37, & 40, which have the maximum density besides a few species e.g. Sl. Nos. 10, 16, 17, 26 & 27.

There is one lacuna which is due to elimination of the species, comprising 57.490 stems per hectare and remaining unidentified; yet most of the plants of the above list are of medicinal value.

Detailed information on Density Status of Tree Vegetation is available in the following forests of various States
[Source: Forest Survey of India]

Andhra Pradesh – Mahaboob Nagar Forests, East Godavari Forests, Adilabad Forests.

Arunachal Pradesh – Kameng and Subansiri Forests, Lohit and Tirap forests, Lohit and Debang forests.

Assam– Nowgong and Karbi Hoglong forests, North Cachar hills forests.

Bihar– Central Bihar forests, Bhagalpur forests, West Champoran forests, Singbhum forests.

Goa– Goa forests.

Gujarat – Surat Circle forests.

Himachal Pradesh – Rajgarh & Nahan forests, Chirbell forests, Kulu seray Kotgarh forests, Simla, Rohhrm, Chopal forests.

Jammu & Kashmir –Kashmir valley forests, Udampur-Kalhna & Rajourie forests.

Karnataka – Kodaga forests, Chikmangalur-Hassem forests, Simogu forests, Chitradurga - Tumkur - Kalar forests, Mandaya forests, Kodaga forests.

Madhya Pradesh – Forests of Narayanpur (Bastar), Balaghat, Mandla, Seoni, Indore, Mandsaur, Ratlam, Khandua, Khargone, Shajapur Rajgarh, Jhabua, Ujain, Dhar, Raigarh, Rajnandgaon, Durg, Bilaspur, Raipur forest.

Maharashtra – Forests of Ballarshah, Chanda, Sondad, Bhandora, Chanda, Melghat, Mewasi, Dhulia, Raigad.

Manipur – Entire State.

Meghalaya – Entire State.

Mizoram– Entire State.

Nagaland – Entire State.

Orissa – Forests of Phulbani, Koraput and Kalahandi.

Punjab and Haryana –Forests of Shiwalik region.

Rajasthan – Forests of Udaipur.

Sikkim– Entire State. Forests

Tripura – Entire State. Forests

Uttar Pradesh – Forests of Bhagirathi, Bilangana, Yamuna, Tehri, Garwal, Alaknanda, Almora, Nainital, Pithorgarh, Chamoli, Dheradun.

West Bengal – Forests of Buxa, Singalila, Purulia, Bankura, Midnapur

Andaman and Nicobar Islands – Forests of Little, Middle and North Andamans.

Dadra and Nagar Haveli – Forests of Dadra & Nagar Haveli.

Nepal and Bhutan – Forests of Entire Bhutan & Forests of Western Nepal.

2.9 MEDICINAL PLANTS : LACUNAE IN CONSERVATION AND PRESERVATION

Herbarium for preservation of species

The time is ripe to consider if there should be a number of herbaria where most of the species of medicinal plants will be preserved in sheets and listed. Various herbaria viz. Sibpur Botanical Garden at Howrah under B.S.I., Indian Council of Forest Research and Education, Dehra Dun; Central Drug Research Institute, Lucknow; Drug Research Laboratory in Jammu and Kashmir have a good collection of medicinal plants of India. Fruits and seeds of all species are to be properly preserved in formalin and in other chemicals. This needs proper planning. The regional herbaria of B.S.I. can be a good resource base, as correct identification of species is of prime importance.

It is worthwhile in this connection to mention about vast literature and preserved materials and manuscripts writen on palm leaves in the past which were mutilated by the invadors. The conquerors have introduced their own system of medicines as they were hostile towards the systems by the alien rulers. For about a thousand years Indian research, as a result, suffered a set back.

Plant species estimation

India's bulk evergreen, moist, wet-mixed, temperate and alpine forests are situated in the Himalayas, Western Ghats and in north-eastern India where most of the medicinal plant species are available. Bulk area of the country has dry deciduous forests and consists of a lesser number of species. A small number of species also occurs at stray places like village fringes, scrub lands, fallow lands and on degraded lands. The Government of India has created substantial number of parks for conservation of forest and faunal resources, but no extra importance has been extended towards the conservation of medicinal plants.

India has approximately more than 1300 species of trees, about 4000 species of shrubs and about 10,000 species of herbs. Herbs are the principal sources of medicinal plants. Of these 15,300 species, about 50% have some grades of medicinal properties; but only about 600-700 species are generally considered as recognised medicinal plants. So there cannot be any definite number to pin- point. The Sidha System of medicine uses 600 while Ayurveda - 700, Amchi - 600 and Unani - 700 species. This confirms that efficacious medicinal plants are of limited number of species.

Besides the aforesaid number of species India has more than 2,500 species of Algae, 23,000 species of fungi, 16 species of Lichens, 1800 species of bryophytes, 1,000 species of ferns and 30 million species of micro-organisms (China, Indonesia, Maxico, Venizuala and Several other countries have much richer flora); but their medicinal values are yet to be determined by the researchers.

Current Identification and Collection Methods of Plant Parts

Various reserarch organisations and business farms, dealing with medicinal plants collect plant materials by engaging contractors. These contractors engage local labourers on daily/monthly rates for such collections. The collectors, being poor adopt easiest means to collect as much bulk material as expeditiously as possible. They don't spend time to select plant parts; instead they uproot the entire plant alongwith roots to facilitate the speedy collection. As a result, most of the species are facing depletion or extinction. For instance, when the flowers and fruits are needed for collection, the collectors either lop the branches or fell the tree while collecting. They collect the bark upto cortex level and practically blaze the entire tree damaging cambium layer, resulting in death of plants. The most damaging is the collection of root or root bark, where the plants suffer most. When entire roots are taken out the plant is lost; in case of bulb, rhizome or corm, when the entire underground part is collected without leaving some portion for future regeneration the plant is lost.

Herbal venders get supply of medicinal plants from "Big Crude Drug Dealers"; the others get materials in the local market (village) and from local contractors. But main collectors are called 'Base level Collectors', who are forest dwellers and forest fringe people.

It is true that the drugs used in the indigenous system of medicine, whose therapeutic value have been investigated on scientific line and drugs of established therapeutic value used in pharmacopias of different countries used to grow in great abundance in nature in many parts of India.

Research and Conservation

The Council of Scientific and Industrial Research has several centres where drug research is carried out. These centres are at Jorhat (Assam), Bhubaneshwar (Orissa), Lucknow (U.P.) and Indian Council of Forestry Research and Education, Dehradun.

I.C.A.R. has also the following research areas.

- National Bureau of plant Genetic Research.
- Indian Horticultural Research Institute, New Delhi.
- Central Arid Zone Research Institute, Jodhpur.
- Central Plantation Crop Research Institute, Kesoragod.

- Universities, involved in drug research are :-
  - G.B. plant University of Agriculture & Technology, Palne.
  - Punjab Agricultural University, Ludhiana.
  - Tamil Nadu agricultural University, Coimbatore.
  - Gujarat Agricultural University, Anand.

Medicinal Plant Directorates of West Bengal and Tamilnadu carry out research.

The research should pinpoint the following points:

- Which part or parts are used :- Is it root or root bark, leaf, flower, fruit or any other part.
- Which time of the year such parts are to be collected to get best efficacy.
- To what extent these are to be collected from a plant, so that the plants do not die out or eradicated?

While collecting medicinal plants, plant parts are to be collected most carefully so that the following precautions are taken :

- Not to collect entire quantity of seed or fruit from a plant.
- Very small part of the plant stem bark has to be stripped.
- Only a small portion, it at all, of the root bark needs to be collected.

Most important point is to collect these parts when the active principle is at its best in order to get maximum efficacy from plant parts (for this purpose itensive laboratory research is necessary).

In this context some salient features are listed below in respect of plants of medicinal value :

- All plants in part of full have some medicinal values.
- A very high percentage of plants, marked as medicinal plants by pharmacologists, have a wide range of usages, with no definite impact on any particular disease.
- No plants can be identified that do not have some medicinal values.
- Root-bark has been found to be very effective for a good number of diseases. This makes such medicinal plants extremely vulnerable for extinction as the collectors uproot the entire plant for collection of root-bark.
- Very few pharmacologists mention about the season of collection of plant parts. Efficacy of the plant parts used in various diseases varies widely from season to season.
- Many plants have almost similar and a number of medicinal properties for cure of various diseases. No mention has been made clearly by the pharmacologists which plants are more effective than the other for any particular disease.

Cultivation of Medicinal plants

"India is a veritable emporium of medicinal plants. Nearly three-fourth of the drugs mentioned in the British and other pharmacopocias grow here in a state of nature - Says R.N. Chopra in his work 'Indigenous Drugs of India (1958). Besides medicinal plants many kinds of perfumes and species grow in this country.

Medicinal plants mostly occur in remote forest areas and occur sporadically, which make collection difficult and transport cost effective. The Indian Council of Agricultural Research, Council of Medical Research and Industrial Research and Indian Council of Medical Research and other organizations have not given as much attention as it is necessary in the cultivation of medicinal plants. Chopra admits what the present authors say - (i) drug plants occur sporadically (ii) transportation is difficult and (iii) ignorance of the collectors lead to admixture of genuine plants with spurious ones.

Cultivation of Digitalis, Cinchona, Ipecaunha, Pyretherum, Dioscorea, Rauwolfia, Podophyllum, Soessaurea, Piper etc. were being done on a small scale.

In 1941. The Committee of Medical Research Council, Great Britain felt the urgent need for cultivation of medicinal plants and the following drugs were recommended for production (Chopra) :-

(1) Acidum tartaricum, (2) Agar, (3) Balsamum, tolutatum (4) Benzoin, (5) Calumba, (6) Camphor, (7) Cantharids, (8) Cascara sograda, (9) Chrysaroben, (10) Cinchona (11) Coca (12) Colophony, (13) Creosote, (14) Datura, (15) Derris, (16) Ephedrine (17) Ergot, (18) Gentian, (19) Glycyrrhiza, (20) Hammamelis, (21) Hyocyamus muticus, (22) Ipecacuanha, (23) Jaborandi, (24) Mnameria (25) Lobelia, (26) Menthol, (27) Mylabris (28) Oleum cadinum (29) Oleum amygcale, (30) Oleum anisi (31) Oleum Chenopodii (32) Oleum Menthae, (33) Santonin, (34) Scilla (35) Storax, (36) Thymol, (37) Tragacanth.

However, the following few species of medicinal plants are being cultivated in various parts of the country.

Pipen longum, Cardamon, Clove, natmeg, Ginnamom, Ginger, Termuric, Hentha arvensis, Mentha piperata cymbopogon in imtranus, C. Mastini, Vetiveria zizamoides Eucalyptus cotriodora, Santalum album, Oseimum, gratissimum, Jasminum sp., Geraneum sp., Saussurea sp. Cyperus rotundus, Skimia laurala etc.

Burning of Dhup, Agarbsati, Havan Samagri during festive occasions of 'Hindus' were in vogue since early civilization. Fragrant flowers were used to be offered to the deities. Indians also used Sandal wood, basil, roses, mint, ajwan and fannei.

Research and Development

Following are R & D centres in India :

1. Council of Scientific and Industrial Research :
   - (a) R.R.L. at Jorhat.
   - (b) R.R.L. at Bhubaneswar.
   - (c) Central Institute of Medicien and Aromatic plants at Lucknow.
   - (d) Central Botanical Research Centre at Lucknow.
2. I.C.A.R. Research :
   - (a) National Bureau of Plant Genetic Research.
   - (b) Indian Horticultural Research Institute.
   - (c) Central Arid Zone Research Institute, Jodhpur.
   - (d) Central Plantation Crop Research Institute, Kasaragod.
3. Universities :
   - (a) G.R. Plant University of Agriculture and Technology,
   - (b) Punjab Agricultural University, Ludhiana.
   - (c) Tamilnady Agricultural University, Coimbatore.
   - (d) Kerala Agricultural University, Oddakali.
   - (e) Gujarat Agricultural University, Anand.
4. Others :
   - (a) Indian Council of Forestry Research and Educagtion : Dehra Dun.
   - (b) Cinchona Departments.

Kirtikar and Bose's list

Ferns 27 species have been mentioned. Of these some common genera are - Pleopeltis, Lygodium, Equisitum,

Adiantum, Stenocloena, Drynaria, Marsilea, Salvinia and Azolla.
Scattered genera (not common) are - Asplenium, Athyrium, Aspidium, Ophioglossum, Botrychium and Cheilanthus.

In temperate forests, where most of ferns occur various species are being depleted due to grazing, fire and extention of cultivation. Some areas of Darjeeling hills which were once home of ferns are now barren.

Lichens About Lichens they mention are : *Parmelia perforata, P. perlata* and *P. kamstchadatisa.*

Gymnosperms

(a) Rare Species are - Cicas rumphii,
C. revoluta
Ephedra gerardiana.
Gnetum scadens.

(About Ferns, Mosses, Liver worts, Fungi and Lichen)
Source : 'Indian Medicinal Plants' – Kirtikar & Bose.

(b) Localised but rare species are - Taxus baccata,
Juniperus communis.
J. recurva.
J. macropoda.

(c) In temperate hills (lower & Upper) :
Common species are -
Pinus wallichiana,
Cedrus deodora
Abies spectabilis (webbiana)
Cupressus sempervirens.

(d) At lower tropical & Temperate belts :
Pinus roxburghii.

Herbal Medicines

India is passing through a period of 'Herbal crisis' which the demand for herbal medicines is blooming across the World.

Primitive people of antiquity have been using herbal

medicines for ages prior to the use of moderned analgesic, antipyretic, antiemetic, antitissive antiseptic, antibiotic and diuretic drugs.

It is said the Sumerians used (3500 B.C.) Atropha belladonna (deadly night shade) against pain and Sickness. The Assyrians (4000 BC) used Ergot (Claviceps purpurea) for control of bleeding. The Egyptians used Sena leaves (Cassia angustifolia). The Chinese used Ephedra in 2000 B.C. for cure of respiratory infections Many more examples may be cited.

The vedic and post-vedic periods since 3500 B.C. saw some celebrated Indian Physicians e.g. Atreya, Mahabarati, Nagarjun Sagar, Vaghbhata, Sushruta, Charaka who were legendry figures of Indian medicine.

Although the traditional Medicines had been replaced to a great extent by chemical and synthetic medicines, herbal medicines are once again gaining ground as the synthetic one are costly and found toxic and proved to be health hazardous.

People suffering from chronic diseases are willing to use traditional medicines than the synthetic ones. Several countries also are looking back towards the 'Traditional Folklore Medicines'. China resisted all imports of modern medicines by use of more and more traditional ones. So herbal wealth of India needs conservation so that traditional species could be used more and more.

In 1948, Col. R.N. Chopra submitted a report recommending revival of Traditional Indian Medicine and its synthesis with modern medicines. Such integration were also advocated by the Government of India.

The World Health Organisation (WHO) also gave gormal recognition to traditional medicines and emphasized global researches into medicinal herbs and to associate the traditional healers in the National Health Care Programme long ago.

Revival of cultivation of medicinal plants — In recent years several states and private organisations have start cultivation and conservation of medicinal plants in gardens and Arbourata. One such endeavour adopted by Orissa State (The Statesman New Service, Feb. 6, 2001) is worth mentioning.

The first phase of an ambitious "Medicinal Garden" was launched here recently with the District Environment Society,

Khurda, making a modest beginning by planting 50 different species of medicinal plants near patrapara village.

The species planted in the first phase area of 40 acres have been procured from the nearby Brebera forests and from the silviculturist, Orissa.

Funded by the forest department, district rural development agency and by a Rajya Sabha MP Mr. BJ Panda, the plan is to develop Medicinal Garden over a total area of 250 acres for plantation in three phases. This area happens to be a degraded Sal forest and is close to the state capital forest department officials said.

Emphasising on the need to recognize, preserve and develop such plants, they said 8000 species of medicinal plants are used by different medicine systems in the country.

Ayurveda uses over 1769 variety of plants while folk medicine depends on over 4600 such plants and Unani on 751.

Yet the forestry sectory is unaware of 90 percent of the Country's medicinal flora that are found in forests, they observed.

In Orissa, the depletion of medicinal plants is primarily due to unscientific collection from forests. In fact there is no database on the number of varieties available in the forests of Orissa, they added.

The officials listed 11 medicinal plant species which were classified as either "critically endagnered" or "endangered". Export of some of these varieties have been benned, said forest officers while highlighting the importance of the project.

They also referred to the Kerala experience where the Girijan Cooperative society is managing over 120 non-timer forest produce, including a sizeable number of medicinal plants.

The Medicinal Garden project has started with an initial allocation of Rs. 13 lakh. It's first phase is spread over 40 acres, the second phase will be of 60 acres with bigger plots for each species and a nursery, and the third phase of 150 acres will be developed. Priority will be given to plants available in Berbera forests.

2.10 MEDICINAL PROPERTIES OF SOME SELECTED PLANT SPECIES

This subject has been dealt with very elaborately by all the front line authorities on the subject. As such the information presented in their work is not being repeated. Sibakali Bhattacharyya has dealt with various aspects of each sp. In his work (in 10 volumes) entitled "Chiranjib Vanusodhi', which seems to cover this subject elaborately. Work of other authority also give similar information.

In order to have an broad idea about the subject the present authors have enumerated some information in diagrams & otherwise for easy access to the subject by ordinary readers. These are some diagramatic and some non-diagramatic presentations of several very important and common plants.

There is perhaps a lacuna in the work of the frontline workers which is the omission in their work in mentioning "remedy of a disease by a specific plant" (Mention has been made of several plants as medicinal agents for particular disease). So as patient does not know what plant would be best to remedy the disease they are suffering from. The present authors are not sure if a plant or plants can be pinpointed to remedy a particular diseases.

Some common Plants and their use in various Diseases

Scientific name (1)	Common Name (2)	Medicinal Uses (3)
Bombax ceiba	Semul	Dried stem and bark used in impotency, Gastro-intestinal disorder, Cough and Female diseases.
Butea monosperma	Palash	Seeds and gum used in eradication of intestinal worm, sexual impotency, restoration of mentrual cycle, prevent pregency.
Mimosa pudica	Lajwati	Seed used in sexual impotency, Checking flow of seman through urine, strong antihelmentic.
Helecteres isora	Marorphali	Whole part used in Flatulence, Gastro intestinal complaint, indigestion and constipation.

(1)	(2)	(3)
Mucuna pruriens	Kiwanch	Seeds are used in treatment of sterility in male, increase sexual power, vigour and strength. When used in combination with Aswagandha best result is obtained.
Tribulus terrestris	Gokhura	Fruit used in urinary disease and nerve tonic.
Piper longum	Pipal	Used in Chronic bronchitis, Cough Cold, Asthma.
Holarrhena antidysentrica	Kurchi	Seeds used in treatment of worms.
Terminalia chebula	Harra	Fruit used in stomach and intestinal diseases in combination with and Bahera.
Centella asiatica	Thankuni	Leaves for stomach disorder and improve mental power.
Evolvulus alsinoides	Gurmarbuti	Leaves a wonder drug in treatment of disbetes.
Cocculus villosus	Jaljamni	Leaves in treatment of venetral disease and urinary disorder.
Cassia angustifolia	Sena	Leaves in treatment of Gastro-intestinal disorder, removes constipation, restore stomach function.
Myristica fragrans	Jaiphal	A nutritive fruit used in treatment of debility and loss of strength.
Astercantha longifolia	Talima-khana	Seeds used in treatment of Gonorrhoea and Jaundice.
Lannea coromendelica	Jhingan	Gum used in body pain.
Rubia cordifolia	Manjisla	Dried roots used as blood purifier.
Gardeina gummifera	Bhusur	Gum used in chronic pain like arthritis
Amaranthus spinosus	Kantanote	Laxative, diuretic, stomachic, antipureetic, appetiser, blood disease lucorrhoea, snake bite.
Spinacia oleracia	Spinach	Cooling, laxative, disease of blood, asthma, Leprosy, antipyretic.
Basella rubra	Puisag	Soporific, narcotic, aptrodisaic, laxative, dysentery, tonic, gonorrhoea.
Ricinus communis	Castor oil	Purgative, Skuidiasese, burns, liver trouble, lumbago, boil, pile, liver trouble, ringworm asthma rheumatism.
Acorus calamus	-	Gastric and repiratory trouble, dyspepsia, dysentery, epilepsy, ulcer, hysteria.
Allium cepa	Pyaj	Bronchial complaint, piles, hysterical feats epileptic fits.

(1)	(2)	(3)
Allium sativum	Rasun	Dyspepsia, hiecough, infantile corvulsions, nervous affections.
Basella rubra	Pui Sag	Genitourinary tract, headache, insomnia.
Rauwolfia serpentina	-	In bites of poisonous reptiles and insects, corneal opacity, Cholders in painful bowlel affection , insanity, high blood pressure, insomnia.
Piper nigram	Golmorich	Anthelmentic, asthma, throat disease, piles night blindness, carminative, aphrodisaic purgative, toothache.
Euphorbia hirta	-	Dysentery, colic pain, bowel complaint, increase flow of mith in women, diarrhoea.
Emblica officinalis	Amloki	Carminative, Laxative, tonic, antipyretic, biliousness, urinary discharge, leprosy, discharge, leprosy, anaema, opthalmia, aphrodasiae asthma, bronchitis.
Jatropha curcas	Bharenda	Anthelmentic, Chronic dysentery, Urinary discharge, anaema, fistula, bad gum, eczema Wing worm.
Cynodon dactylon	Durba graze	Dysentery, Wound, Pyrrhoea, Piles.
Phoenix dactylifera	Khejur	Bronchitis, Haemoptysis, worm, Cough.
Piper cubeca	Kababchini	Dysentery, Asthma, Gont
Aegel marmelos	Bel	Cough, Cold, Dysentery, Scurvy, Typhoid, Constipation.
Santalum album	Chandan	Fever, cough, Gonorrhoea Epilepsy, Pox, Whooping cough.
Andrographis paniculata	Kalmegh	Debility, Dysentery, dyspepsia, febrifuge, Stomachic, tonic, anthelmintic, Kalazar.
Justicia adhatoda	Vasaka	Bronchitis, leprosy, blood purifier, heat trouble, asthma, vomitting, loss of memory, leucoderma, diuretic, bronchitis, asthma, gonorrhoea.
Vitex negunda	Nisindha	Astingent, Stomachic, Anthelmintic, for growth of hair, eye disease, leucoderma, bronchitis, asthma.
Ocimum basilicum	Bhuitulsi	Stomachic, anthelmenlic, cough, gonorrhoea, diarrhoea, scorpeon sting, snake bite.
Ocimum sanctum	Tulsi	Stomachic, antipyretic, heart disease, leucoderma, bronchitis, expectorant.
Curculigo orchioides	Shyam mushli	Sexual impotency, Veneral and urinary diseases, asthma, jaundice, (Dried tuberous roots)

LOSS OF QUALITIES OF SOME MEDICINAL PLANTS OF INDIA
(DIAGRAMMATIC PRESENTATION)

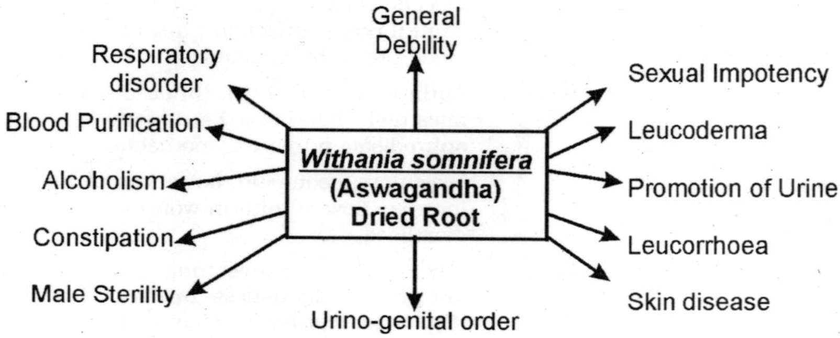

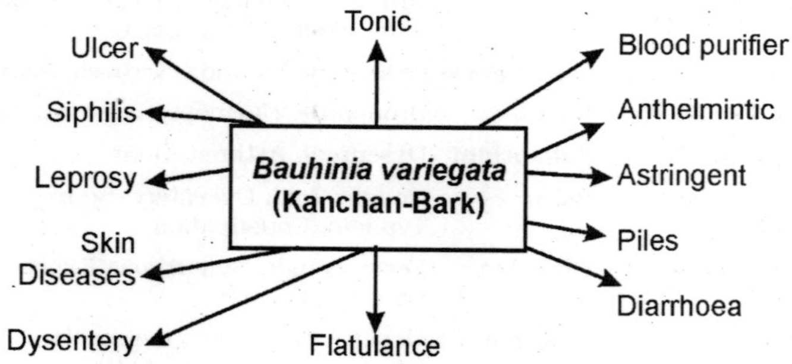

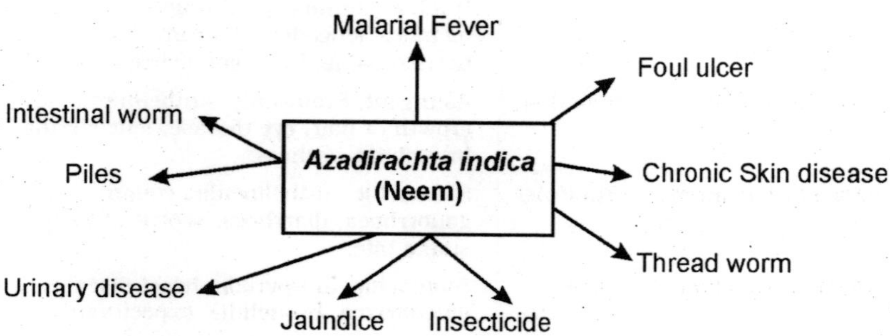

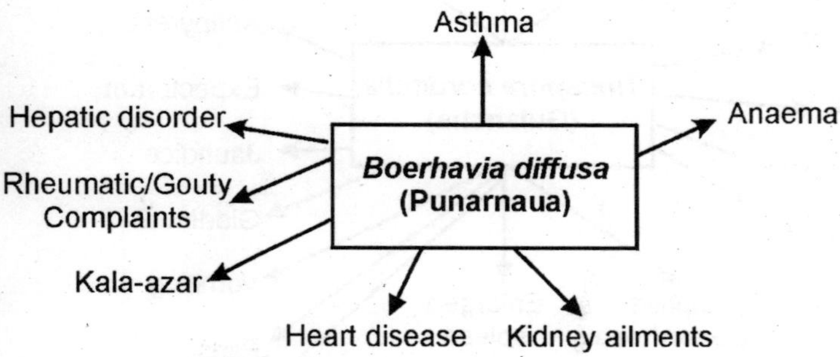
Asthma
Hepatic disorder
Anaema
Boerhavia diffusa
(Punarnaua)
Rheumatic/Gouty
Complaints
Kala-azar
Heart disease
Kidney ailments

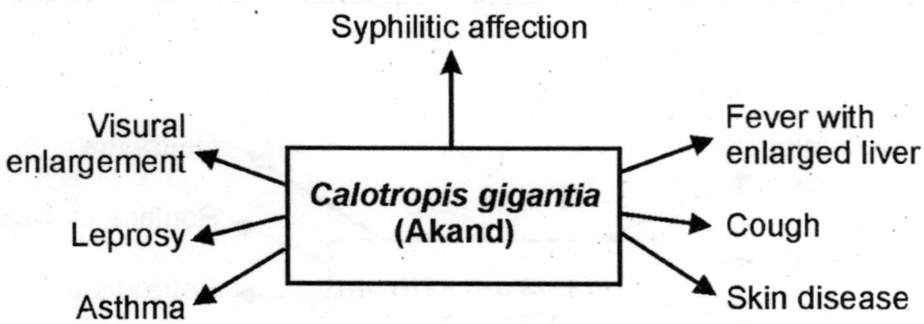
Syphilitic affection
Visural
enlargement
Fever with
enlarged liver
Calotropis gigantia
(Akand)
Leprosy
Cough
Asthma
Skin disease

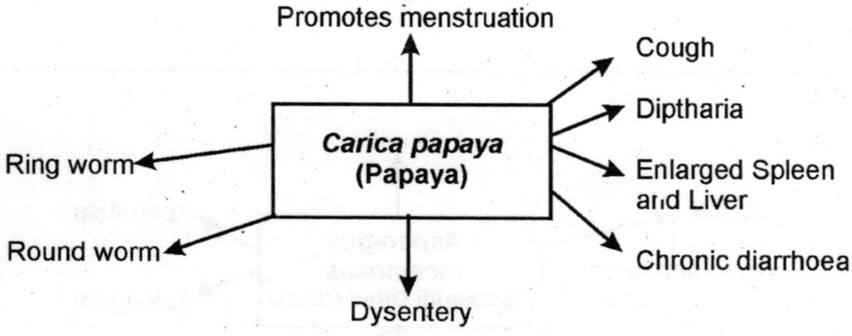
Promotes menstruation
Cough
Diptharia
Carica papaya
(Papaya)
Ring worm
Enlarged Spleen
and Liver
Round worm
Chronic diarrhoea
Dysentery

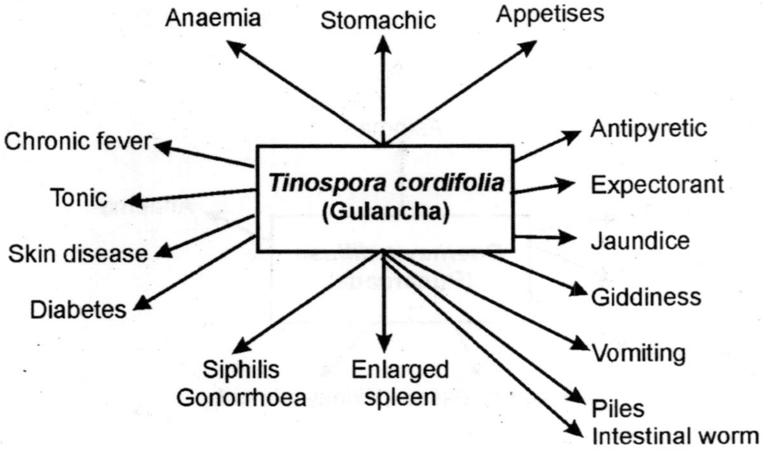
Anaemia
Stomachic
Appetises
Chronic fever
Tonic
Skin disease
Diabetes
Tinospora cordifolia
(Gulancha)
Antipyretic
Expectorant
Jaundice
Giddiness
Vomiting
Piles
Intestinal worm
Siphilis
Gonorrhoea
Enlarged
spleen

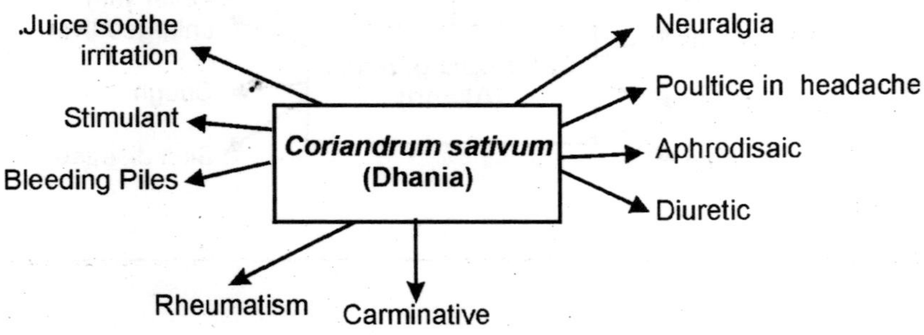
.Juice soothe
irritation
Stimulant
Bleeding Piles
Coriandrum sativum
(Dhania)
Neuralgia
Poultice in headache
Aphrodisaic
Diuretic
Rheumatism
Carminative

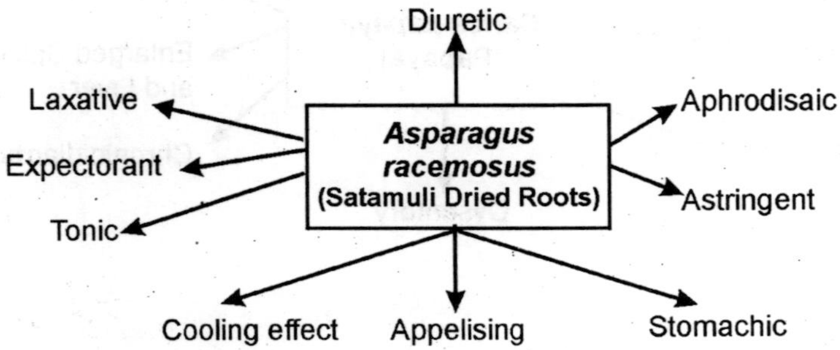
Diuretic
Laxative
Expectorant
Tonic
Asparagus
racemosus
(Satamuli Dried Roots)
Aphrodisaic
Astringent
Cooling effect
Appelising
Stomachic

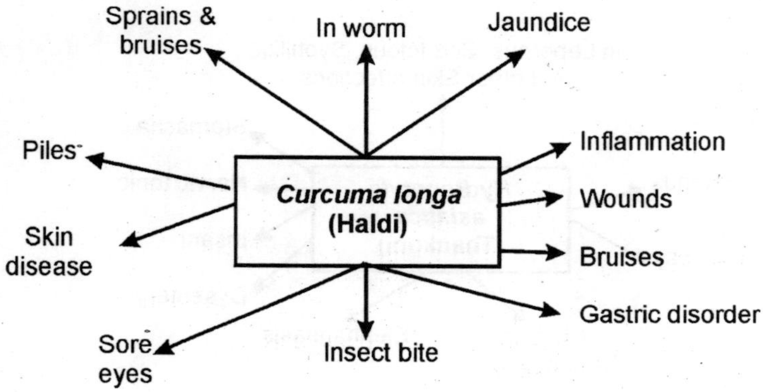
Sprains &
bruises
In worm
Jaundice
Piles
Inflammation
Curcuma longa
(Haldi)
Wounds
Skin
disease
Bruises
Gastric disorder
Sore
eyes
Insect bite

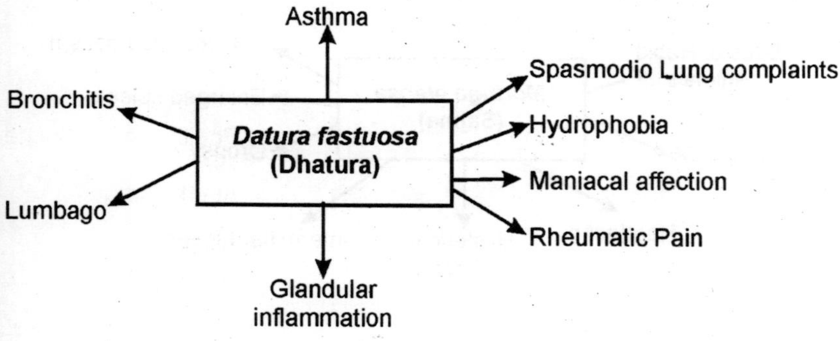
Asthma
Spasmodio Lung complaints
Bronchitis
Hydrophobia
Datura fastuosa
(Dhatura)
Maniacal affection
Lumbago
Rheumatic Pain
Glandular
inflammation

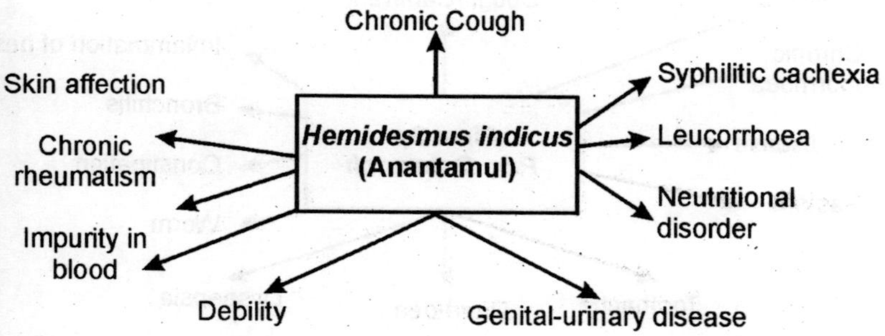
Chronic Cough
Syphilitic cachexia
Skin affection
Hemidesmus indicus
(Anantamul)
Leucorrhoea
Chronic
rheumatism
Neutritional
disorder
Impurity in
blood
Debility
Genital-urinary disease

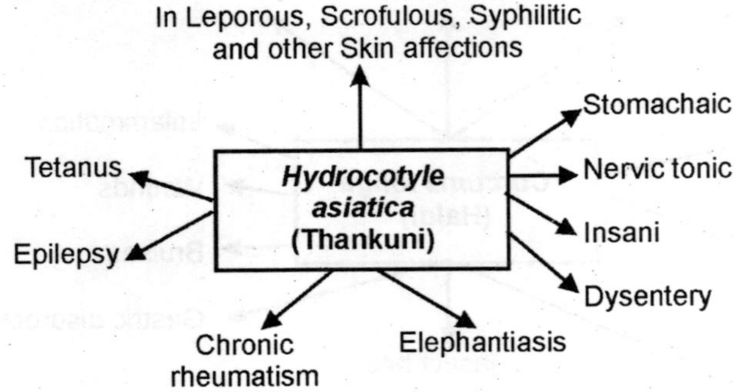
In Leporous, Scrofulous, Syphilitic and other Skin affections
Stomachaic
Nervic tonic
Insani
Dysentery
Elephantiasis
Chronic rheumatism
Epilepsy
Tetanus
Hydrocotyle asiatica (Thankuni)

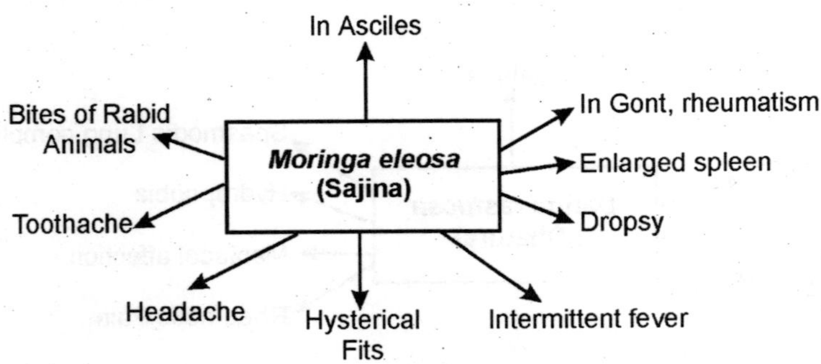
In Asciles
In Gont, rheumatism
Enlarged spleen
Dropsy
Intermittent fever
Hysterical Fits
Headache
Toothache
Bites of Rabid Animals
Moringa eleosa (Sajina)

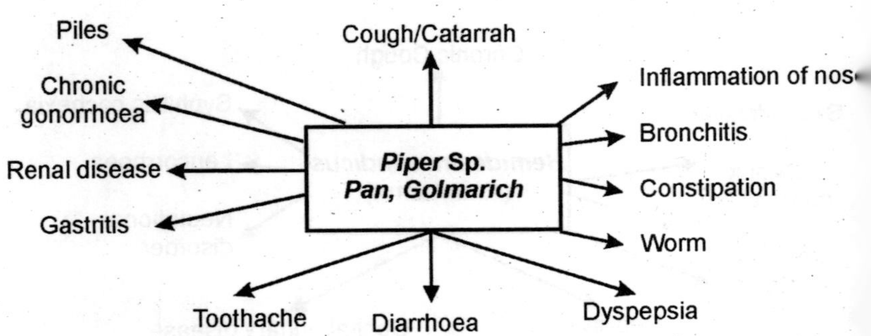
Cough/Catarrah
Inflammation of nos
Bronchitis
Constipation
Worm
Dyspepsia
Diarrhoea
Toothache
Gastritis
Renal disease
Chronic gonorrhoea
Piles
Piper Sp. Pan, Golmarich

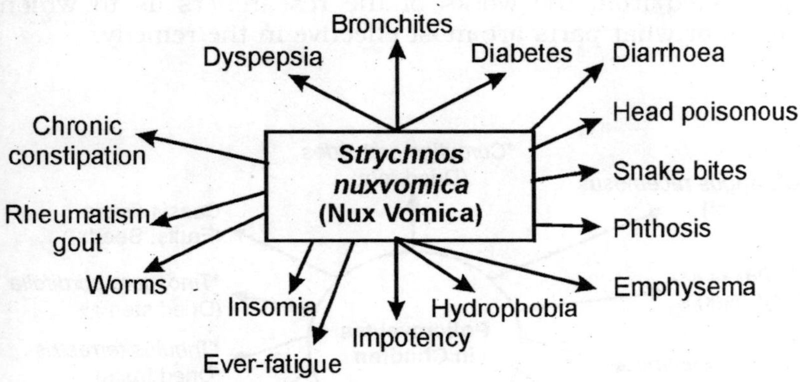
Bronchites
Dyspepsia
Diabetes
Diarrhoea
Head poisonous
Chronic constipation
Strychnos nuxvomica
(Nux Vomica)
Snake bites
Rheumatism gout
Phthosis
Worms
Emphysema
Insomia
Hydrophobia
Impotency
Ever-fatigue

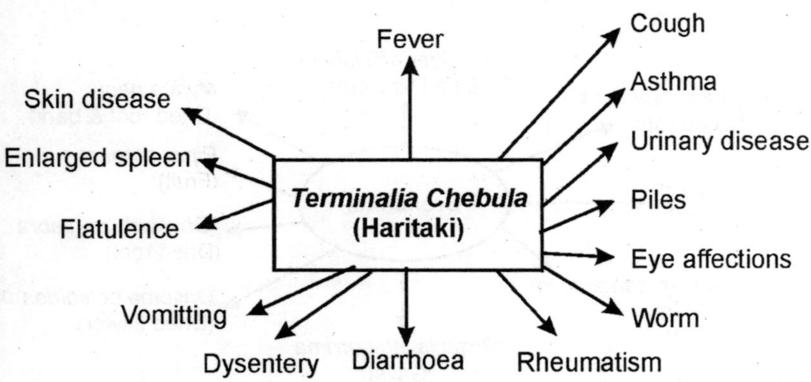
Cough
Fever
Asthma
Skin disease
Urinary disease
Enlarged spleen
Terminalia Chebula
(Haritaki)
Piles
Flatulence
Eye affections
Vomitting
Worm
Dysentery
Diarrhoea
Rheumatism

Some common diseases and their remedy from plants

Many plants and their parts are source for remedy of a single disease and also many more diseases. It could not be ascertained from the works of the researchers as to which species or what parts are most effective in the remedy.

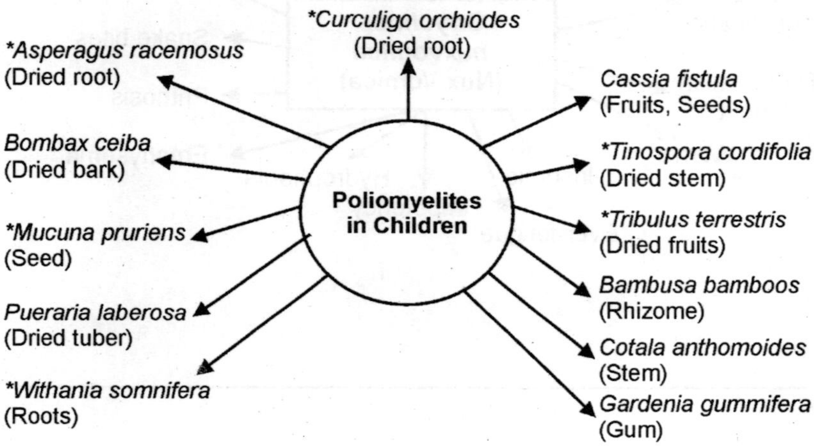

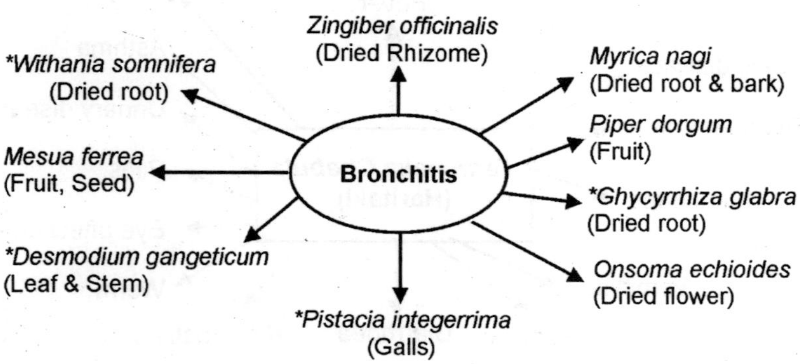

* Not common, rare

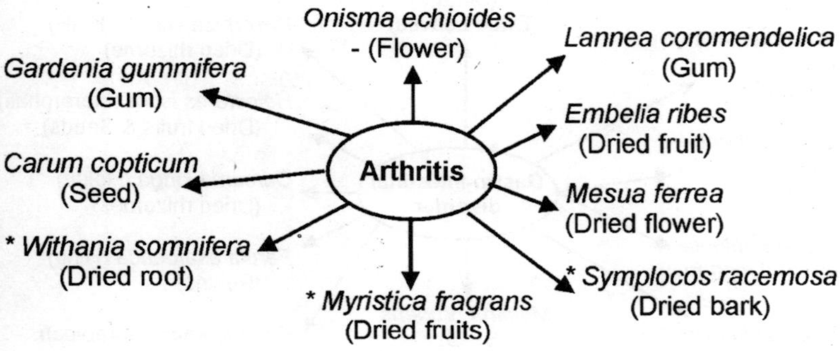

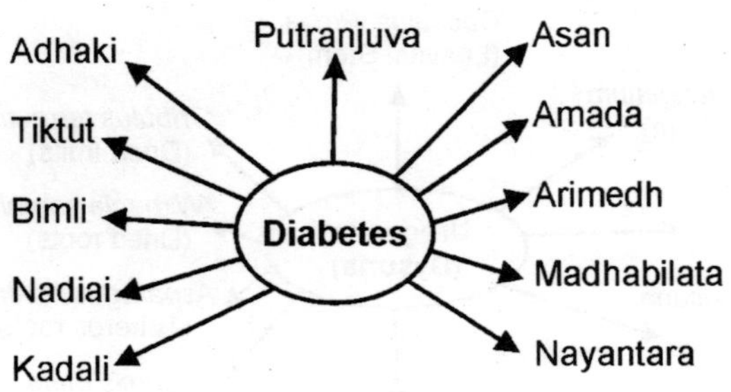

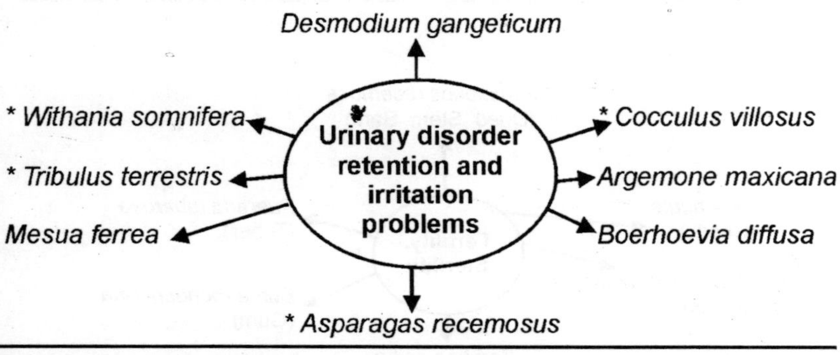

*Not common, but rare

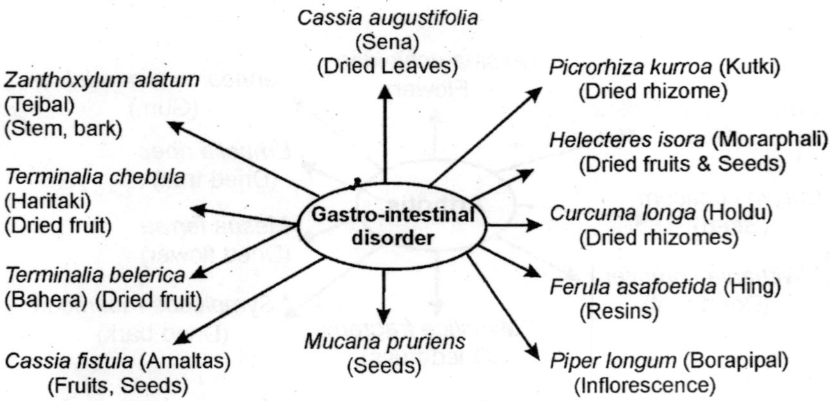

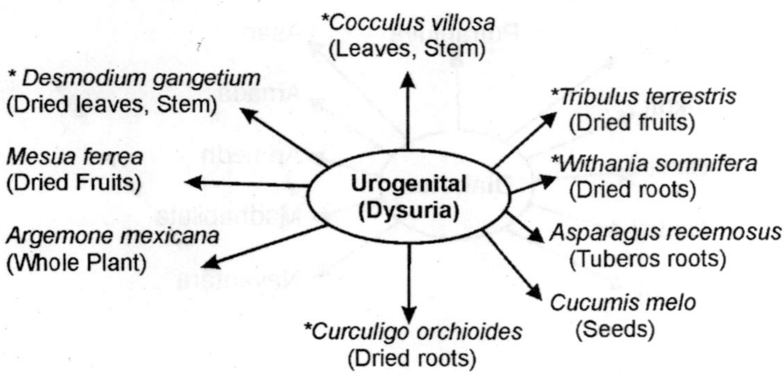

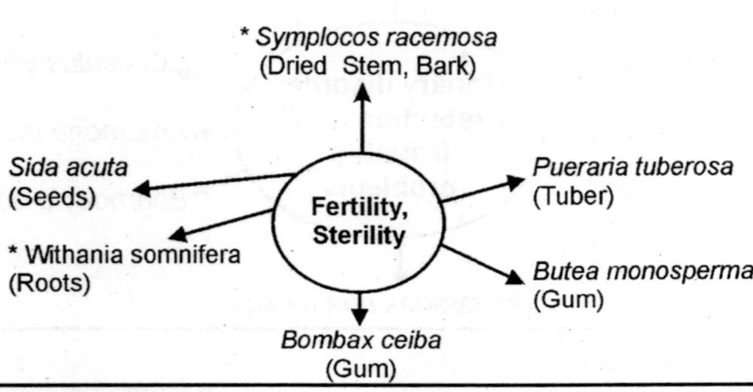

* Not common, but rare

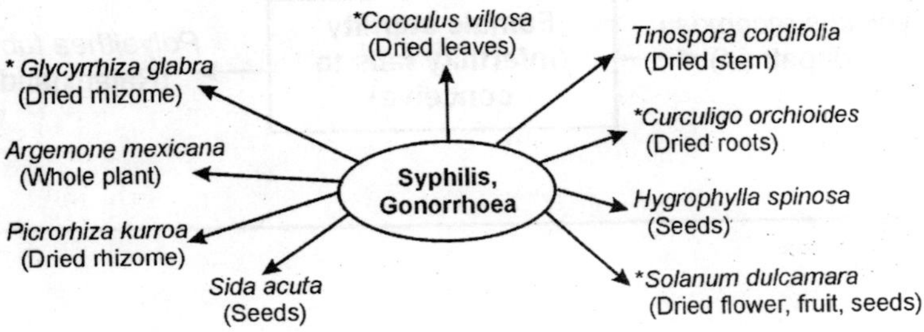

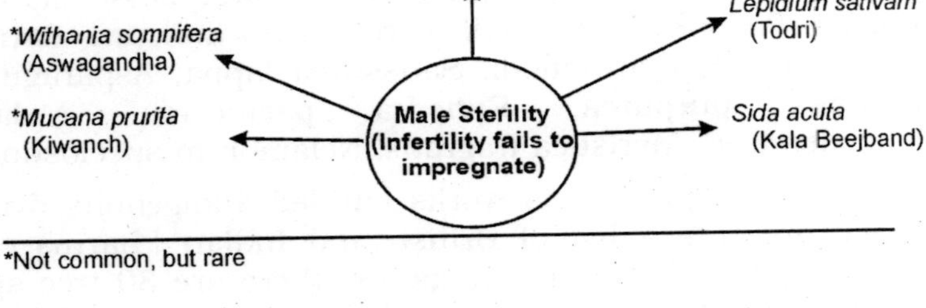

*Not common, but rare

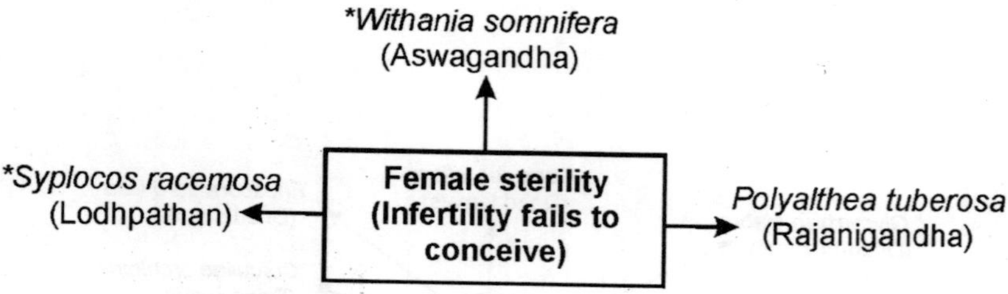

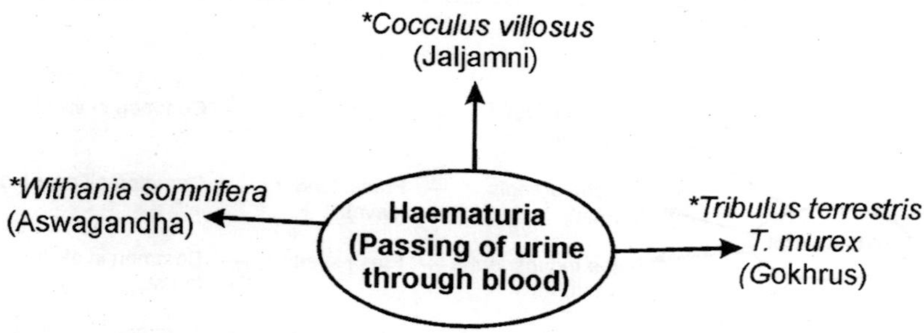

*Not common, but rare

Of the entire list of 550 species, some localised endemic and rare plants listed are : Aconitum sp., Aloe sp., Balsamodendron sp., Bixa orellana, Bryonia epigoena, Caryota urens, Coptis teeta, Gaultheria fragantissima, Glycyrrhiza glabra, Hyoscyamus niger, Nardostachys jatamansi, Picrorrhiza kurroa, Podophyllum emodi, Pyrenthrum indicum, Rauwolfia serpentina, Rheum emodi, Saussurea lappa, Asparagus sp., Digitalis purpurea, Ephedra pachiclada, Melaleuca leucodendron, Myristica fragrans, Nelambium speciosum etc.

R. N. Chopra in his works entitled 'Indigenous drugs of India' presents a list of British and Indian Pharmaceutical Drugs, growing in India. In the list there are 30 tree sp., 13 shrubs, 5 climbers or trailers and balance 81 are herbs. Of the herbs the uncomon genera are Aconitum, Gaultheria, Glycrrrhiza, Hyoscyamus, Picrorhiza, Podophyllum, Rauwolfia, Rheum, Saussurea, Swertia etc.

Some more diseases and plants used as remedy

Anti-coagulatant : *Justicia adhatoda, Allium sativum, Azadirachta indica, Camelia sinensis.*

Cardiovascular (Hypertensive and Cardio tonics) : Aegle marmelos (root, bark), Allium sepa, Allium sativum, Arachis hypogea, Asclepias curassavica, Cururbita maxima, Eclipta alba, Piper aurantiacum, Rauwolfia serpentina (root), Terminalia arjuna (bark).

Aphrodisiac : Allium sepa, Allium sativum, Asparagum racemosus, Glycyrrhiza glabra, Hibiscus-rosa-sinensis, Myristica fragrans, Mucuna pruriens, Strychnos nux-vomica, Tinospra cordifolia.

Antiarthritis, Anti inflamatory : Acorus calamus, Allium sativum, Boswellia serrata, Calophyllum innophyllum, Capsicum sp., Curcuma longa, Echlipta alba, Solanum nigrum, Withania somnifera, Xanthium Strumerium, Zingiber officinale.

Antiasthmatic : Acalypha indica, Acorus calamus, Justicia adhatoda, Albizzia lebbek, Ephedra sp., Ocimum sanctum, Vitex negundo, Terminalia belerica, Piper longum, Saussurea lappa, Picrorhiza Kurroa, Lobelia sp., Hedycium sp.

Antidiabetic : Aegle marmelos, Allium sativum, Anona squamosa, Asteracantha longifolia, Azadirachta Indica, Blumea sativum, Cinnamomum tamala, Curcuma longa, Emblica officinalis, Grewia sylvestre., Momordica charantia, Zingiber officinale, Swertia chirata, Zyzygium cumini (seed), Musa paradisiaca.

Bronchitis (Respiratory disorder) : Picrorhiza kurroa (Kutki)-dried rhizome; Zingiber officinale (Adrak) - dried rhizome, Myrica nagi (Kaifal) - dried root & bark; piper longum (bari pipal) - Fruits; Glycyrrhiza glabra - (Muleltee) - dried roots; Pistacia integerrima (Kakrusungi) - Galls; Desmondium gangeticum (Salparni) - Leaf, stem; Mesua ferrea (nag-Kashar)-Fruit, Seed; Withania somnifera (Anwagandha) - Dried roots.

Liver disease : Acacia catechu, Azadirachta indica, Andropogon paniculata, Asteracantha, longifolia, Andrographis paniculata, Allium sativum.

Boerhaavia diffusa, Cynodon dectylon, Capparis spinosa (Climbing shrub); Emblica officinalis.

Memory stimulant : Acorus calamus, Bacopa monrieri, Centrella asiatica, Celastrus paniculata (Climber), Glycyrrhiza glabra.

Hyper acidity : Acorus calamus, Asparagus racemosa, Emblica officinalis, Glycyrrhiza glabra, Zingiber officinale.

Viral Diseases : Among the viruses that infect human beings, most common are HIV (Human Immunodeficiency Virus), HBV (Hepatitis-B Virus), IV (Influenza virus), PV (Polio Virus), MV (Measles virus) and some other viruses. Of these HIV cause AIDS (Acquired Immuno Deficiency Syndrome), appears lethal. An HIV attacks four lymphocytes of the human immune system. Plants kill HIV directly are papaver (Opium), wild snake, Periwinkle, Croton, Anemone, Gourd, Licorice (Jastimadhu, Castor (Aranila), Pine cone, King Bitter (Kalmegh) etc. Plants having components to interfere replications.

Galactagogue :- Plants containing several compounds which induce milk production in greater volume and secretion of the same in the human beings are termed galactagogues.

Number of plants species recognised as experimentally proven species are Alstonia Scholaris (Chattni), Oat, Fig (Dumoor), Fenugreek (Methi), Lettuce (Jeera), Barley, Onion, Garlic etc.

2.11 SOME POTENTIAL DRUG PLANTS OF INDIA (List prepared by R.N. Chopra)

Chopra remarks "the drug resources of India are vast and inhaustible" . . . . .and that "India could supply the whole of the civilised work with medicinal herbs".

The present authors do not agree with the statement as the present scenario is completely different.

Chopra has prepared a list of British and Indian Pharmacopoeal Drugs that grow in India which is reproduced below with remark on the present status.

The symbols used are :

D = Depleted; PS = Presently safe; C = cultivated; S = Sporadic (vulnerable) throughout the list.

Similar lists have been prepared by Shibakali Bhattacharya and Kirtikar & Bose.

(NB. Present status of species indicated by the authors).

2.11.1 Some Potentical drug Plants of India (From the work of Chopra and Charaka)

List of British and Indian Pharmacopoeial Drugs Growing in India (Present Status has been indicated by the present authors)

Scientific name	Status	Local Name/Parts used
Abroma augusta	S	Abroma bark
Acacia arabica	PS	Indian acacia
Acacia catechu	PS/S	Black catechu
Acalypha indica	S	Indian acalypha
Aconitum chasmanthum	D	Aconite
Adhatoda vasica	PS	Vasaka
Aegle marmelos	PS	Bael, Bela fructus
Allium sativum	PS	Garlic allium
Alpinia barbadensis	C	Aloes
Alpinia officinarium	S	Alpinia, Galangal.
Alstonia scholaris	S	Alstomia bark, Dita bark
Andrographis paniculata	D	Kalmegh
Anethum graveolens	D	Dill
Anethum sowa	C	Sowa
Arachis hypogea	C	Groundnut
Areca catechu	C	Betel nut
Aristolochia indica	D	Aristolochia
Artcmisia marifima	S	Santonin, Artcmisia santonica
Astragalus strobiliforus	D	Tragacanth
Atropa acuminata	D	Indian belladonna
Atropa belladonna	D	Belladonna
Bacopa monnicra (Herpestis monnicra)	D	Herpestis
Berberis aristata	S	Berberis root
Barrhaavia repens	S	Punarnaba
Brassica integrifolia	C	Sinapis
Brassica juncea	C	Brown mustard

S = Sporadic (vulnerable), PS = Presently safe,
D = Depleted species, C = Cultivated.

Scientific name	Status	Local Name/Parts used
Butea monosperma	PS	Butea seed
Calatropis procera	PS	Calatropis
Camellia sincnsis	PS	Tea plant
Cannabis sativa	PS	Cannabis
Capsicum frutescens	PS	Capsicum
Carica papaya	PS	Papaya
Carum carvi	C	Caraway
Cassia angustifolia	D	Senna
Cassia fistula	PS	Cassia fruit
Centella asiatica	PS	Hydrocotyle
Cephaelis pecacuanha	D	Ipecae
Chenopodium album	PS	Wormseed
Chenopodium ambrosioides var. *anthelminicum*	PS	American worm seed
Chrysanthcemum cinerariafolium	C/D	Pyrethrum
Cinchona ledgeriana, C.Succirubra and other species and hybrids	C	Cinchona
Cinnamomum camphora	C	Camphor
Cinnamomum zeylanicum	C	Cinnamon
Cissampelos parcira	D	Cissampelos
Citrullus colocynthis	C	Colocynth
Citrus aurantium	C	Bitter-orange peel
Citrus medica var. *limon*	C	Lemon peel
Cocos nucifera, C. butyraceae	C	Coconut
Coffea arabica	C	Coffee plant
Colchicum luteum	D	Colchicum corm and seed
Coriandrion sativum	C	Coriander
Crocus sativa	D	Saffron
Cuminum cyninum	C	Cumin
Curcuma longa	C	Turmeric
Cymbopogon flexuosus	PS	Lemon grass
Datura fastuosa	PS	Datura
Datura metal	PS	"

Scientific name	Status	Local Name/Parts used
Datura stramonium	PS	"
Derris ferruginca	PS	Derris, Tuba root
Digitalis lanata	D	Digoxin
Digitalis purpurea	D	Digitalis leaf
Dryopteris flix mas	D	Male fern. Aspidium
Elettaria cardamomum	C	Cardamom fruit
Ephedra gerardiana, *E. nebrodensis*	D	Ephedra
Eucalyptus globulus	PS	Eucalyptus
Eugenia caryophyllus	C	Clove
Eupatorinon ayapana	C	Ayapana
Ferula narthex (Ferula factida)	C	Asaioetida
Facniculum vulgare	C	Fennel
Gaultheria jragrantissima	D	Wintergreen
Glycyrrhiza glabra	D	Liquorice
Hemidesmus inducus	D	Indian sarsaparilla
Holarrhena antidysenterica	PS	Kurchi bark
Hydnocarpus wightiana	D	"
Hyoscyamus muticus	D	Hyoscyamus
Hyoscyamus niger	D	Hyoscyamus leave
Ipomaa hederacca	C	Kaladana
Ipomaa turpethum	C	Turpeth
Juniperus macropoda	D	Juniper
Limon usitatissimon (Lobelia nicotiata)	C	Linseed
Lobelia nicotiatafolia		Lobelia
Melia azadirachta	PS	Nim
Mentho arvensis	C	Peppermint
Mentho piperita	C	"
Moringa olcifera	C	Moringa
Myristica fragrans	D	Nutmeg
Papaver somniferum	C	Opium
Picrana quassioides	D	Quassia
Picrorhiza kurrooa	D	Picrorhiza
Pimpincll anisum	C	Anise

Scientific name	**Status**	**Local Name/Parts used**
Pinus excelsa	PS	Colophony
Pinus Khasya	PS	"
Pinus longifolia	PS	"
Piper cubela	C	Betel
Piper ovata	C	Isabgul
Podophyllum hexandrum (Podophyllum emodi)	D	Indian podophyllum
Polygala chinensis	S	Indian senega
Prunus amygdaius	S	Almond oil
Psoralea corylifolia	S	Babchi
Ptcrocarpus marsupium	D	Kino
Rauwolfia serpentina	D	Rauwolfia
Rheum emodi; R. webbianum,	D	Rhubarb
Ricinus communis	PS	Castor oil
Rosa damascena	PS	Rose
Rosmarimus officinalis	PS	Rosemary
Santichom album	PS	Sandal wood
Saraca indica	C	Asoka bark
Saussurea lappa	D	Saussurea
Sesamum indicum	C	Sesame oil
Strophantinus Kombi	D	Strphanthus
Strycinos nux-domina	*D*	*Nux vomica*
Swertia chirata	D	Chiretta
Terminalia chebula	S	Myrobalam
Thymus vulgaris	C	Thyme
Tinospora cordifolia	S	Tinospora
Trachyspermum animi (Carum copticum)		Ajowan
Tylophora asthmatica	S	Antamul
Urginea indica	S	Indian squill
Valeriana wallichii and other species	D	Valerian
Vernonia anthelmintica	D	Vernonia
Vitex peduncularis	D	Vitex leaf
Withania somnifera	S	Ashwagandha
Zingiber officinale	C	Ginger

2.11.2 List of medicinal plants prepared by Dr. Shibakali Bhattacharya.

Botanical names	**Status**
Abies webbiana	PS
Abroma augusta	S
Abrus precatorius	S
Abutilon indicum	S
Acacia arabica	PS
Acacia catechu	PS
Acacia farnesiana	PS
Acacia leucophloea	PS
Acacia suma	PS
Acalypha indica	S
Achyranthes aspera	PS
Aconitum chasmanthum	
Aconitum elwessii	
Aconitum falconeri	
Aconitum ferox	D
Aconitum heterophyllum	
Aconitum	
Aconitum napellus	
Acorus calamus	D
Adenanthera pavonina	S
Adhatoda vasica	PS
Adiantum capilus	PS
Adiantum caudatum	PS
Adiantum lunulatum	PS
Adina cordifolia	PS
Aegle marmelos	PS
Aeschynomene aspera	D
Aeschynomene indica	D
Aganosma caryophyllata	C
Aganosma dichotoma	S

Legends
PS = Presently safe   S = Sporadic (Threatened)
D = Depleted   C = Cultivated   NK = Not known

Botanical names	Status
Aglaia roxburghiana	D
Ailanthus excelsa	PS
Alangium lamarchii	PS
Alangium salvifolium	PS
Albizzia lebbek	PS
Allium cepa	C
Allium sativum	C
Alocasia indica	C
Aloe indica	C
Aloe vera	C
Alpine galanga	S
Alstonia scholaris	PS
Philoxeroides	PS
Alternanthera sessilis	PS
Altingia excelsa	PS
Amaranthus airopurpureus	PS
Amaranthus blitum var. *oleracea*	PS
Amaranthus gangeticus	PS
Amaranthus lividus	PS
Amaranthus spinosus	PS
Amaranthus viridis	PS
Amomum aromaticum	C
Amomum subulatum	C
Amorphophallus campanulatus	C
Anacardium occidentale	C
Ananas comosus	C
Andrographis paniculata	S
Andropogon jwarancusa	S
Anogeissus latifolia	S
Anona reticulata	C
Anona squamosa	C
Anthocephalus chinensis	PS
Aphanamixis polystachya	PS
Aquilaria agallocha	D
Areca catechu	PS

Botanical names	Status
Areca triandra	PS
Argemone mexicana	PS
Argyreia speciosa	S
Aristolochia indica	S
Artemisia maritima	PS
Artemisia vulgaris	PS
Artocarpus heterophyllus	C
Artocarpus lakoocha	S
Arum trilobatum	PS
Arundo donax	PS
Asparagus racemosus	D
Asterocantha longifolia	S
Astragalus candolleanus	S
Astragalus leucocephalus	S
Averrhoa carambola	C
Azadirachta indica	C
Bacopa monniera/monnieri	S
Balanites aegyptiaca	D
Baliospermum axillare	D
Baliospermum montanum	S
Balsamo dendron mukul	D
Bambusa arundinacea	PS
Bambusa balcooa	PS
Bambusa bambos	PS
Bambusa spinosa	PS
Bambusa tulda	PS
Bambusa vulgaris	PS
Barleria cristata	S
Barleria prionitis	PS
Barleria strigosa	PS
Barringtonia acutangula	PS
Barringtonia racemosa	PS
Basella alba	PS
Basella rubra	PS
Bassia longifolia	PS

Botanical names	Status
Bauhinia purpuria	PS
Bauhinia racemosa	PS
Bauhinia variegata	PS
Berberis aristata	S
Berberis asiatica	S
Betula utilis	D
Bixa orellana	C
Bilepharis edulis	S
Blumea lacera	PS
Boerhaavia differ	PS
Bombax ciba	PS
Borasus flabellifer	PS
Boswellia serrata	PS
Brassica campenstris	C
Brassica juncea	C
Brassica napus	C
Bryonia laciniosa	D
Buchanania lanzan	PS
Butea monosperma	PS
Caesalpinia bonducella	S
Caesalpinia pulcherrima	C
Cajanus indicus	C
Calamus viminalis	S
Callicarpa macrophylla	PS
Calophyllum inophyllum	S
Calotropis gigantia	PS
Calotropis procera	PS
Camellia drupifera	S
Camellia japonica	S
Camellia kissi	S
Camellia sinensis	C
Camellia thea	C
Camellia theifera	C
Canabis sativa	PS
Canscora decussata	S

Botanical names	Status
Capparis decidua	S
Capparis zeylanica	S
Capsicum annum	C
Capsicum frutescens	C
Cardiospermum halicacabum	S
Careya arborea	S
Carica papaya	C
Carissa carandas	C
Carthamus tinctorius	C
Carum carvi	C
Carum copticum	C
Cassia alata	PS
Cassia fistula	PS
Cassia occidentalis	PS
Cassia sophera	PS
Cassia tora	PS
Catharanthus roseus	C
Cadrella toona	PS
Cedrus deodara	PS
Ceiba pentandra	PS
Celastrus paniculatus	PS
Celosia argentea	C
Celosia cristata	C
Centella asiatica	PS
Cephalandra indica	PS
Chenopodium album	C
Chenopodium ambrosioides	C
Cicer arietinum	C
Cimicifuga foetida	C
Cinnamomum camphora	C
Cinnamomum tamala	C
Cinnamomum zeylanicum	C
Cissampelos pareira	S
Cissus quadrangularis	S
Citrullus vulgaris	C

Botanical names	Status
Citrus aurantifolia	C
Citrus decumana	C
Citrus limetoides	C
Citrus limon	C
Citrus maxima	C
Citrus medica	C
Citrus reticulata	C
Citrus sinensis	C
Cleome pentaphylla	S
Cleome viscosa	S
Clerodendrum indicum	PS
Clerodendrum viscosum	PS
Clerodendrum serratum	PS
Clitoria ternatea	PS
Coccinia indica	S
Cocculus hirsutus	PS
Cochlospermum religiosum	PS
Cocos nucifera	PS
Coix aquatica	PS
Coix gigantea	PS
Coix lachryma - Jobi	PS
Commelina benghalensis	S
Commelina salicifolia	S
Convolvulus arvensis	S
Coptis teeta	D
Corchous capsularis	C
Cordia dichotoma	S
Coriandrum sativum	C
Costus speciosus	S
Crocus sativus	S
Crotalaria juncea	C
Croton tiglium	S
Cucumis melo	C
Cucumis sativa	C
Cucumis utilissimus	C

Botanical names	**Status**
Cuminum cyminum	C
Curculigo orchioides	S
Curcuma amada	S
Curcuma aromatica	C
Curcuma domestica	C
Curcuma longa	C
Curcuma zedoaria	PS
Cuscuta reflexa	S
Cymbopogon citratus	S
Cymbopogon jwarancusa	C
Cymbopogon schoenanathus	C
Cynodon dactylon	PS
Cyperus rotundus	PS
Daemia extensa	S
Dalbergia sissoo	PS
Datura metal	PS
Daucus carata var. *sativa*	C
Delonix regia	PS
Delphinium zalil	D
Dendrocalamum hamiltonni	PS
Dendrocalamus strictus	PS
Dendrophthoe falcata	PS
Desmodium gangeticum	S
Desmodium triflorum	S
Desmostachya bipinnata	PS
Dillenia indica	PS
Dioscorea bulbifera	S
Dioscorea jacquemontii	S
Dioscorea pentaphylla	S
Dioscorea triphylla	S
Diospyros peregrina	S
Dolichos biflorus	S
Dolichos lablab	C
Drynaria quercifolia	PS
Echinochloa frumentacea	PS

Botanical names	Status
Eclipta alba	PS
Elaeocarpus ganitrus	D
Elaeocarpus tuberculatus	S
Elephantopus scaber	S
Elettaria cardamomum	C
Embelia ribes	S
Emblica officinalis	PS
Enhydra fluctuans	PS
Ephedra gerardiana	D
Ephedra vulgaris	D
Eriodendron anfractuosum	S
Ervatamia coronaria	PS
Erythrina variegata	PS
Eupatorium ayapana	C
Eupatorium triplinerve	PS
Euphorbia hirta	PS
Euphorbia neriifolia	PS
Euphorbia prostrata	PS
Evolvulus alsinoides	PS
Evolvulus nummularius	PS
Feronia elephantum	PS
Feronia limonia	PS
Ferula foetida	C
Ficus bengalensis	PS
Ficus carica	PS
Ficus cunia	PS
Ficus heterophylla	PS
Ficus hispida	PS
Ficus infectoria	S
Ficus racemosa	S
Ficus religiosa	PS
Ficus rumphii	S
Flacourtia cataphracta	S
Flacourtia indica	S
Flacourtia jangomas	S
Fleurya interrupta	S

Botanical names	**Status**
Foeniculum vulgare	C
Fritillaria cirrhosa	D
Fritillaria roylei	D
Fumaria indica	S
Fumaria parviflora	S
Garcinia indica	S
Garcinia mangostana	S
Garcinia morella	S
Garcinia pedunculata	S
Garcinia tintoria	S
Garcinia xanthochymus	S
Gardenia gummifera	S
Gardenia lucida	S
Gloriosa superba	S
Glycosmis pentaphylla	PS
Glycyrrhiza glabra	S
Gmelina arborea	PS
Gomphrena globota	PS
Gossypium herbaceum	PS
Grangea maderaspatana	S
Grewia asiatica	PS
Grewia tillaefolia	S
Gymnema sylvestre	S
Gynandropsis gynandra	PS
Gynocardia odorata	S
Habenaria edgeworthii	S
Habenaria latilabris	S
Helicteres isora	PS
Heliotropium indicum	S
Hemidesmus indicus	PS
Hemigraphis hirta	PS
Hibiscus abelmoschus	PS
Hibiscus esculentus	C
Hibiscus mutabilis	C
Hibiscus rosa-sinensis	C
Hiptage benghalensis	C

Botanical names	Status
Hiptage madablota	C
Holarrhena antidysenterica	PS
Hordeum vulgare	C
Hydnocarpus kurzii	S
Hydnocarpus wightiana	S
Hygroryza aristata	PS
Hyoscyamus niger	C
Ichnocarpus frutescens	S
Imperata cylindrica	PS
Indigofera tinctorea	S
Ipomoea bataus	C
Ipomoea quamoclit	C
Ipomoea reptans	C
Jasminum auriculatum	C
Jasminum grandiflorum	C
Jasminum heyneana	C
Jasminum multiflora	C
Jasminum pubescens	C
Jasminum sambac	C
Juglans regia	D/C
Juniperus communis	D
Juniperus macropoda	D
Jussiaea repens	PS
Justicia gendarussa	PS
Kalanchoe pinnata	C
Laccifer lacca	C
Lagenaria vulgaris	C
Lannea grandis	PS
Laportea crenulata	PS
Lathyrus sativus	C
Lawsonia inermis	C
Leucas cephalotes	S
Leucas lavandulaefolia	S
Lilium polyphyllum	C
Lilium tigrinum	C
Limnanthemum cristatum	PS

Botanical names	**Status**
Linum usitatissimum	C
Lippia nodiflora	PS
Lobelia inflata	C
Loranthus falcatus	PS
Loranthus longiflorus	PS
Luffa acutangula	C
Luffa amara	C
Luffa echinata	C
Madhuca indica	PS
Mallotus philippinensis	PS
Malaxis acuminata	D
Malaxis muscifera	D
Mangifera indica	PS
Marsilea minuta	PS
Marsilea quadrifolia	PS
Melia azedarach	PS
Melilotus indica	PS
Melothria heterophylla	S
Mentha spicata	C
Merremia tridentata	PS
Mesua ferrea	PS
Michelia champaca	PS
Mimosa pudica	PS
Mimusops elengi	PS
Mimusops hexandra	S
Mirabilis jalapa	C
Momordica charantia	C
Momordica cochinchinensis	C
Momordica muricata	C
Mollugo hirta	S
Mollugo sperula	S
Morinda citrifolia	PS
Morinda tinctoria	PS
Moringa oleifera	C
Morus alba	C
Morus atropurpurea	S

Botanical names	**Status**
Morus indica	S
Morus laevigata	S
Mucuna prurita	S
Murraya Koenigii	PS
Murraya paniculata	PS
Musa paradisiaca	PS
Musa sapientum	PS
Myrica nagi	C
Myristica fragrans	D
Myristica malabarica	S
Nardostachys jatamansi	D
Nauclea cordifolia	S
Nelumbo nucifera	C
Nephelium longana	C
Nerium indicum	C
Nerium odorum	C
Nicotiana plumbaginifolia	PS
Nicotiana tabacum	C
Nigella sativa	C
Nyctanthes arbortristis	C
Nymphaea alba	C/D
Nymphaea nouchali	C/D
Nymphaea rubra	C/D
Nymphaea stellata	C/D
Ochrocarpus longifolius	S
Ocimum basilicum	C
Ocimum kilimandscharicum	C
Ocimum sanctum	C
Oldenlandia corymbosa	S
Oroxylum indicum	S
Oryza fatua	C
Oryza sativa	C
Ougeinia oojeinensis	PS
Oxalis acetosella	PS

Botanical names	Status
Oxalis corniculata	PS
Paederia foetida	S
Pandanus odoratissimus	S/C
Pandanus tectorius	S
Panicum frumentacea	S
Paris polyphylla	D
Paspalum scrobiculatum	PS
Pergularia extensa	S
Periploca aphylla	S
Phaseolus radiatus	C
Phaseolus radiatus var. *attrea*	C
Phaseolus radiatus var. *grandis*	C
Phaseolus sublobatus	C
Phalogacanthus thyriflorus	S
Phoenix aculis	PS
Phoenix dactylifera	PS
Phoenix patudosa	PS
Phoenix sylvestris	PS
Phragmites karka	PS
Phyllanthus niruri	S
Phyllanthus simplex	S
Physalis minima	S
Picrorhiza kirroa	D/C
Pinanga gracilis	D
Pinus longifolia	PS
Piper aurantiacum	C
Piper betle	PS/C
Piper cubeba	C
Piper longum	C
Piper nigrum	C
Pistacia integerrima	S
Pisum sativum	C
Plumbago zeylanica	S
Plumeria acutifolia	PS
Plumeria rubra	PS
Podophyllym emodi	D

Botanical names	**Status**
Podophyllum hexandrum	D
Poincianaa pulcherrima	PS
Poinciana regia	PS
Polyanthes tuberosa	C
Polyatthia longifolia	PS
Polyalthia suberosa	S
Polygonatum cirrhifolium	D
Polygonatum oppositifolium	D
Polygonatum verticillatum	D
Polygonum hydropiper	PS
Polygonum orientate	PS
Polypodium quercifolium	PS
Pongamia pinnata	PS
Portulaca oleracea	PS
Portulaca quadrifida	PS
Potamogeton indicus	PS
Premna integrifolia	PS
Premna latifolia	PS
Prunus amygdalus	S
Prunus cerasoides	S
Prunus communis	C
Prunus persica	C
Prunus puddum	C
Psidium guyava	C
Psoralia corylifolia	S
Pterocarpus marsupium	PS
Pterocarpus santalinus	PS
Pterospermum suberifolium	PS
Pueraria tuberosa	PS
Punica grantum	C
Putranjiva roxburghii	PS
Pyrus communis	C
Quamoclit pinnata	PS
Quisqualis densiflora	PS
Quisqualis indica	PS
Randia dumetorum	PS

Botanical names	**Status**
Ranuculas sceleratus	PS
Raphanus sativus	C
Rauwolfia canescens	D
Rauwolfia serpentina	D
Rauwolfia tetraphylla	D
Rheum emodi	D
Rhododendron arboreum	S
Rhus succedanes	PS
Ricinus communis	C
Rosa damascena	C
Roscoea purpurea	S
Rubia cordifolia	PS
Saccharum benaglense	PS
Saccharum officinarum	C
Saccolabium papillosum	S
Salvadora oleoides	PS
Salvadora persica	PS
Salvinia cucullata	PS
Sansevieria roxburghiana	PS/C
Santalum album	PS
Sapindus trifoliatus	S
Saraca indica	C
Sarcosttemma acidum	S
Saussurea appa	D
Schleichera oleosa	PS
Schleichera trijuga	PS
Shcrebera swiettenioides	S
Scindapsus officinalis	PS
Semecarpus anacardium	PS
Sesamum indicum	C
Sesbania grandiflora	C
Sesbania sesan	C
Setaria italica	C
Shorea robusta	PS
Sida cordifolia	S
Sida rhombifolia	S

Botanical names	**Status**
Sida rhomboidea	S
Smilax glabra	S
Smilax indica	S
Solanum indicum	S
Solanum khasianum	S
Solanum melongena	C
Solanum nigrum	S
Solanum torvum	S
Solanum xanthocarpum	S
Sorghum vulgare	C
Soymida febrifuga	PS
Sphaeranthus indicus	S
Spinacia oleracea	C
Spondias dulcis	S
Spondias mangifera	PS
Stephania glabra	S
Stephania hernandifolia	S
Stereospermum suaveolens	PS
Stereospermum tetragonum	PS
Streblus asper	PS
Strychnos nux-vomica	PS
Strychnos potatorum	PS
Swertia chirata	D
Symplocos laurina	D
Symplocos racemosa	D
Syzygium aromaticum	PS
Syzygium cumini	PS
Syzygium fruticosa	S
Syzygium operculatum	S
Tabernaemontana coronaria	PS
Tacca aspera	S
Tacca integrifolia	S
Tamarindus indica	PS
Tamarix aphylla	PS
Tamarix articulata	PS
Tamarix dioica	PS

Botanical names	Status
Tamarix gallica	PS
Tamarix indica	PS
Tamarix troupii	PS
Taxus baccata	D
Tectona grandis	PS
Tephrosia purpurea	PS
Terminalia arjuna	PS
Terminalia belerica	PS
Terminalia chebula	PS
Terminalia citrina	PS
Thalictrum foliolosum	D
Thespesia populnea	PS
Thevetia nerifolia	C
Thevetia peruviana	C
Tinospora cordifolia	D
Tinospora malabarica	D
Tinospora tomenstosa	D
Trachyspermum ammi	D
Tragia involucrata	S
Trapa bispinosa	PS
Trewia nudiflora	PS
Trianthema mongyna	PS
Trianthema portulacastrum	PS
Tribulus terrestris	D
Trichosanthes anguina	C
Trichosanthes bracteata	C
Trichosanthes dioica	C
Trichosanthes palmata	C
Trigonella corniculata	C
Trigonella foenum-graecum	C
Triticum aestivum	C
Tylophora asthmatica	D
Tylophora indica	D
Typha elephantina	PS
Typha latifolia	PS
Typhonium trilobatum	PS

Botanical names	Status
Uraria hamosa	S
Uraria lagopoides	S
Uraria picta	S
Valeriana hardwickii	D
Valeriana officinalis	D
Vallaris solanacea	S
Vanda roxburghii	D
Vanda tessellata	D
Vangueria spinosa	S
Vateria indica	S
Vernonia cinerea	PS
Vetiveria zizanioides	PS
Viburnum foetidum	S
Viburnum prunifolium	S
Vigna cylindrica	S
Vinca pusilla	S
Vinca rosea	C
Viscum album	S
Viscum heyneanum	S
Viscum indicum	S
Viscum monoicum	S
Vitex negundo	PS
Vitex peduncularis	D
Vitex pedata	PS
Vitex vinifera	C
Wedelia calendulacea	S
Withania somnifera	D/C
Woodfordia fruticosa	PS
Wrightia tinctoria	PS
Wrightia tomentosa	PS
Zanthoxylum acanthaopodium	PS
Zanthoxylum alatum	PS
Zanthoxylum rhetsa	PS
Zea mays	C
Zehneria umbellata	S
Zingiber officinale	C
Ziziphus jujuba	PS
Ziziphus nummularia	PS
Ziziphus oenoplia	PS

2.11.3 List of Medicinal Plants (Kirtikar and Bose)

Scientific Name	Status
RANUNCULACEAE	
Clematis	
napaulensis	
triloba	
smilacifolia	very sporadic
gouriana	
graveolens	almost depleted
Anemone	
obtusiloba	-do-
Thalictrum	
foliolosum	-do-
Adonis	
aestivalis	NK
Nigella	
sativa	NK
Ranunculus	
trichophyllus	NK
lingua	
sceleratus	PS
pensylvanicus	
arvensis	NK
muricatus	
falcatus	
Caltha	
palustris	D
Coptis	
teeta	D
Delphinium	
denudatum	
caeruleum	sporadic
elatum	
brunonianum	
Actaea	
spicata	NK
Cimieifuga	
foetida	Very sporadic
Paeonia	
emodi	Depleted
Aconitum	
lucidum	
chasmanthum	
violaceum	Depleted
heterophyllum	
palmatum	
deinorrhizum	
balfourii	
falconeri " var. I	
spicatum	
laciniatum	
ferox	
elwesii	
lethal	
napellus	
DILLIENIACEAE	
Dillenia	
indica	PS
MAGNOLIACEAE	
Michelia	
champaca	PS
nilagirica	
montana	S
Illicium	
griffithii	S
ANNONACEAE	
Uvaria	

LEGEND : C = Cultivated; P.S. = Presently Safe; S = Sporadic (threats)
D = Depleted; N K = Not Known

Scientific Name	Status
narum	S
Artabotrys	
odoratissimus	C
Canangium	
odoratum	NK
Annona	
squamosa	
reticulata	C
Sageraea	
laurifolia	NK
Polyalthia	
longifolia	
simiarum	PS
thwaites	NK
MENISPERMACEAE	
Tinospora	
malabarica	S
cordifolia	sporadic
Anamirta	
cocculus	NK
Tiliacora	
acuminata	S
Cocculus	
hirsutus	
pendulus	
laurifolius	S
macrocarpus	
Pachygone	
ovata	NK
Pericampylus	
glaucus	NK
Stephania	
hernandifolia	S
glabra	
Cissampelos	
pareira	S
Tinospora sinensis	
Stephania japonica	S

Scientific Name	Status
BERBERIDACEAE	
Berberis	
petiolaris	
aristata	
lycium	D
asiatica	
Mahonia	
napaulensis	S
Podophyllum	
emodi	D
NYMPHAEACEAE	
Nymphaea	
alba	
rubra	Depleted/C
pubescens	
stellata	
Euryale	
ferox	Depleted/C
Nelumbo	
nucifera	-do-
Nymphaea nouchali	D
PAPAVERACEAE	
Argemone	
mexicana	PS
Meconopsis	
aculeata	
napaulensis	D
Papaver	
somniferum	C
FUMARIACEAE	
Corydalis	
govaniana	S
ramosa	
CRUCIFERAE	
Nasturtium	
palustre	
indicum	PS

Scientific Name	Status
montanum	
Cardamine	
impatiens	NK
Capsella	
bursa-pastoris	PS
Lepidium	
sativum	S
draba	
crassifolium	
latifolium	NK
ruderale	
perfoliatum	
Cochlearia	
flava	NK
Crambe	
cordifolia	NK
Raphanus	
sativus	C
Brassica	
oleracea	C
campestris " var. rapa	C
nigra	C
CAPPARIDACEAE	
Cleome	
monophylla	
brachycarpa	
icosandra	PS
felina	
heptaphylla	
chelidonii	
gynandra	
Macrua	
arenaria	NK
Crataeva	
murvala	PS
Cadaba	
farinosa	
trifoliata	NK
Capparis	

Scientific Name	Status
spinosa	
heyneana	PS
decidua	PS
grandis	
sepiaria	
zeylanica	
RESEDACEAE	
Reseda	
odorata	
Oligomeris	NK
subulata	
Ochradenus	
baccatus	
VIOLACEAE	
Viola	
serpens	S
odorata	
cinerea	
diffusa	
patrini	NK
tricolor	
biflora	
sylvestris	
Ionidium	
enneaspermum	S
COCHLOSPERMACEAE	
Cochlospermum	
gossypium	PS
BIXACEAE	
Bixa	
orellana	C
FLACOURTIACEAE	
Flacourtia	
cataphracta	S
ramontchi	
sepiaria	

Scientific Name	Status
Cynocardia	
odorata	D
Hydnocarpus	
wightiana	D
venenata	
anthelmintica	NK
Taraktogenos	
kurzii	D
PITTOSPORACEAE	
Pittosporum	
napaulense	NK
POLYGALACEAE	
Polygala	
crotalarioides	NK
Chinensis	S
telephioides	
glomerata	NK
sibirica	
FRANKENIACEAE	
Frankenia	
pulverulenta	NK
CARYOPHYLLACEAE	
Saponaria	
vaccaria	
Polycarpea	S
corymbosa	
Vaccaria pyramidata	
PORTULACACEAE	
Portulaca	
oleracea	
quadrifida	PS
tuberosa	
TAMARICACEAE	
Tamarix	
troupii	
dioica	PS

Scientific Name	Status
aphylla	
gallica	
Myricaria	
elegans	
germanica	NK
ELATINACEAE	
Bergia	
odorata	NK
IIYPERIACACEAE	
Hypericum	
patulum	
perforatum	
humifusum	
sampsoni	
japonicum	S
chinense	
GUTTIFERAE	
Garcinia	
mangostana	C
indica	NK
morella	
xanthochymus	S
dulcis	NK
cornea	
cowa	S
heterandra	NK
Ochrocarpa	
longifolius	S
Calophyllum	S
inophyllum	S
apetalum	
elatum	NK
Mesua	
ferrea	PS
TERNSTROEMIACEAE	
Schima	
wallichii	PS

Scientific Name	Status
Camellia	
thea	PS
Gordonia	
obtusa	S
DIPTEROCARPACEAE	
Dipterocarpus	
turbinatus	
tuberculatus	PS
alatus	
pilosus	
Shorea	
robusta	PS
tumbuggaia	NK
Hopea	
odorata	S
Vateria	
indica	S
MALVACEAE	
Althaea	
officinalis	NK
rosea	
ludwigii	
Malva	
sylvestris	
rotundifolia	S
parviflora	
verticillata	
Malvastrum	
coromandelianum	S
Sida	
veronicaefolia	
rhomboidea	
spinosa	
acuta	S
rhombifolia	
cordifolia	
Abutilon	
indicum	
hirtum	
theophrasti	S
glaucum	
asiaticum	
Malachra	
capitata	S
Urena	
lobata	
sinuata	S
repanda	
Pavonia	
odorata	
zeylanica	NK
Hibiscus	
furcatus	
micranthus	
cannabinus	
sabdariffa	
abelmoschus	
esculentus	PS/C
tiliaceus	
rosa-sinensis	
lampas	
trionum	
surattensis	
manihot	
mutabilis	
Thespesia	
populnea	PS
Gossypium	
herbaceum	
arboreum " var.	PS
barbadense	
hirsutum	
Kydia	
calycina	PS
BOMBACACEAE	
Adansonia	
digitata	PS

Scientific Name	Status
Bombax	
ceiba	PS
Ceiba	
pentandra	PS
STERCULIACEAE	
Pterygota	
alata	PS
Sterculia	
foetida	
urens	
balanghas	S
rubiginosa	
Pterocymbium	
javanicum	S
Scaphium	
affine	
wallichii	NK
Helicteres	
isora	PS
Peterospermum	
suberifolium	
acerifolium	PS
heyneanum	
Pentapetes	
phoenicea	NK
Eriolaena	
quinquelocularis	S
Melochia	
corchorifolia	S
Abroma	
augusta	PS
Guazuma	
tomentosa	PS
Theobroma	
cacao	C
Buettneria	
herbacea	NK

Scientific Name	Status
TILIACEAE	
Grewia	
tiliaefolia	
asiatica	
sclerophylla	
villosa	
hirsuta	S
tenax	
umbellata	
paniculata	
microcos	
Triumfetta	
bartramia	S
semitriloba NK	
Corchorus	
capsularis	
olitorius	
trilocularis	C
fascicularis	
depressus	
ELAEOCARPACEAE	
Elaeocarpus	
ganitrus	D
oblongus	
serratus	S/D
tuberculatus	
LINACEAE	
Linum	
usitatissimum	C
perenne	NK
strictum	NK
Reinwardtia	
trigy	S
Hugonia	
mystax S	
ERYTHROXYLACEAE	
Erythroxylon	

Scientific Name	Status
monogynum	
coca	S
MALPIGHIACEAE	
Hiptage	
benghalensis	C
ZYGOPHYLIACEAE	
Tribulus	
terrestris	
alatus	D
Zygophyllum	
simplex	
coccineum	NK
Fugonia	
cretica	NK
GERANIACEAE	
Geranium	
wallichianum	
napalense	
robertianum	
ocellatum	
lucidum	S
molle	
pratense	
pusillum	
rotundifolium	
sibiricum	
OXALIDACEAE	
Oxalis	
corniculata	
acetosella	PS
Biophytum	
sensitivum	PS
Averrhoa	
carambola	C
bilimbi	

Scientific Name	Status
BALSAMINACEAE	
Impatiens	
balsamina	
tripetala	S
chinensis	
RUTACEAE	
Evodia	
lunur-ankenda	
fraxinifolia	PS
rutaecarpa	
Ruta	
graveolens	NK
tuberculata	
Peganum	
harmala	NK
Dictamnus	
albus	NK
Zanthoxylum	
alatum	
acanthopodium	S
oxyphyllum	
hamiltonianum	
budrunga	
Toddalia	
asiatica	S
Vepris	
bilocularis	NK
Skimmia	
laureola	D
Glycosmis	
cochinchinensis	S
Acronychia	
laurifolia	S
Murraya	
koenigii	PS/C
paniculata	
Clausena	
excavata	

Scientific Name	Status
pentaphylla	S
wampi	
Limonia	
crenulata	S
Luvunga	
scandens	S
Paramignya	
monophylla	S
longispina	
Atalantia	
monophylla	D
Citrus	
medica	
medica var. proper	
medica var. limonum	
medica var. acida	
medica var. limetta	C
aurantium	
aurantium var. proper	
aurantium var. bigaradia	
aurantium var. bergamia	
maxima	
Feronia	
elephantum	S
Aegle	
marmelos	PS
SIMAROUBACEAE	
Ailanthus	
glandulosa	S
excelsa	C
malabarica	S
Samadera	
indica	NK
indica var. lucida	
Picrasma	
quassioides	D
javanica	D
Brucea	
sumatrana	NK
Eurycoma	
longifolia	NK
Balanites	
aegyptiaca	
roxburghii	D
OCHNACEAE	
Ouratea	
angustifolia	NK
Ochna	
pumila	S
squarrosa	
BUSERACEAE	
Boswllia	
serrata	PS
glabra	
Garuga	
pinnata	PS
Commiphora	
mukul	
agallocha	D
stocksiana	
Canarium	
commune	
strictum	Very sporadic
bengalense	
MELIACEAE	
Turracea	
villosa	
Naregamia	
alata	NK
Azadirachta	
indica	PS
Melia	
azedarach	PS
composita	
Dysoxylum	
hamiltonii	D
malabaricum	C

Scientific Name	Status
Sandoricum	
indicum	NK
Aglaia	
odoratissima	S
roxburghiana	
Aphanamixis	
polystachya	PS
Amoora	
cucullata	D
Walsura	
piscidia	S
heynea	
trijuga	S
Carapa	
moluccensis	
moluccensis	S
obovata	
Soymida	
febrifuga	S
Chukrasia	
tabularis	PS
toona atach	PS
Chloroxylon	
swietenia	PS
ICACINACEAE	
Ximenia	
americana	NK
OLACACEAE	
Olax	
scandens	
nana	NK
Sarcostigma	
kleinii	NK
CELASTRACEAE	
Euonymus	
tingens	
javanicus	S
Kokoona	
zeylanica	NK
Celastrus	
paniculata	S
Gymnosporia	
spinosa	S
royleana	
Elaeodendron	
glaucum	S
Salacia	
oblonga	
reticulata	NK
RHAMNACEAE	
Ventilago	
maderaspatana	S
talyculata	
Berchemia	
lineata	S
Zizyphus	
mauritiana	
trinervia	PS
nummularia	
sativa	NK
rugosa	PS
oenoplia	PS
Rhamnus	
dahuricus	
wightii	
purpureus	NK
triqueter	
nipalensis	S
Gonania	
leplostachya	PS
VIITACEAE	
Vitis	
quadrangularis	
adnata	
latifolia	
vinifera	

Scientific Name	Status
indica	
setosa	
carnosa	PS
araneosa	
pedata	
tomentosa	
repens	
pallida	
Leea	
macrophylla	
crispa	
indica	PS
robusta	
aequata	
SAPINDACEAE	
Cardiospermum	
halicacabum	S
Aesculus	
indica	S
hippocastanum	S
Allophylus	
serratus	S
Schleichera	
trijuga	PS
Sapindus	
trifoliatus	S
mukorossi	
Litchi	
chinensis	C
Nephelium	
longana	
lappaceum	C
Acer	
pictum	S
Dodonaea	
viscosa	PS
ANACARDIACEAE	
Rhus	

Scientific Name	Status
parviflora	
semialata	
wallichii	D
insignis	
succedanea	
Pistacia	
integerrima	D
Mangifera	
indica	S
caesia	
Anacardium	
occidentale	PS
Buchanania	
Ianzan	PS
Melanorrhoea	
usitata	
wallichii	S
curtisii	
Lannea	
coromendelica	S
Semecarpus	
anacardium	PS
Holigarna	
arnottiana	S
longifolia	
Spondias	
pinnata	PS
CORIARIACEAE	
Coriaria	
nepalensis	D
MORINGACEAE	
Moringa	
oleifera	C
concanensis	NK
CONNARACEAE	
Connarus	
monocarpus	NK
Rourea	

Scientific Name	Status
santaloides	NK
bofalaria	
burhia	
prostrata	
albida	
verrucosa	
juncea	S
medicaginea	
trifoliastrum	
retusa	
sericea	
Trigonella	
occulta	
foenum-graecum	S
polycerata	
corniculata	
Melilotus	
indica	
officinalis	S
alba	
Cyamopsis	
tetragonoloba	S
Indigofera	
linifolia	
glandulosa	
enneaphylla	
áspalathoides	
aspalathoides	Very sporadic
trifoliata	
oblongifolia	
tinctoria	
pulchella	
trita	
articulata	
glabra	
Pseudarthri	
viscid	NK
Uraria	
picta	

Scientific Name	Status
lagopoides	S
hamosa	
Alysicarpus	
longifolius	S
Arachis	
hypogaea	C
Ougenia	
oojeinensis	S
Desmodium	
tiliaefolium	
gangeticum	
polycarpum	
trifolium	S
lasiocarpum	
retroflexum	
pulchellum	
Abrus	
precatorius	S
Cicer	
arietinum	C
Lathyrus	
sativus	
aphaca	C
pratensis	
Pisum	
sativum	C
Glycine	
soja	C
Teranmus	
labialis	NK
Mucuna	
monosperma	
gigantea	S
prurita	
Erythrina	
indica	PS
Psoralea	
corylifolia	S
Colutea	

Scientific Name	Status
nepalensis	S
Mundulea	
suberosa	S
Tephrosia	
purpurea	
villosa	S
petrosa	
Glycyrrhiza	
glabra	S
Millettia	
auriculata	
pachycarpa	S
Adinobotrys	
atropurpureus	NK
Sesbania	
aegyptiaca	C
aculeata	
grandiflora	
Astragalus	
tribuloides	
hamosus	S
multiceps	
strobiliferous	
Taverniera	
cuneifolia	NK
Lens	
esculenta	C
Alhagi	
camelorum	NK
Zornia	
diphylla	S
Smithia	
conferta	S
sensitiva	
Ornocarpum	
sennoides	
lunatus	
aconitifolius	NK
adenanthus	
catiang	
Clitoria	
ternatea	C
Dolichos	
biflorus	
lablab	C
Rhynehosia	
minima	NK
Cajanus	
cajans	C
Atylosia	
searabaeoides	S
Cylista	
seariosa	
Flemingia	
strobilifera	
chappar	S
grahamiana	
congesta	
tuberosa	
nana	
Dalbergia	
sisso	
multiflora	
lanceolaria	
volubilis	S
spinosa	
tamarindifolia	
latifolia	
Pterocarpus	
santalinus	
indicus	D
marsupium	
Pongamia	
pinnata	PS
Derris	
scandens	
elliptica	PS
uliginosa	

Scientific Name	Status
Sophora	
tomentosa	
mollis	S
griffithii	
" *var.* roxburghi	
suberosa	
Butea	
monosperma	PS
superba	
Canavalia	
virosa	
ensiformis	C
Pueraria	
tuberosa	C
Phaseolus	
trilobus	
radiatus	C
mungo	
CAESALPINIACEAE	
Caesalpinia	
crista	S
jayabo	
nuga	NK
sapan	
pulcherrima	C
sepiaria	
digyna	S
coriaria	
Delonix	
elata	C
Wagatea	
spicata	NK
Cassia	
fistula	PS
occidentalis	PS
sophera	PS
obtusifolia	
auriculata	S
obovata	

Scientific Name	Status
alata	PS
glauca	
absus	
mimosoides	
pumila	Very sporadic
angustifolia	
javanica	
tora	
Cynometra	
mimosoides	NK
cauliflora	
Hardwickia	
pinnata	PS
Saraca	
indica	C
Ceratonia	
siliqua	NK
Haematoxylon	
campechianum	NK
Tamarindus	
indica	C
Humboldtia	
vahliana	S
Bauhinia	
tomentosa	
racemosa	
retusa	
vahlii	PS
purpurea	
variegata	
malabarica	
macrostachya	
MIMOSACEAE	
Neptunia	
oleracea	NK
Xylia	
dolabriformis	S
Entada	

Scientific Name	Status	Scientific Name	Status
scandens	S	cerasus	
Adenanthera		avium	C/S
pavonia	D	cerasoides	
Prosopis		communis	
spicigera	PS	domestica	
Dichrostachys		insititia	
cinerea		cornuta	
Leuceena		mahaleb	
glauca	PS	undulata	NK
Mimosa		triflora	
Pudica		*Prinsepia*	
rubicaulis	PS	utilis S	
Acacia		*Rubus*	
farnesiana	D	moluccanus	
nilotica	PS	saxatilis	S
leucophloea	D	fruticosus	
catechu	PS	*Geum*	
ferruginea		urbanum	NK
senegal		alatum	
modesta	D	*Potentilla*	
rugata		nepalensis	
caesia		supina	
pennata		anserina	S
suma		fruticosa	
Albizzia		reptans	
lebbeck		fragarioides	
odoratissima		kleiniana	
julibrissin	PS	sericea	
amara		*Agrimonia*	
procera		eupatoria	S
stipulata		*Rosa*	
Pithecellobium		*damascena*	
bigeminum	PS	centifolia	
dulce		gallica	
ROSACEAE		alba	C/S
Prunus		indica	
amygdalus		banksiae	
persica		multiflora	
armeniaca		*Cydonia*	

Scientific Name	Status
vulgaris	
Pyrus	
malus	C
communis	
Eriobotrya	
japonica	C
Cotoneaster	
nummularia	S/D
bacillaris	
microphylla	
var. buxifolia	
SAXIFRAGACEAE	
Bergenia	
ligulata	
Dichora	
febrifuga	
Ribes	
orientale	S
nigrum	
Kalanchoe	
pinnata	
spathulata	
laciniata	
Sedum	
asiaticum	S
multicaule	
DROSERACEAE	
Drosera	
lunata	
indica	D
burmanni	
HAMAMELIDACEAE	
Exbacklandia populnea	S/C
RHIZOPHORACEAE	
Rhizophora	

Scientific Name	Status
mucronata	PS
Ceriops	
candolleana	PS
Kandelia	
candel	D
COMBRETACEAE	
Terminalia	
catappa	
belerica	PS
chebula	D
citrina	D
arjuna	PS
tomentosa	PS
paniculata	
pyrifolia	
bialata	D
oliveri	
myriocarpa	PS
coriacea	
pallida	D
Calycopteris	
floribunda	D
Anogeissus	
latifolia	PS
Quisqualis	
indica	C
LECYTHEDACEAE	
Barringtonia	
racemosa	
acutangula	PS
speciosa	
Careya	
arborea	PS
MYRTACEAE	
Myrtus	
communis	NK
Melaleuca	
leucadendron	C

Scientific Name	Status
Eucalyptus	
citriodora	PS
globulus	
Psidium	
guyava	PS
Jambosa	
jambos	PS
Syzygium	
operculatum	
cumini	
spicatum	
hemispherican	
Pimenta	
acris	NK
MELASTOMACEAE	
Memecylon	
umbellatum	
amplexicaule	S
angustifolium	
Melastoma	
malabathricum	
decemfidum	S
Osbeckia	
nepalensis	S
crinita	
Ammannia	
baccifera	
auriculata	S
LYTHRACEAE	
Woodtordia	
fruticosa	S
Lawsonia	
inermis	PS
Lagerstroemia	
speciosa	
indica	
Sonneratia	
caseolaris	D

Scientific Name	Status
PUNICACEAE	
Punica	
granatum	C
ONAGRACEAE	
Jussieua	
suffruticosa	S
repens	
Trapa	
bispinosa	S
SAMYDACEAE	
Casearia	
graveolens	
esculenta	S
tomentosa	
CARIACEAE	
Carica	
papaya	C
PASSIFLORACEAE	
Passiflora	
foetida	C
edulis	
Adenia	
palmata	C
CUCURBITACEAE	
Trichosanthes	
palmata	
cordata	
dioica	
nervifolia	C
cucumerina	
anguina	
wallichiana	
Gymnopetalum	
cochinchinense	
Lagenaria	
vulgaris	C

Scientific Name	Status
Luffa	
aegyptiaca	
acutangula	C
"*var.* amara	
echinata	
Benincasa	
hispida	C
cerifera	
Momordica	
charantia	
balsamina	
dioica	C
cochinchinensis	
tuberosa	
Cucum	
trigonus	
melo	C/PS
" var. momordica	
" var: utilissimus	
prophetarum	
sativus	
Citrullus	
colocynthis	
vulgaris	
" var fistulosus	C
Coccinia	
indica	S
Cucurbita	
maxima	
pepo	C
Bryonopsis	
laciniosa	C
Melothria	
maderaspatana	
perpusilla	S
heterophylla	
Blastania	
garcini	
Kedostris	

Scientific Name	Status
rostrata	NK
Corallocarpus	
epigaeus	
Zanonia	
indica	
BEGOPNIACEAE	
DATISCACEAE	
Datisca	
cannabina	NK
CACTACEAE	
Opuntia	
coccinellifera	
monacantha	
stricta	S
nigricans	
dillenii	
FICOIDACEAE	
Trianthema	
portulacastrum	
pentandra	S
decandra	
Mollugo	
hirta	
oppositifolia	
pentaphylla	
cerviana	PS
nudicaulis	
Giesekia	
pharmacoides	NK
UMBELLIFERAE	
Hydrocotyle	
asiatica	PS
javanica	
Eryngium	
caeruleum	NK
Bupleurum	

Scientific Name	Status
falcatum	NK
jucundum	
Apium	
graveolens	
Carum	
carvi	
bulbocastanum	C
roxburghianum	
copticum	
Pimpinella	
heyneana	C
sexifraga	
diversifolia	
stocksii	
Seseli	
indicum	NK
Pycnocycla	
aucheriana	NK
Foeniculum	
capillaceum	C
Prangos	
pabularia	S
Angelica	
glauca	
Ferula	
narthex	C
jaeschkeana	
toctida	
Peucedamum	
graveolens	
grande	
dhana	NK
aucheri	
nagpurense	
Heracleum	
wallichii	NK
Zosimia	
orientialis	
Coriandrum	
sativum	C
Nothopanax	
fruticosum	
Hedera	
helix	C
Cuminum	
cyminum	NK
Daucus	
carota	C
Psammogeton	
biternatum	
ALANGIACEAE	
Alangium	
lamarckii	S
CAPRIFOLIACEAE	
Sambucus	
ebulus	
javanica	S
Viburnum	
foetidum	
Lonicera	
glauca	
RUBIACEAE	
Sarcocephalus	
missionis	NK
cordatus	
Anthocephalus	
chinensis	PS
Adina	
cordifolia	PS
Nauclea	
sessilifolia	PS
Mytragyna	
parvifolia	
Uncaria	
gambir	
Hymenodictyon	
excelsum	PS

Scientific Name	Status
Paederia	
foetida S	
Pavetta	
indica	PS
Randia	
dumetorum	PS
uliginosa	
Rubia cordifolia	S
Vangueria	
spinosa	S
Morinda	
citrifolia	PS
Ophiopogon	
serpentinum	S
Cuises	
officinalis	C
Oldenlandia	
corymbosa	S
Psychotria	
ipecacuana	S
Ophiorrhiza	
mungo's	S
Ixora	
parviflora	PS
coceinea	
Mussaenda	
frondosa	PS
VALARIANACEAE	
Nardostachys	
jatamansi	D
Valeriana	
hardwickii	D
officinalis	
ASCLEPIACACEAE	
Hemidesmus	
indicus	D
Cryptolepis	
buchanani	S

Scientific Name	Status
Cryptostegia	
grandiflora	S
Periploca	
aphylla	S
Secamone	
emetica	S
Glossonema	
varians	S
Oxystelma	
esculentum	S
Calotropis	
gigantea	PS
procera	
Asclepias	
curassavica	PS
Pentatropis	
cynanchoides	
microphylla	
Pergularia	
extensa	S
Daemia	
cordata	
Holostemma	
annulare	
Cynanchum	
arnottianum	
Sarcostemma	
brevistigma	
brunonianum	
intermedium	
stocksii	
Gymnema	
sylvestre	
Marsdenia	
roylei	S
volubilis	
Leptadenta	
reticulata	
Tylophora	

Scientific Name	Status
fasciculata	
asthmatica	
tenuis	
Cosmostigma	
racemosum	
Dregea	
volubilis	
Ceropegia	
bulbosa	NK
tuberosa	
Caralluma	
edulis	NK
Boucerosia	
aucheriana	NK
LOGANIACEAE	
Fagraea	
racemosa	S
Strychnos	
colubrina	S
nux-vomica	D
potatorum	D
bourdilloni	
cinnamomifolia	S
Cyrtophyllum	
peregrinum	
GENTIANACEAE	
Exacum	
tetragonum	S
bicolor	
pedunculatum	S
lawii	
Enicostemma	
littorale	S
Erythraea	
roxburghii	
Hoppea	
dichotoma	S
Canscora	
diffusa	
decussata	
Gentiana	
tenella	
kurroo	
decumbens	S
dahurica	
Swertia	
purpurascens	
paniculata	
chirata	
angustifolia	D
angustifolia Vr	
decussata	
alata	
lawii	
Limnanthemum	
nymphaeoides	PS
Menyanthes	
trifoliata	NK
HYDROPHYLLACEAE	
Hydrolea	
zeylanica	
BORAGINACEAE	
Cordia	
obliqua	PS
wallichii	
rothii	PS
vestita	
macleodii	
Ehretia	
aspera	SP
microphylla	
Coldenia	
procumbens	NK
Rotula	
acquatica	PS
Heliotropium	

Scientific Name	Status
eichwaldi	
tuberculosum	
strigosum	PS
brevifolium	
indicum	
Trichodesma	
indicum	
africanum	PS
zeylanicum	
Cynoglossum	
glochidiatum	PS
Macrotomia	
benthami	
perennis	
Onosma	
echioides	NK
bracteatum	
Caccinia	
glauca	
Lithospermum	
officinale	
arvense	

CONVOLVULACEAE

Scientific Name	Status
Argyreia	
speciosa	
fulgens	S
Lettsomia	
aggregata	S
Calonyction	
bona-nox	NK
muricatum	
Quamoclit	
pinnata	PS
coccinea	
vulgaris	
Erycibe	
paniculata	NK
Rivea	
ornata	PS
Ipomoea	
hederacea	
uniflora	NK
paniaelata	
batatas	
pes-tigridis	PS
reniformis	
obscura	NK
sepiaria	
reptans	C
campanulata	
pes-caprae	PS
dissecta	
tuberosa	NK
dasysperma	
hispida	
Operculina	
turpethum	
Merremia	
vitifolia	PS
tridentata	
Convolvulus	
arvensis	PS
glomeratus	
spinosus	
Evolvulus	
alsinoides	PS
Cressa	
cretica	NK
Cuscuta	
reflexa	PS
hyalina	
chinensis	

SOLANACEAE

Scientific Name	Status
Solanum	
nigrum	PS
dulcamara	D
spirale	
verbascifolium	PS

Scientific Name	Status
ferox	
indicum	
melongena	S
xanthocarpum	
trilobatum	
gracilipes	
torvum albicaule	
incanum	
sarattenae	
Physalis	
minima	
minima var. indica	PS
angulata	
peruviana	
Capsicum	
frutescens	
annuum	PS
minimum	
Withania	
somnifera	D
coagulans	
Nicandra	
physaloides	D
Lycium	
barbarum	D
ruthenicum	
Atropa	
belladonna	D
Datura	
stramonium	
fastuosa var al.	C/S
metal	
Scopolia	
lurida	D
Physoclaina	
praealta	D
Hyoscyamus	
niger	
muticus	D

Scientific Name	Status
reticulatus	
Nicotiana	
tabacum	S
rustica	
SCROPHULARIACEAE	
Torenia	
asiatica	PS
Vandellia	
pysidaria	PS
pedunculata	
Bonnaya	
reptans	PS
Scoparia	
dulcis	PS
Picrorrhiza	
kurrooa	D
Verbascum	
thapsus	S
Celsia	
coromandeliana	S
Linaria	
ramosissima	PS
Schweinfurthia	
sphaerocarpa	NK
Lindenbergia	
urticaefolia	PS
Stemodia	
viscosa	NK
Limnophila	
gratissima	
gratioloides	S
roxburghii	
Moniera	
cuneifolia	
Artanema	
sesamoides	NK
Curanga	
amara	S
Veronica	

Scientific Name	Status
anagallis	S
beccabunga	
Striga	
lutea	S
orobanchoides	
Sopubia	
delphinifolia	S
Pedicularis	
pectinata	S
siphonantha	
OROBANCHACEAE	
Cistanche	
tubulosa	D
Orobanche	
aegyptiaca	D
LENTIBULARIACEAE	
Utricularia	
bifida	S
BIGNONIACEAE	
Oroxylum	
indicum	PS
Tecomella	
undulata	C
Dolichandrone	
spathacea	
falcata	
Heterophragma	
roxburghii	S
Stereospermum	
Chelonoides	PS
suaveolens	PS
Radermachera	
xylocarpa	
Amphicome	
emodi	NK
Tecoma	

Scientific Name	Status
stans	C
Crescentia	
cujete	C
PEDALIACEAE	
Martynia	
annua	
Pedalium	S
murex	
sesamum	
indicum	C
ACANTHACEAE	
Auriculatus	
ciliatus	C
Blepharis	
edulis	S
sindica	
Acanthus	
ilicifolius	PS
Barleria	
prionitis	
noctiflora	S
cristata	
cristata var. dichotoma	
strigosa	
courtallica	
longiflora	
Neuracanthus	
sphaerostachys	NK
Andrographis	
paniculata	S
echiodes	
Haplanthus	
verticillaris	NK
tentaculatus	
Gymnostachyum	
febrifugum	NK
Phlogacanthus	
thyrsiflorus	PS

Scientific Name	Status
Crossandra	
undulaefolia	NK
Asystasia	
gangetica	NK
Lepidagathis	
cristata	
trinervis	S
hamiltoniana	
Justica	
gendarussa	PS
procumbens	
Adhatoda	
vasica	C
Rhinacanthus	
communis	
nasuta	NK
Ecbolium	
linneanum	NK
Graptophyllum	
pictum	NK
Rungia	
repens	S
parviflora	
Dicliptera	
roxburghiana	S
Peristrophe	
bicalyculata	
Cardanthera	
uliginosa	S
Asteracantha	
longifolia	S
Ruellia	
prostrata	S
suffruitcosa	
Daedalacanthus	
roseus	S
Strobilanthes	
callosus	S

Scientific Name	Status
VERBENACEAE	
Lantana	
indica	PS
aculeata	
Lippia	
nodiflora	PS
Verbena	
officinalis	C
Callicarpa	
arborea	
lanata	S
macrophylla	
cana	
Stachytarpheta	
indica	S
Tectona	
grandis	PS
Premna	
integrifolia	
tomentosa	
latifolia	PS
esculenta	
herbacea	
Gmelina	
arborea	PS
asiatica	
Vitex	
trifolia	
negundo	
peduncularis	PS
glabrata	
agnus-castus	
pubescens	
leucoxylon	
Clerodendrum	
inerme	
phlomidis	C/S
serratum	
viscosium	

Scientific Name	Status
siphonanthus	
Avicennia	
officinalis	PS
tomentosa	
LABIATAE	
Ocimum	
canum	
basilicum	C
gratissimum	
sanctum	
Geniosporum	
prostratum	NK
Orthosiphon	
stamineus	S
Coleus	
amboinicus	C
Anisochilus	
carnosus	NK
Lavandula	
bipinnata	NK
Pogostemon	
plectranthoides	
purpurascens	S
parviflorus	
Colebrookea	
oppositifolia	S
Mentha	
viridis	
piperita	
sylvestris	C
arvensis	
Lycopus	
europaeus	NK
Origanum	
majorana	NK
vulgare	
Thymus	
serpyllum	S
Hyssopus	
officinalis	NK
Micromeria	
capitellata	S
Calamintha	
clinopodium	S
Melissa	
parviflora	S
Perowskia	
abrotanoides	NK
atriplicifolia	
Meriandra	
strobilifera	NK
bengalensis	
Salvia	
moorcroftiana	
lanata	
plebeia	
aegyptiaca	
cabulica	S
spinosa	
officinalis	
Nepeta	
elliptica	
ciliaris	
ruderalis	S
glomerulosa	
Dracocephalum	
moldavicum	NK
Lallemantia	
royleana	NK
Brunella	
vulgaris	S
Marrubium	
vulgare	NK
Anisomeles	
indica	S
malabarica	
Stachys	
parviflora	NK

Scientific Name	Status
Leonurus	
sibiricus	S
Roylea	
elegans	NK
Otostegia	
limbata	NK
aucheri	
Leucas	
cephalotes	
zeylanica	
aspera	PS
linifolia	
urticeafolia	
stelligera	
Leonotis	
nepetaefolia	
Eremostachys	
vicaryi	NK
acanthocalyx	
Hyptis	
suaveolens	S
Ajuga	
bracteosa	
Zataria	
multiflora	
Ziziphora	
clinopodioides	NK
tenuior	
Hymenocrater	
sessilifolius	
Tenucrium	
stocksianum	
scordium	
PLANTAGINACEAE	
Plantago	
major	
lanceolata	S
amplexicaulis	
ovata	

Scientific Name	Status
psyllium	
ciliata	NK
lagocephala	
NYCTAGINACEAE	
Boerhavia	
diffusa	PS
repens	
Pisonia	
aculeata	S
morindaefolia	
Mirabilis	
jalapa	S
AMARANTHACEAE	
Celosia	
argentea	S
" var. cristata	
Digera	
arvensis	S
Amaranthus	
spinosus	
paniculatus	PS
gangeticus	
" var. tristis	
virdis	
blitum	
" var. oleracea	
Aerva	
tomentosa	PS
lanata	
Achyranthes	
aspera	PS
bidentata	
Alternanathera	
sessilis	PS
CHENOPODIACEAE	
Chenopodium	
album	PS
botrys	

Scientific Name	Status
ambrosioides	
Beta	
vulgaris	C
Spinacia	
oleracea	C
Kochia	
indica	
sedoides	S
scoparia	
Arthrocnemum	
indicum	
Salicornia	
brachiata	PS
Suaeda	
fruticosa	PS
monoica	
Salsola	
kali	
foetida	
Basella	
rubra	C
Haloxylon	
salicornicum	
recurvum	NK
PHYTOLACCACEAE	
Phytolacca	
acinosa	S
POLYGONACEAE	
Calligonum	
polygonoides	S
Pteropyrum	
olivierii	NK
Polygonum	
aviculare	
plebejum	
viviparum	
glabrum	
persicaria	

Scientific Name	Status
barbatum	PS
hydropiper	
punctatum	
molle	
chinense	
ientale	
virginianum	
sphaerostachyum	
serrulatum	
Fagopyrum	
cymosum	
esculentum	PS
talaricum	
Rheum	
spiciforme	NK
emodi	D
webbianum	NK
nobile	D
Oxyria	
digyna	NK
Rumex	
maritimus	
dentatus	
nepalensis	
vesicarius	PS
acetosella	
acetosa	
scutatus	
Rumex maritimus L.	PS
Rumex vesicarius L.	
ARISTOLOCHIACEAE	
Bragantia	
wallichii	
tomentosa	NK
Aristolochia	
bracteata	
indica	S
tagala	

Scientific Name	Status
PIPERACEAE	
Piper	
longum	
cluba	
sylvaticum	
betle	PS
nigrum	
attenuatum	
sarmentosum	
aurantiacum	
CHLORANTHACEAE	
Chloranthus	
officinalis	
brachystachys	NK
MYRISTICACEAE	
Myristica	
malabarica	
fragrans	PS
LAURACEAE	
Cinnamomum	
tamala	
obtusifolium	C/S
iners	S
zeylancum	C
macrocarpum	NK
glanduliferum	S
parthenoxylon	NK
camphora	C
cassia	
pauciflorum	NK
javanicum	
Machilus	
macrantha	S
Actinodaphne	
hookeri	S
Litsea	
chinensis	
polyantha	S
stocksii	
Lindera	
neesiana	S
Cassytha	
filiformis	PS
Litsea glutinosa Lot. Pers.	PS
Litsea monopetala (Rc)	
ELAEAGNACEAE	
Elaeagnus	
hortensis	
umbellata	S
latifolia	
Hippophae	
rhamnoides	S
salicifolia	
LORANTHACEAE	
Dendrophthoe falcata cochinchinensis	PS
Loranthus	
elasticus	
falcatus	PS
Viscum	
album	
monoicum	S
orientale	
articulatum	
SANTALACEAE	
Santalum	
album	PS
Osyris	
arborea	NK
EUPHORBIACEAE	
Euphorbia	
hypericifolia	
hirta	
thymifolia	

Scientific Name	Status	Scientific Name	Status
microphylla		rhamnoides	S
tirucalli		patens	
neriifolia		*Putranjiva*	
nivulia		roxburghii	PS
antiquorum		*Antidesma*	
royleana	C/PS	bunius	S
thomsoniana		zeylanicum	
jhelioscopia		*Jatropha*	
dracunculoides		glandulifera	
longifolia		nana	
granulata		multifida	PS
sanguinea		curcas	
turcomanica		gossypifolia	
Buxus		*Aleurites*	
sempervirens	D	moluccana	C
Bridelia		*Bischofia*	
retusa	S	Javanica	S
montana		*Aporosa*	
Cleistanthus		lindleyana	S
collinus	S	*Croton*	
Andrachne		reticulatus	
cordifolia	NK	oblongifolius	PS
Phyllanthus		caudatus	
reticulatus		tiglium	
maderaspatensis		*Chrozophora*	
urinaria	S	rottleri	PS
simplex		prostrata	
niruri		*Acalypha*	
Cicca		fruticosa	
disticha	S	indica	PS
Clochidion		hispida	
hohenackeri	S	paniculata	
zeylanicum		*Trewia*	
Flueggea		nudiflora	PS
virosa	S	*Mallotus*	
leucopyrus		philippinensis	PS/S
Sauropus		*Macaranga*	
quadrangularis	S	peltata	
Breynia		indica	S

Scientific Name	Status
Homonoia	
riparia	S
Ricinus	
communis	C
Baliospermum	
montanum	S
axillare	
Tragia	
involucrata	S
Sapium	
indicum	
insigne	S
sebiferum	
Excoecaria	
agallocha	
acerifolia	PS
Sebastiania	
chamélaea	NK
Huru	
crepitans	NK
Manihot	
utilissima	C
Hippomane	
mancinella	NK
URTICACEAE	
Hunulus	
lupulus	NK
Cannabis	
sativa	C
Streblus	
asper	S
Morus	
indica	
alba	C
nigra	
Ficus	
gibbosa	
bengalensis	
benjamina	
retusa	
rumphii	
religiosa	
lacor	
heterophylla	
asperrima	
hispida	PS
cunia	
ribes	
palmata	
glomerata	
carica	
arnottiana	
dalhousiae	
talboti	
tsiela	
Holoptelea	
integrifolia	PS
Celtis	
australis	S
cinnamomea	
Terma	
orientalis	PS
Gironniera	
reticulata	NK
Girardinia	
zeylanica	NK
Pouzolzia	
indica	PS
Antiaris	
toxicaria	D
Artocarpus	
hirsuta	
integrifolia	C
lakoocha	
Urtica	
parviflora	
dioica	S
pilulifera	

Scientific Name	Status
Laportea	
crenulata	PS
Boehmeria	
nivea	PS
PLATANACEAE	
Platanus	
orientalis	S
JUGLANDACEAE	
Juglans	
regia	C/D
MYRICACEAE	
Myrica	
Myrica esculenta	D
CASUARINACEAE	
Casuarina	
equisetifolia	C
CUPULIFERAE	
Betula	
utilis	S
alnoides	
Quercus	
incana	
lamellosa	S
pachyphylla	
Corylus	
colurna	
Quercus infectoria Olivei	S
SALICACEAE	
Salix	
tetrasperma	
acmophylla	
caprea	S
alba	
babylonica	
Populus	
nigra	
ciciata	S
CERATOPHYLLACEAE	
Ceratophyllum	
demersum	S
Elettaria	
cardamomum	C
Alpinia	
galanga	
alihugas	
calcarata	S/PS
malaccensis	
speciosa	
Maranta	
arundinaceae	C
Canna	
indica	C
Musa	
sapientum	
textilis	C
HAEMODORACEAE	
Sansevieria	
roxburghiana	C
IRID	
Iris	
ensata	
nepalensis	C
kumaonensis	
soongarica	
Crocus	
sativus	C
Belamecanda	
chinensis	C
AMARYLLIDACEAE	
Agave	
americana	
angustifolia	
vera-cruz	

Scientific Name	Status
Curculigo	
orchioides	S
Crinum	
asiaticum	
latifolium	
defixum	
Polianthes	
tuberosa	S
TACCACEAE	
Tacca	
pinnatifida	S
aspera	
BROMELIACEAE	
Ananas	
sativus	C
DIOSCOREACEAE	
Dioscorea	
pentaphylla	
oppositifolia	
bulbifera	S
triphylla	
alata	
var. globosa	
sativa	S
LILIACEAE	
Smilax	
glabra	
lanceaefolia	
zeylanica	S
prolifera	
merophylla	
Asparagus	
filicinus	
racemosus	
adscendens	S
gonoclados	
officinalis	

Scientific Name	Status
Yucca	
gloriosa	
aloifolia	C
Aloe	
vera	C
Polygonatum	
multiflorum	
Asphodelus	S
tenuifolius	
Chlorophytum	S
arundinaceum	
Allium	
ascalonicum	
cepa	
sativum	C
schaenoprasum	
tuberosum	
ampeloprasum	
Urginea	S
indica	
coromandeliana	
Scilla	
indica	C
Lilium	
giganteum	C
wallichianum	
Fritillaria	
imperialis	C
roylei	
cirrhosa	
Colchicum	
luteum	C
Gloriosa	
superba	C
Polyanthes	
tuberosa	C
PONTEDERIACEAE	
Monochoria	
vaginalis	PS

Scientific Name	Status
XYRIDACEAE	
Xyris	
indica	
anceps	PS
pauriflora	
COMMELINACEAE	
Commelina	
obliqua	
suffruticosa	
nudiflora	PS
benghalensis	
salicifolia	
Aneilema	
scapiflorum	
Cyanotis	
tuberosa	
axillaris	
Floscopa	
scandens	
FLAGELLARIACEAE	
Flagellaria	
indica	PS
JUNCACEAE	
Luzula	
campestris	PS
PALMAE	
Areca	
catechu	PS
nagensis	
Loxococcus	
rupicola	
Pinanga	
dicksonii	PS
Irachis	
Arenga	
saccharifera	
obtusifolia	PS
Nallichia	
disticha	D
Caryota	
urens	D
mitis	
Phoenix	
dectylifera	PS
sylvestris	
pusilla	
Nannorphos	
ritchieana	NK
Copernicia	
cerifera	NK
Corypha	
umbraculifera	D
Borassus	
flabellifer	
Lodoicea	
seychellarum	PS
Elaeis	
guineensis	
Cocos	
nucifera	
schizophylla	PS
yatai	
Calamus	
rotang	
travancoricus	D
rheedii	
ziminalis	
Nipa	
fruticans	D
PANDANACEAE	
Pandanus	
tectorius	D
tascinalatus	

Scientific Name	Status	Scientific Name	Status
TYPHACEAE		aromatica	
Typha		rubescens	
angustata		*Scindapsus*	
elephantina	PS	officinalis	
laxmanni		*Rhaphioophora*	
ARACEAE		pertusa	PS
Cryptocoryne		*Lasia*	
spiralis	PS	heterophylla	
Pistia		*Pothos*	
stratiotes	PS	scandens	PS
Lagenandra		cathecarti	PS
ovata	PS	*Acorus*	
Arisaema		calamus	S
speciosum		gramineus	D
tortuosum	PS	ALISMACEAE	
leschenaultii		*Sagittaria*	
Sauromatum		sagittifolia	PS
guttatum		HYDROCHARITACEAE	
Typhonium		*Vallisneria*	
trilobatum	PS	spiralis	PS
Amorphophallus		ORCHIDACEAE	
campanulatus		*Desmotrichum*	
prainii		fimbriatum	
Synantheras		*Dendrobium*	
sylvatica		ovatum	
Plesmonium		*Eulophia*	
margaritiferum		campestris	
Remusatia		nuda	
vivipara		*Cymbidium*	
Colocasia		aloifolium	D
esculenta		*Vanda*	
antiquorum		roxburghii	
Alocasia		spathulata	
indica		tessellata	
macrorrhiza	PS	*Saccolabium*	
montana		papillosum	
denudata		*Acampe*	
Homalomena		wightiana	

Scientific Name	Status
Zeuxine	
strateumatica	
ORCHIDACEAE	
Orchis	
latifolia	
Habenaria	
commelinifolia	D
cultivated	
Currcuma	
angustifolia	
aromatica	
zedoaria	
caesia	S/D
amada	C
longa	C
Kaempferia	
galanga	
angustifolia	S/D
rotunda	
Gastrochilus	
pandurata	
Hedychium	
spicatum	S
Amomum	
xanthioides	
subulatum	C
aromaticum	S
costatum	
Zingiber	
officinale	C
zerumbet	
cassumunar	S/D
Costus	
speciosus	S/D
PLUMBAGINACEAE	
Plumbago	S
zeylanica	
rosea	S

Scientific Name	Status
indica	
SAPOTACEAE	
Aerus	
sapota	S/C
Madhuca	
indica	
longifolia	S/PS
Mimusops	
elengi	
hixandra	PS/C
Mamikara	
kanki	S
Sarcos stoma	
brevistina	S
ERENACEAE	
Diospyros	
embryopleis	C/S
SYMPLOCACEAE	
Symplocos	
racemosa	S
STYRACEAE	
Styrax	
benzoin	S
OLEACEAE	
Jasminum	
arborescens	
grandiflorum	C/S
sambac	
multiflorum (su. pubescens)	C/S
humilis	
Nyctanthes	
arbor-tristis	C/S
Schrebera	
swietenoides	S

Scientific Name	Status
SALVADORACEAE	
Azima	
tetracantha	S
Salvadora	
persica	PS
ASTERACEAE	
Sonchus	
arvensis	S
brachyotes	S
Pluchia	
lanceolata	S
Vernonia	
cinera	PS
anthelmentica	PS
Elephantopus	
scaber	PS
Gnangea	
maderaspetara	S
Eupatorium	
ayapana	C
triplinerve	S
Blumea	
lacera	S
Anacyelus	
pyrethrum	S
Artimesia	
vulgaris	S
Carthamus	
tinctorius	C
Crysanthemum	
coronarium	C
Eclipta	
alba	PS
prostrata	
Enhydra	
fluctuans	PS
Saussurea	
lappa	D
Guisotia	
abisinica	PS
Xanthium	
strumarium	PS
Wedelia	
calendulacea	PS
Sphaeranthes sp.	PS
AGOCYNACEAE	
Carissa	
carandus sp.	PS
Aganosma	
caryophyllata	C
dichotoma	C
Alstonia	
scholaris	PS
Ichnocarpus	
fruitescens	PS
Holarrhene	
antidysenterica	PS
Rauwolfia	
seropentina	D/C
Nerium	
odorum	C
indicum	C
Wrightia	
tomentosa	PS
tinctoria	S
Thevetia	
neriifolia	C
Vallaris	
haynei	PS
solanea	PS
Plumeria	
acutifolia	C
rubra	C
Ervatania	
coronaria	C

Scientific Name	Status
POLYPODIACEAE	
Cibotium	
barometz	
Stenoloma	
chinensis	
Adiantum	S
lunulatum	
caudatum	
capillus-veneris	
aethiopicum	
venustum	
pedatum	
flabellulatum	
Cheilanthes	
tenuifolia	S/D
Pteris	
aquilina	
Asplenium	
adiantum-nigrum	
ruta-muraria	S/P
trichomanes	
falcatum	
Athyrrium	
filix-foemina	
Actiniopteris	
dichotoma	
Aspidium	
polymorphum	
Drynaria	
quercifolia	PS
Pleopeltis	
lanceolata	
Lygodium	
flexuosum	PS
japonicum	
OSMUNDACEAE	
Osmunda	
regalis	D

Scientific Name	Status
OPHIOGLOSSACEAE	
Ophioglossum	
vulgatum	
Helminthostachys	
zeylanica	D
Cotrychium	
lunaria	D
ternatum	
EQUISETACEAE	
Equisetum	
debile	D
POLYPODIACEAE	
1. *Actiniopteris australis* (l.f.) L.	
2. *Adiantum capillus* veneris L.	
3. *Adiantum incisum* Forsk.	
syn	D
Adiantum philippense	
Adiantum venustum	
POLYPODIACEAE	
Drynaria quercifolia	
MARSILEACEAE	
Marsilea minata	PS
SALVINIACEAE	
Azolla pinnata	
Salvinia cucullata Roxb.	
FUNGI	
Agaricus	
campestris	
ostreatus	PS
igniarius	
Polyporus	
anthelminticus	
officinalis	PS
Boletus	
crocatus	

Scientific Name	Status
Mylitta	
lapidescens	
Auricularia	
sambucina	PS
LICHEN	
Parmelia	
kamstchadalis	
perlata	PS
perforata	
CYPERACEAE	
Kyllinga	S
triceps	
monocephala	
Fimbrystylis	
junciformis	
Juncellus	
inundatus	
Cyperus	
scariosus	
rotundus	
esculentus	PS
longus	
articulatus	
iria	
Scripus	PS
grossus	
articulatus	
kysoor	
maritimus	
POACEAE	
Oryza	
sativa	
Hygroryza	
aristata	
Coix	
lachryma-jobi	
Polytoca	
barbata	

Scientific Name	Status
Zea	
mays	
Saccharum	
officinarum	
arundinaceum	
munja	
spontaneum	
Manisuris	
granularis	
Vetiveria	
zizanioides	
Amphilophis	
odorata	PS
Cymbopogon	
jwarancusa	
schoenanthus	
nardus	
citratus	
Heteropogon	
contortus	
Avena	
fatua	
sativa	
satvia var. orientalis	
Desmostachya	
bipinnata	PS
Cynodon	
dactylon	
Eleusine	
coracana	
indica	
GYMNOSPERMAE (CONIFERAE)	
Cycas	
rumphii	D
revoluta	D
Ephedra	
gerardiana	D/C
Gnetum	

Scientific Name	Status
scandens	D
Cupressus	
sempervirens	PS
Juniperus	
communis	D
recurva	D
macropoda	D
Taxus	
baccata	D
Pinus	
longifolia	
gerardiana	PS
excelsa	
Cedrus	
deodara	
Abies	
webbiana	PS
POACEAE	
Phragmites	
maxima	PS
Dactyloctenium	
aegypticum	PS
Triticum	
aestivum	C
durum	C
spelta	C
amyleum	C
Hordeum	
vulgare	C
Paspalum	
scrobiculatum	PS
Pennisetum	
spicatum	PS
compressum	PS
Thysanolaena	
maxima	PS
Panicum	
miliaceum	PS
miliare	PS
antidotale	PS
Echinochloa	
colona	PS
frumentacea	PS
crus-galli	PS
Setaria	
italica	PS/C
plicata	PS
viridis	PS
Sorghum	
halepense	C
vulgare	C
Bambusa	
arundinacea	C/PS
Dendrocalamus	
stricutus	C/PS

CHAPTER - 3

THE THREATENED WILDLIFE OF INDIA

3.1 INDIA'S WILDLIFE SITUATION IS UNDER SEVERE THREAT

A well-known authority on Indian Animals – S.H. Pratar observed (1928) : "in the distribution of its wildlife few regions of the earth exceed India in its interest".

The wildlife of India derived partly from originally indigenous species, and partly from species that migrated from elsewhere. It is an admixture of Indian, Malayan, African and European elements. The main peninsular region of India, however, remained the home of the true Indian fauna.

P.D. Stracey of the Indian Forest Service while reviewing the status of wildlife (1959) remarked : "Our bountiful wildlife has reached a sorry plight through excessive destruction both of numbers and of habitat, the former as the result of the growing number of weapons and the army as potential killers of animals and birds, and the latter as the result of the urge to many more land under cultivation". Alas the situation depicted by Stracey sharply deteriorated in the last 40 years aggravated by population explosion.

According to a report of the Zoological Survey of India, the mammalian fauna is represented by 372 species, belonging to 169 genera and 45 families which is 9 per cent of the total world's mammalian faunda. Out of these, only 10 genera are endemic but their present status needs careful examination.

There must not be any misconception about the vulnerable status of the fauna as more than 80 species have been considered to be threatened to extinction and included in Schedule I of the Wildlife (Protection) Act of 1972. The Zoological survey of India consideres 25 of these 80 species to be highly endangered, and considers a general deadline on the population of each species.

In this connection it is worth mentioning that several species of edible frogs, especially Rana trigina, R. hexadactyla, R. crassa and odd number of frog were killed in the years 1973-1984. One could imagine the impact of this on the balance of insect pest (prey) in our agricultural field vis-a-vis the likely dearth of food for reptilian species. But this figure of the killing of frogs may be several times higher now. It is a telling example of how various species are facing negative biotic impact.

- India has one-sixth of world human population, highly disproportionate to its land area and available resources.
- Population has had a leap from 238 million in 1901 to 1000 million in 2000 which means a rise of about 300 percent in 100 years.
- India has problems of a serious magnitude : food, shelter, and various other necessities of life.
- Due to rise in population the requirement of various food materials will rise and the consequent demand for birds, reptiles and and other animals will increase.
- India has half the world's buffalo population and one-seventh of cattle and goat population which put a tremendous pressure on the natural ecosystem.
- Vast genetic heritage, the animal wealth in India, itself is being lost at an alarming rate.

The breakthrough in biotechnology in general and genetic engineering in particular has created some awareness of livelihood and also some new avenues of employment for the growing population. It is now possible to transfer genes from their wild relations, to domesticated plants and animals. Large scale environmental disturbances have thrown out of gear the evolutionary processes of nature which shapes new species of organisms.

Also steps are being taken to keep extensive protected areas for conservation of flora and fauna especially those are under threat.

The Zoological Survey of India report reveals that India is losing at an alarming rate its vast genetic hertiage, the animal wealth of the country. Useful efforts have been made to save and protect some endangered species, particularly some mammals like tiger, lion, deer, sambar, elephant and birds. But the concept of conservation should have a holistic approach which involves protection of habitat and biodiversity as every species is interlinked with one another, has coevolved and remains tied together in a web of life. The biological diversity and its components have ecological, genetic, scientific, aesthetic, recreational, cultural, educational, social and economic value which has been depicted in some charts.

The Zoological Survey of India has prepared a list of the estimated number of invertebrate and vertebrate fauna of India broadly indicating their status, although research in many field is still in the initial stage.

The countrywide endangerment of flora and fauna is a phenomenon of the day. It is a product of the continuing use of various natural resources by vast and growing human and cattle population with a growing needs of people for food, fuel and shelter, there is less room for wildlife. Most wildlife species are getting endangered primarily due to loss of habitat and proaching though there are other causes of endangerment.

It is wellknown that flightless birds, small reptiles and amphibians are often easily eliminated by people. Trade in animal products such as rhino horns, crocodile skins, in live animals such as parrots, peacock pigeons, partridges is often responsible for reducing species number. The demand for fish and fish products and the modern harvesting technology has reduced abandance of many species, besides industrial pollution, poisoning of waterways and aquatic life. Dams have cut off migration access to breeding or feeding areas. Alien species introduced to new habitats without controls choke out or eliminate local species by occupying the habitat, competing with them; eating them or by bringing new diseases to which the local species are susceptible.

About birds, 9,700 world species, 1100 are endangered or threatened, but a substantial other species are being depleted. Only the public and private conservation initiatives can save them from being depleted further.

Dr. G. Bertrand of M.A. Society, USA records the decline of tiger population of a few thousands now from 1000000 at the turn of the century, Bertrand feels that the internet is an instant source of information on environmental matters, and a way of sharing strategies for saving the balance wildlife population. Information and databases shared by governments and private organizations are allowing us to approach problems in a unified way.

In 1983, the conservation movement received an enormous boost from the Convention on Trade of endangered species by Wild Fauna and Flora (CITES). The renewed commitment of governments to the recent Biodiversity convention and the global environmental fund and the growth of non-profit conservation organization worldwide will also help.

In the words of Dr. G. Bertrand Massachusetts Anderbor Society -

> "In the end, wild endangered species are no more endangered than we are ourselves by our own population growth, pollution, and the hostilities and conflict brought on by resource competition. Endangered wildlife and our survival are interwined - a fact we are still learning".

India's natural vegetation has undergone drastic changes by various forms of agriculture, forestry, and urbanization, although some species are also threatened by natural causes as landslides, floods & droughts. The rate of deforestation has been estimated at 1350 sq. km per annum out of a total of 460,440 sq. km although only just over half of this can be considered adequately stocked forest lands. Moist forests are degraded rapidly. The widespread effects of this degradation include reduction in rainfall in some areas and flooding in others. At Mahabaleswar region the rain fall decreased from 1000 cm a year to just over 600 cm.

Podophyllum emode, Rauwolfia serpentina, Paphiopedilum drury, Dendrobium pauciflorum, Paphiopedilum micranthum etc are being over exploited.

Over-exploitation of garden plants such as Anemone japonica, Peulownica tomentosa, Jasminum rudiflorum, Kerria japonica, Lonicera nitida, Crotoneaster frigidus, Gentiana farrei, Magnolia Campbelli, Meconopsis, betonicifolia are being made.

3.2 HUMAN POPULATION EXPLOSION VIS-A-VIS WILDLIFE

A Z.S.I. report enumerated the following (for detail refer Chapter 1) :

- India has one-sixth of World human population highly disproportionate to its land area and available resources.
- Population has had a leap from 238 million in 1901 to 1000 million in 2000 which mean a rise of about 300 percent in 100 years.
- India has its problems of serious magnitudes - for food, shelter, and various other necessities of life.
- Due to rise in population the demand for various food materials will rise and the use birds, reptiles and other animals will increase.
- India has half the worlds buffalo population and one-seventh of cattle and goat population which put on tremedous pressure on natural ecosystem.
- Vast genetic heritage, the animal wealth in India is being lost at an alarming rate.

The break-through in the biotechnology in general and genetic engineering in particular has created some new avenues of employment for the growing population. It is now possible to transfer genes from wild relations to domesticated plants and animals. Large scale environmental disturbances has thrown out of gear, the evolutionary processes of nature which shape new species of organisms.

Also steps are being taken to keep extensive protected areas for conservation of flora and fauna especially those are under threat.

The Z.S.I. reveals that India is losing at an alarming rate its vast genetic heritage, the animal wealth of the country. Useful efforts have been made to save and protect some endangered species, particularly some mammals like tiger, lions, deer, sambar, elephant and birds. But the concept of conservation should have a holistic approach which involves protection of habitat and biodiversity as every species is interlinked with each other, has coevolved and are tied together in a web of life. The biological diversity and its components have ecological, gentic, scientific, aesthetic, recreational, cultural, educational, social and economic value which have been clearly shown in some charts.

The Zoological Survey of India has prepared list of estimated number of invertebrate and vertebrate fauna of India which broadly depicts their status, although research in many field are still in the initial stage.

(a) Threatened wildlife

Out of innumerable cases of prevailing stress (threatening) on wildlife only a few are mentioned here under due to brevity of space.

S.C. Sharma, Addl. Inspector General of Forests and CHES Management Authority of India writes (1998) about the vulnerable status of Wildlife as,

> "During recent years, there has been a lot of concern about the heavy toll on wild animals being taken by the poachers and persons involved in illegal trade. The State Governments are now convinced that their efforts need to be augmented by the Wildlife enthusiasts outside the Government System".

Vivek Menon et al., in his book "Wildlife Crime" (1998) write in the Introduction of the book.

"India is an importer, exporter and a conduct of Wildlife that enters the $ billion annual global trade. In response to this, the Government of India has set its policy and made its laws quite unambiguously. *Most wildlife species in India are protected* and it is a crime to kill any of them, with the exception of rats, mice, crows and fruit bats. Many plants are also protected by law. Export and Import of all wild animals and their parts and products (except shed peacock feathers) is prohibitaed and the same is true for more than forty species of plants " Despite all these laws and policies the illegal trade in Wildlife continues to flourish. Just as mere laws do not bring down the incidence of heinous crimes in society, the poaching of animals, importing of plants and their subsequent trade has also to be dealt with firmly. We firmly believe that the time has come to recognise the gravity of situation and try to arrest the cataclysmic decline of species".

(b) Wildlife Crime (Source : Menon *et al.*, 1998)

Several species are mentioned hereunder which are in trade and are now threatened :

Wild ducks, geese, assorted water fowl, jungle fowl, partridges, quails, pheasants, doves, pigeons, rose-ringed parakeet, alcxandrina parakeet, blossom-headed parakeet, red-breasted parakeet, spotted munia, white throated munia, white rumped munia, hornbills, flamingoes, storks, pheasants, cranes, ducks, hawks, falcons, merlin; the source also mentions that between 1970-1976, about 13 million birds were exported from India (80% of which were parakeets and munias). Export was banned in 1990 and domestic trade was banned in 1991 of Rana hexadactyla, R. crassa and R. tigrina the legs of which are protected under Schedule - IV of the Wildlife (Protection) Act, edible nest of Swiftlet (Collocalia, fuciphaga, C. uniculor) and trade in antlers of Chital, Sambar, Swamp deer, Barking deer, Hog deer.

The illegal trade in Wildlife and its products continues to flourish despite wide range of existing laws. Mere laws cannot contain this illegal trade unless the provision are rigidly enforced. Militants and Wildlife Crimes are leading to depletion of fauna. Extremist groups have been found to poach and trade as a means to accumulate cash for arms. Several project tigers reserves have this problem besides other protected areas. Wildlife smugglers and drug barons often operate hand in hand. Even snakes are used for 'couriering drugs internationally. Vivek Memon *et al.*, in his book Wildlife Crime (1998) write, "In June 1993, a Shipment of (103 boa snakes from Colombia to USA) was detected by X-ray to have been stuffed with Cocaine-filled Condoms.

The Himalayan black bear and Sloth bear are killed for their gall bladder. Musk deer is heavily poached in the Himalayas for Kasturi. Elephant is heavily poached for ivory. Rhino horns are poached for horns said to be curing diseases. Tiger and leopard skin, bone, animal fur (The Tibetan antelope or Chiru), Skins of rat-snakes, Cobra, saw scaled viper, Russell's viper, Keelbacks etc. are in trade. All crocodiles, pythons and the yellow monitor lizards are given maximum protection by the CITES. Indian reptiles in trade are - Vipera russelli, Xenochrophis piscator, Ptyas mucosus, Naja naja, Ophiophagus hannah, Python nalurus, Varanus bengalensis, Varanus flavescens, Varanus salvator, Star Tortoises (Geochelone elegans), roofed turtle (Kachuga tecta), pond turtle (Melanochelys trijuga), Spiny-tailed lizard (Uromastyre hardwickii) etc.; of the primates rhesus (Macaca mulatta), bonnet macaques (Macaca radiata), Common langur (Presbytes entellus), hoolock gibbon (Hylobates hoolock), Slender loris

(Nycticebus coucang and Loris tardigradus), pig-tailed macaque (Macaca nemestrina), etc.

All these are being poached for trade purposes; it is therefore, very convincing that Indian Wildlife is facing a cricis for existence.

(c) Threatened waterbirds of Assam

(Media report Telegraph dated Dec. 20, 2000)

To highlight the destruction of migratory birds in this country a report is reproduced entitled "Killing of Birds" at Jhatinga in Cachar, Assam.

Hundreds of thousands of rare birds dot the landscape, making for a mesmerising sight. But beneath the facade lie ugly nets cast by hunters.

The scene is the same in marshlands across Upper Assam, which plays host to an astonishing variety of migratory birds every winter.

Wildlife officials admit that hundreds of migratory birds are caught and smuggled out of Upper Assam everyday. However, no one is sure where these avian visitors are taken to and whether they end up as meat on dining tables or trophies on living room walls.

Accompanied by members of a non-government organisation, Jorhat wildlife warden Santa Sarma recently raided several *beels* (lagoon) along the Brahmaputra and seized a few hunting nets. However, not a single poacher could be apprehended.

Sarma told The Telegraph that "informers" tipped her about the presence of poachers in almost all the beels along the Brahmaputra, but all of them fled before she reached the avian habitats.

"Birds are usually caught after midnight (between 2 am and 3 am), making it difficult for the authorities to catch anyone redhanded", she said.

Sarma said only a fraction of the birds trapped in nets were sold in local markets. "Poachers know that they can be caught if they sell birds openly. So they smuggle them out of the state." she added.

An organised gang of poachers is believed to be behind the racket, but Sarma does not rule out the involvement of forest personnel, too.

"We did not find a single forest department employee at the Kokliamukh beat office during our visit. There should be at least four personnel present at the beat office at all times," she said.

Divisional forest officer of Jorhat circle, M.N. Durarh, said he had received a complaint, but was yet to conduct a survey.

"I took up this assignment recently and am yet to visit the areas where poaching is allegedly rampant," He said.

However, Durarh clarified that the forest department weas not taking things lightly. "We are definitely worried and are determined not to spare the people involved in smuggling of rare birds," he said.

The forest official said people living in the vicinity of beels could play a major role in protection of migratory avian species. "We plan to organise public awareness camps soon," he added.

Sarma echoed Duarah's view that villagers should be invloved in the campaign to protect birds. But she said the onus was on the forest department to show the way.

"All beels with a sizeable population of migratory birds should be taken over by the forest department. I have already forwarded my personal suggestions to the forest department," she said.

Eminent zoologist P.C. Bhattacharjee, who has been invited to the 2002 International Ornithological Congress in Beijing, believes Assam is a "unique bird-watching site" as two "flyways" - the east Asian and the central Asian routes - pass through it.

"No other state in the country has two flyways used by migratory birds", he told The Telegraph.

According to conservationist Anwaruddin Choudhury, 290 species of migratory birds visit Assam almost every year. Of these, waterfowls comprise the bulk.

(d) Vanishing vultures fuel funeral rethink

No screeching, no sound of their flapping of wings. An ominous quiet hangs over the Towers of Silence, the Parsi tomb for the dead, at the heart of the financial capital, where fewer and fewer vultures are flying in to feast on the bodies.

With the bird's population taking a nosedive in Mumbai and in many other areas in the country, the community is searching for ways to dispose of the rotting bodies without the help of nature's most efficient clean-up bridge.

Behram Contractor, a veteran journalist from the community, said Zoroastrians prohibits burial or cremation of bodies.

"But still, many Parsis may prefer to be buried after death as an alternative; though it may not strictly have the approval of the religion," Contractor, the editor of abloid, *Afternoon Despatch*, said.

"It does not really matter what happens to your body once you are gone and burying the body in any case is better than leaving it out to rot," he said.

Instead of burying or cremating their bodies, devout Zoroastrians leave their bodies out for vultures at the open-air amphitheatres nestled in 20 hectares of forest on upscale Malabar Hill.

"The community is facing an unprecedented crisis in its 2000 - year old history," B.K. Karanjia, a Parsi writer of repute said. "There are very few vultures left to do the job," he added.

A 1999, study by the Bombay Natural History Society reported an alarming drop in the vulture population in the country. Some ornithologists attributed this decline to an unidentified virus sweeping across South Asia.

"You could see hordes of vultures circling over the Towers of Silence on Malabar Hill a few years ago, but now they are all gone," director of the society, R. Rahmani said.

"No one knows for sure what's happened to them, but they may have become victims or the mysterious disease killing the vultures in other parts of the country," he said.

Alarmed by the rotting and stinking corpses piling high in the towers, some senior members of the community recently met the Parsi panchayat, the community's custodian, and offered several alternatives including the setting up of a crematorium or a burial ground for its approval. Contractor said there was enough land for a graveyard in the towers if the Parsi panchayat gave the green light.

Karanjia said several members of the community were in favour of cremation after Dastur Daboo, a reformist, was cremated. "But the problem is with the priests who generally refuse to perform the last rites."

He said though the Towers of Silence, *dakhmas* to the Parsis, had enough land, the government might not give the go- ahead for a crematorium in the heart of the city.

(Media report Dec. 17, 2000)

(e) Calcutta - bound turtles rescued

Two truckloads of Gangetic turtles, an endangered species under the Wildlife Protection Act, escaped hungry human palates when they were rescued on their way to Calcutta from Kannauj and put back into Gomti river today.

In the biggest ever catch in the state, forest department officials, along with the railway police, nabbed two trucks with 46 quintals of smuggled turtles, on their way to markets in Bengal and Bangladesh.

However, forest officials are worried that if the smuggling and poaching of the delicious turtles, which is on an unprecedented rise, is not halted immediately, it would spell irreparable doom for the Ganges.

Troubled by the "disturbing trend", chief wildlife warden R.L. singh said : "The demand for such turtles, which is believed to have aphrodisiacal properties, is on an increase and the market has spread from Bengal and Bangladesh to other states also."

What is worrying Singh and Wildlife experts is the fact that killing these turtles would upset the river eco-system as they feed on carcass and other pollutants in the river.

The population of this endangered species has nose-dived from 11 lakh in 1990 to a mere four lakh this year.

Following this "alarming" development, the International Union for Nature and Natural Resources, a UN body, and the Convention on International Trade of Endangered Species have declared these turtles an endangered species.

Singh said the steep hike in the smuggling of these animals is because of the belief that its meat increases sexual potency. "Though this has not been proved scientifically, we are aware that it sells for as much as Rs. 300 a kg in some places because of this myth," he added.

The threat to the Ganges eco-system has forced the wildlife department to request the DG for help last year in manning certain points in the river and its tributaries so that poachers could be caught.

They have also stationed personnel at railway stations. The department has also started a mission to sensitive school children about the threat posed to these animals and the river. "School children have to be taught about them" Singh said, adding. "They also double-up as of formers."

(Telegraph December. 2000).

3.3 CONSERVING BIODIVERSITY RESOURCES

India has a very extensive network of protected areas (National parks, Sanctuaries, Tiger reserves etc.) covering more than 1,50,000 sq. km. Most fascinating and prestigeous protected areas are project Tiger areas, Kaelodeo National Park, Kaziranga National Parks and many other such areas.

Yet, the animals are not fully secured inside these protected areas. Financial constraints, crippling shortage of adequate staff, poaching, insurgency, inadequate communication system, lack of judicial support are some deficiencies besides several others.

Some of these detrimental factors are responsible for various animals, insecurity inside the protected areas. The export of tiger and leopard skins, bones, claws are rampant as may be witnessed from frequent illegal traders being held with booty by the customs, forest and police officials.

The forest covering 6,00,000 sq. km., are under control of Forest Department, a quarter area of it is under the umbrella of Protected areas. The very net-work that preserved the green cover and its denizens over many decades is now falling apart.

Sometime past about half a million mature sal tree were felled in Madhya Pradesh ostensibly to prevent spread of an insect epidemic. However, the clear cutting was halted by the Court. Similar situation is common in many states.

Even zoo animals have become vulnerable to poachers. The incidence of Hyderabad, Nandankanan are two such examples. Pilferage of bird, animals skin and various other parts of their bodies, even live birds and animals are being detected all over the country. The official registers of the Govt. of India's Export-Import Trade Control offices testify to the type of prohibited flora and fauna detected.

"Two vultures, the white-backed vulture and the long-billed vultures are heading for extinction. There has been 90 percent drop in the population of these vultures over just the past three to five years. In the past Keoladeo National Park (area 29 sq.

km.) supported more than 2000 vultures; it was now having a population of 100 birds. The author mentions about the research investigation of Vibhu Park of Bombay Natural History Museum about the aforesaid and four other potected reserves. Buxa, Jaldapara, Gir, Corbett, Simlipal which had fewer number of vultures. The crisis has galvanised Wildlife organizations both in India and abroad for investigation and they consider a relative of poultry virus may be cause of vulture dealth." *(Source : Sunday Telegraph dt. 26.11.2000 by G.S. Madu and A.S.Gill)*

The genetic resources serve as raw materials for tomorrow's biotechnology. Gene pools in the wild constitute the additional genetic resources for the future improvement of crop plants, forest trees as well as new and novel plants of potential use in agriculture, forestry, medicine and industry. UNCED Earth Summit deals with various aspects concerning conservation and sustainable use of all components of biological diversity.

"India is now facing a severe threat for her rich biodiversity and genetic resources from habitat fragmentation. The only way to avert a mass extinction is 'Conservation' - our evolutionary responsibility". *(Source : M.S. Mondal of B.S.I. in Nal. Symposium of Zoo. Soc. 1995)*

Various issues about wanton destruction of Tropical forests was first brought to public notice by the National Academy of Sciences (NAS, USA) in their report entitled "Research priorities in Tropical Biology". Later NAS published a book entitled 'Biodiversity' where the editor E.O. Wilson (1988) stressed its importance of exploding human populations responsible for degrading the forest resources; he also found that much biodiversity was being lost through extinction caused due to destruction of forests. According to Raven, 1988; Wildon, 1989, Ehrlich and Wilson, 1991 "Hundreds of books, articles, and reports were written, and scores of symposia, workshops, meetings, and seminars were organized to publicize the fact that the Earth's biological diversity was decreasing. Locally and internationally financed programs to conserve biodiversity were implemented. The promise of these new efforts, however, is tampered by the reality of the situation : Between 1950 and 1990 an estimated 30 to 40 percent of the tropical rains forest disappeared, and with the projected disappearance of an equal amount of forests over the next thirty to fifty years; a quarter of the world's species diversity may vanish forever. The situation in India is serious as may be viewed from the charts and tables drawn in various chapters.

3.4 CHECKLIST (TENTATIVE) AND STATUS OF MAMMALIAN FAUNA

(Most of the species are predected under Wildlife (Protection) Act, 1972)

MAMMALS

Local or Common Name	Zoological Name
CATS	
Tiger	*Panthera tigris*
Leopard/Panther	*Panthera pardus*
Fishing Cat	*Felis virerrina*
Leopard Cat	*F. bengatensis hersffieldi*
Jungle Cat	*F. chaus*
Large Indian Civet	*Vevera zibetha zibetha*
Small Indian Civet	*V. malaccensis malacensis*
Indian Palm Civet	*Paradoxurus* sp.
Malayan Palm Civet/ Toddy Cat	*P. hermaphroditus*
The snow Leopard	*Panthera uncia*
The Coluded Leopard	*Neofelis nebulosa*
The Marbled Cat	*Felis marmorata*
The Golden Cat	*F. temmincki*
The Desert Cat	*F. libyca*
The Caracal	*F. caracal*
The Lynx	*F. lynx*
The Pallas's Cat	*Octolobus manul*
The spotted Linsang or Tiger Civet	*Prionodon pardicolor*
The Brown Palm Civet	*Paradoxurus jerodni*
The Himalayan Palm Civet	*Paguma larvata*
The Red Fox	*Vulpes vulpes*
The Dhole	*Cuor alpinus*
The Brown Bear	*Ursus arctos*

Local or Common Name	Zoological Name
The Himalayan Black Bear	*Selenarctos thibetanus*
The Cat Bear or Panda	*Ailurus fulger s*
The Common Otter	*Lutra lutra*
The smooth Indian Otter	*L. perspicillata.*
The clawless Otter	*Aonyx cinerea*
The Beach or Stone Marten	*Martes folna*
The Himalayan Weasel	*Mustela sibrica*
Ermine	*M. erminea*
The Black Buck	*Antelope cervicabra*
Chinkara or Indian Garelle	*Gorella gazella*
The Pale Weasel	*M. altaica*
The Yellow-bellied Weasel	*M. kathian*
The stripe-backed Weasel	*M. strigidorsa*
The Marvelled Pale Cat	*Vormela peregusna*
The Ratel or Honey Badger	*Mellivora capensis*
The Longeared Hedge Hog	*Hemiechinus auritus*
The Eastern Mole	*Talpa micrura*
The Indian Short-tailed Mole	*T. micrura micrura*
While-tailed Mole	*T. micrura leucura*
The Himalayan Marmot	*Marmota bobak*
Long-tailed Marmot	*M. caudata*
The Alitopiral/Gerbilles	*Tatera indica*
The Indian Desert Gerbilles	*Meriones nurrlanae*
Common Dolphin	*Delphinus delphis*
Plumbeous Dolphin	*Sotalia plumbea*
Red-sea Bottle-nosed Dolphin	*Tursiops adeuneus*
Indian Broad beaked Dolphin	*Lagenorhynchus clectra*
Little Indian Perpase Dolphin	*Neomeris phocaenoide*
Short-tailed Mole	*Talpa micrura*
Indian Porcupine	*Hystrix indica*
Hodgson's Porcupine	*Hystrix hodgsoni*

Local or Common Name	Zoological Name
Brush-tailed Porcupine	*Atherurus macrourus*
Indian Hare	*Lepus nigricollis*
Cape Hare	*L. capensis tibetanus*
Arabian Hare	*L. arabicus craspidotis*
Himalayan Mouse Hare	*Ochotone royler*
Wild Goat	*Capra hircus*
Markhor	*C. indica*
Thar (Himalayan Thar)	*Hemitragus jemlahicus*
Thar (Nilgiri)	*Hemitragus hylocrius*

OTHER MAMMALS

Crestless Himalayan Porcupine	*Hystrix hodgsoni*
Common Indian Hare	*Lepus ruficandalus*
Hispid Hare	*L. hispidus*
Indian Elephant *E*	*lephas maximus*
Great Indian One-horned Rhinoceros	*Rhinoceros unicornis*
The Gaur	*Bos gaurus*
The Bengal Barking Deer	*Muntiacus muntijac vaginalis*
The Sambhar	*Cervus unicolor*
Spotted Deer	*Axis axis*
Hog Deer	*Axis porcinus*
Wild Boar	*Sus scrofa cristatus*
Indian Pangolin	*Manis crassicaudata*
Chinese Pangolin	*M. pentadactyla*
Binturong	*Arctitis binturong*
Small Indian Mungoose	*Herpestes auropunctatus*
Common Mungoose	*H. edwardsii*
Crab-eating Mungoose	*H. urva*
Jackal	*Canis aurinus*
Indian yellow-throated Marten	*Martes flavigula*

Local or Common Name	Zoological Name
Brown Ferret Badger H.sp.	*Melogale personata*
Hog Badger	*Arctonyx collaris*
Common otter	*Lutra lutra monticola*
Smooth-coated Indian Otter	*L. perspicillata*
Clawless otter	*L. leptonyx*
Sloth Bear	*Melursus urcinus*
Serow	*Capricornis sumatrensis*
Malayan/Common Tree Shrew	*Tapaia glis*
The Ruddy Mungoose	*Herpetes smithi*
The striped-necked mungoose	*H. vitticollis*
The Brown Mungoose	*H. juscus*
The striped Hyena	*Hyaena hyaena*
The wolf	*Canis lupus*
Goral	*Nemor headus goral*
Takin	*Budorcas taxicolor*
Tibetan Antelope	*Pantholops hodgsoni*
The Dugong	*Dugong dugon*
The Asiatic Wild Ass	*Equns hemionus*
Indian Wild Ass	*E. hemionus khur*
The Banteng/Tsaine	*Bos banteng*
The Yak	*Bos grunniens*
Wild Buffalo	*Bubulus bubalis*
The Shapu/Urial	*Ovis oriculals*
The Nayan/Great Tibetan Sheep	*O. ammon*
Marcopolos sheep	*O. sp.*
The Bharal/Blue Sheep	*Pseudois nabura*
The Ibex	*Capra ibex*
Wild Water Buffalo	*Babulus bulalis*
Pigmy Hog	*Sus salvanius*
Indian Fox	*Vulpes bengalensis*
The Gangetic Dolphin	*Platanista gangetica*

Local or Common Name	Zoological Name
Andaman Wild Pig	*Sus scrofa andamanensis*
Swamp Deer	*Cervus duvauceli*
Thamin	*Cervus eldi*
Desert fox	*Vulpes bucopus*
SQUIRREL	
Kashmir Wooly Flying Squirrel	*Eupetaurus cinereus*
Orange Himalayan Giant squirrel	*Dremonys lokriah*
Himalayan Striped squirrel	*Calloniaurus maccielland*
Irrawati Squirrel (Honey-bellied)	*C. pygerythrus lokroides*
Golden-backed Squireel	*C. caniceps crumps*
Common Giant Flying Squirrel	*Petaurista magnificus*
Lesser Giant Flying Squirrel	*P. elegans*
The larger Brown Flying Squirrel	*P. philippinensis*
Red Flying Squirrel	*P. altiventer*
Grey-headed flying squirrel	*P. caniceps*
Small Travancore Flying Squirrel	*Pelinomys fuscocapillus*
Kashmir Flying squirrel	*Hylopetes fimbriatus*
Particoloured Flying Squirrel	*H. alboniger*
Hairy footed flying squirrel	*Belomys pearsoni*
3-striped palm squirrel	*Funambulus palmarus*
5-striped Palm Squirrel	*F. pennanti*
Giant Squirrel	*Ratufa indica*
Grizzled Giant Squirrel	*R. macrour*
The Malayan Giant Squirrel	*R. bicolor*
Dusky-striped squirrel	*Funambulus bublineatus*

Local or Common Name	Zoological Name
BATS	
Common Flying Fox	*Pteropus giganteus*
Fulvous Fruit Bat	*Rousettus leschensulti leschensulti*
Southern Short-nosed Fruit Bat	*Cynopterus sphinx sphinx.*
Allied Horse-shoe Bat	*Rhinolophus rouxi rouxi*
Common Yellow Bat	*Scotophilus kukli*
Wroughton's Bat	*S. temmineki wroughtoni*
Hairy-winged Bat	*Marpiocephalus harpia lasyunes*
Wall Bat	*Myotis mystacinus muricola*
Tikell's Bat	*Hespirotenus tickelli*
Barlequin Bat	*Scotophilus ornatus*
Painted Bat	*Kerivoula picta*
Bearded Sheath-tailed Bat	*Taphazon melanopogor*
The Great Himalayan Leaff nosed Bat	*Hiaposideras armiger*
Serotine	*Pipistrellus coromandra*
Common Yellow Bat	*Scotophilus heathi*
Dobsonstub-nosed Bat	*Murina Cyclotis*
Himalayan Hairy-winged Bat	*Harlocephalus harpia lasyurus*
Hardaricke Temb Bat	*Taphozons longimanus*
Himalayan Horse shoe Bat	*Rhinolophus affinis*
Eastern Himalayan Little Horse-shoe Bat	*Rhinolophus connutus*
Trifoil Horse-shoe Bat	*R. trifoliatus*
Great-Eastern Horse-shoe Bat	*Myotis siligorensis*
Asian Barbastelle	*Barbastella leucomelas*
Hollons Tube-nosed Bat	*Murina buttoni*
Indian Vampaire Bat	*Megaderma lyra*
Indian Pigmy Pipistrella	*Prepistrellus minus*

Local or Common Name	Zoological Name
RATS & MOUSE	
House Mouse	*Mus musculus*
Edwardis Rat	*Rattus edwardsi*
Little Himalayan Rat	*R. eta*
Chestnut Rat	*R. fulvescens*
White-bellied Rat	*R. niviveater*
Himalayan Rat	*R. nitidus*
House Rat	*Rattus rattus*
Mallards/Large-tothen Rat	*Dacuomys mallardi*
Bay Bamboo Rat	*Cannomys oleracea*
Little Indian Field Mouse (Bush Rat)	*Golunda ellioti*
The Indian Tree Shrew	*Anathana elloti*
Bhutan Duars Rat	*Rattus ratus bhotia*
Bengal Mole Rat	*Bandicoota bengalensis*
Bengal Bandicoot	*Bandicoota sp.*
The Melad/Soffured Field Rat	*Millardia meltada*
Indian Field Mouse	*Mus booduga*
The Spiny mouse	*M. platythrix*
The White-tailed wood rat	*Rattus blanfordi*
Alexandrine Rat	*R. alexandrus*
The Brown Rat	*R. norvegicus*
Royle's vole	*Alticola roylei*
Sikkim vole	*Pitymys sikkimensis*
Murree vole	*Hyperacrius wynnei*
PRIMATES	
The Bonnet Macaque	*Macaca radiata*
The Rhesus Macaque	*M. mulatta*
The Assamese Macaque	*M. assamensis*
The Stump-tailed Macaque	*M. arctoides*

Local or Common Name	Zoological Name
The Pig-tailed Macaque	*M. nemestrina*
The Lion-tailed Macaque	*M. silenus*
The Capped Langur or Leaf monkey	*Presbytis pileatus*
The Nilgiri Langur	*P. Johni*
The Slow Loris	*Nycticebus coucang*
The Slender Loris	*Loris tardigradus*
The Golden Langur	*Presbytis geei*
The common langur	*Semnopethecus entellus*
Barbe's leaf monkey	*Presbytes barbei*
Phayre's Leaf monkey	*Trachipihacus pheyrei*
The Crab-eating Macaque	*Macaca fascicularis*
The Hoolock	*Hylobetes hoolock*

NON-HUMAN PRIMATES

World : 58 Genera

Madagascar Island is richest with 23 Genera.

India has : 15 species (some are endemic over a very small area).

(An overview)

N.E. Assam/Tripura

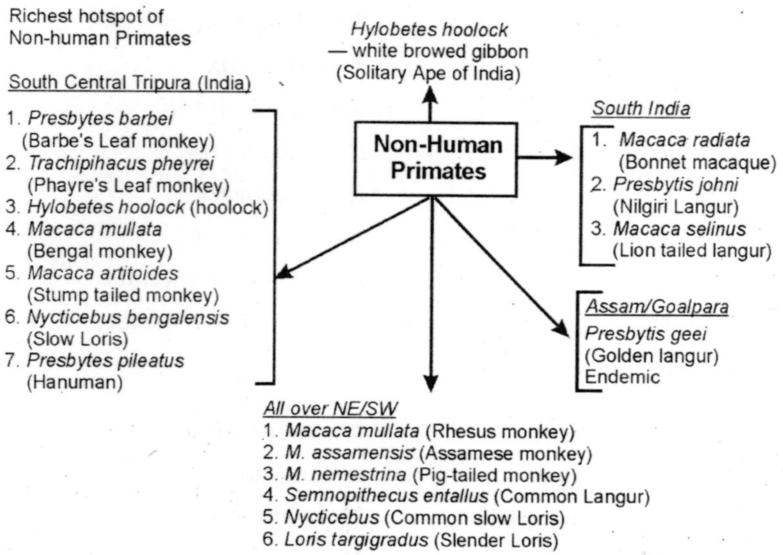

3.5 PRIMATES

The order Primates in India has 3 families, 5 genera and 15 species (world has 12 families, 58 genera and 181 species).

Loriside - 2 sp. of Loris. Slow. loris and Slender loris.

Cercopithecidae - 12 sp. of Macaca and Presbytis.

Hylobatidae - 1 sp. of Hylobatidae.

A check list has been prepared which may be referred to.

India does not have Gorilla or Chimpanzee or Orang Utan, but has one ape, a single species, which is Hoolock or white browed Gibbon. This ape exists in Assam, Tripura and other eastern hilly states of Manipur, Mizoram, Nagaland and Arunachal Pradesh. The distribution of apes, monkeys and lemus are as folows.

- Hoolock - Assam, Tripura, Arunachal Pradesh, Mizoram, Manipur, and Nagaland.
- The bonnet macaque - Peninsular India - near Bombay, Godavari river, Trivancore.
- The rhesus macaque - Himalaays, Assam, Central and Northern India, Tripura.
- The Assamese macaque - Himalayan foot hills from Mussoorie eastward, Assam & North east hilly states, Sunderbans, Tripura.
- The Liontailed macaque - Western Ghat from Canara, Southwards to Kerala, Tamil nadu.
- The Common langur - All over India.
- The Capped langur or leaf monkey - Assam and North east hilly states, Tripura.
- The Golden langur - East of Snakes river at Manas.
- The Nilgiri langur - Western ghat from Coong to South-Nilgiri, Anamalai, Brahmagir, Tinnevally, Palai hills.
- The slow loris - Assam, North east hills, Tripura.
- The Slender Loris - Southern India, Srilanka.
- Bengal Monkey - Tripura and eastern India.
- Stumptailed monkey - Tripura and eastern India.
- Banbe's leaf monkey - Tripura.

Several species of primates are in danger.

A small stretch of forest in Tripura from Cherilam Reserve to Trishna Reserve is perhaps the most interesting and vulnerable stretch as the forests are being biotically very much disturbed. The home of golden langur in Manas- Bhutan once is very much disturbed. Three sp e.g. Bonnet macaque, Liontailed macaque and Nilgiri langur need close protection as these are endemic to the peninsular area. The entire stretch of forests is under heavy depletion.

3.6 PANGOLINS

Sangita Mitra of Z.S.I., Calcutta studied biodiversity aspects of Indian Pangolin (Manis sp.). She stated that 77 mammals, of 372 mammalian species of India, as vulnerable and endangered as per the Indian Red Data Book. Pangolin is one such highly specialised species. It's population is not known, but it is rare in the forests. They are killed for the supposed medicinal value of their scales and meat. Though a protected animal included in Schedule -I of the Indian Wildlife (Protection) Act (1972) and Appendix - II of CITES' the illegal trade continues. Other biotic and abiotic factors are also causes for the depletion of this species.

3.7 STATUS OF REPTILES AND LIZARDS

A Broad view

Biodiversity depletion

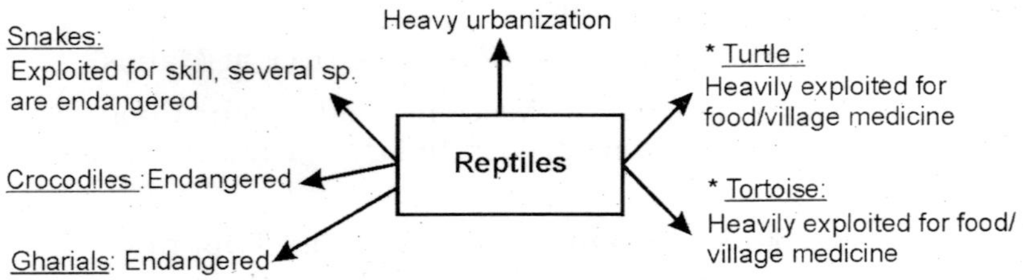

*Many species are endangered

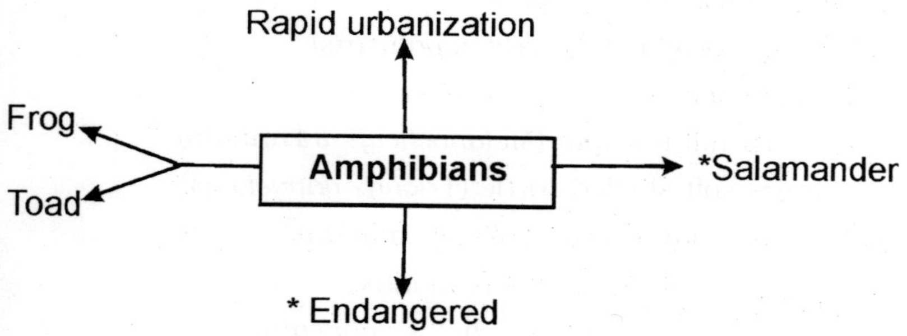

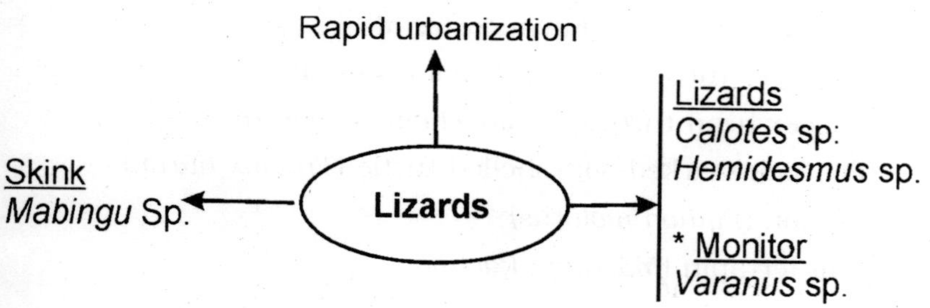

Perilous indication of depletion

Several schedules of Wildlife (Protection) Act, 1972 of India which lists a good number of animals, birds, insects and plants as threatened indicate the sad story of depletion due to various biotic impacts. Some of the species are endangered. More and more species are being enlisted every year that highlights the vulnerability. To justify the authors' findings a small list is reproduced from the schedule of Protected animals as follows:

(i) In Schedule I, Part I.

Water lizard (*Varanus salvator*)

Large Bengal Monitor lizard (*Varanus bengalensis)*

Agra monito (*Varanus griseus)*

Yellow monitor lizard (*Varanus flavescens)*

Salt water crocodile (*Crocodilus pqrosus)*

Gharial (*Gavialis gangeticus)*
Marsh crocodile (*Crocodilus palustris)*
Batagur terrapin (*Batagur baska)*
Eastern hill terrapin (*Melanochelys tricarinata)*
Ganges soft shelled turtle (*Trionyx gangeticus)*
Golden jacko (*Caloductyloides aureaus)*
Green Sea turtle (*Chelonia mydas)*
Hasks bill turtle (*Eretmochelys imbricata)*
Indian Egg eating snake (*Elashistodon westermanni)*
Indian soft shelled turtle (*Lissemys punctata)*
Indian tent turtle (*Kachuga tecta)*
Leathery turtle (*Deermochelys coriarea)*
Lagger-head turtles (*Lepidochetys olivacea)*
Peacock marked soft-shelled turtle (*Trionyx hurun)*
Python (*Python molurus)*
Sail terrapin (*Machuga kachuga)*

(ii) In Schedule - II, Part- I the following have been recorded:

Chameleon (*Chameleon calcaratus)*
Spingy Tailed Lizard or Sanda (*Hromastrix hardwickii)*

(iii) In Schedule - II Part II the following are recorded :

Checked neelback (*Xanochrophis piscator)*
Dhamon or rate snake (*Ptyas mucosus)*
Dog faced water snake (*Cerberus rhynchops)*
King Cobra (*Ophiophagus hannah)*
Indian Cobra (*Naja* sp.)
Olivaceous Keelback (*Artretin mschistosum)*
Russel's viper (*Viper ruselli)*.

REPTILES and LIZARDS (General information)

India is not very rich in these animal species. It has about 52 species of snakes of which 20-30 species may be listed as common, besides another about 30 species of inland turtles and tortoises. About the lizards India has about 158 species. But the status of all these species cannot be easily ascertained as authentic information on individuals of each species is absent.

Only evidence (factual) is available in the records of a Government of India Department, which controls the exports and imports of flora and fauna. The records show how innumerable species of flora and fauna and their parts are being exported either legally or illegally. Large number and quantity of such materials protected under Wildlife (Protection) Act, 1972 are being illegally killed and exported which include mammalian skins and bones (tiger, leopard, Jackal, fox, wolf, hayena, mongoose, otter, porcupine, deer and several other species). Lizard skins of various species including those of Varanus, Living turtles and tortoises, snakes (by thousands) and prohibited bird species ; besides hairs, quills, bones, flowers, fruits, seeds, roots and other floral products.

On December 20, 2000 two truck loads of gangetic turtles (about 46 quintals) were intercepted in Lucknow which were being transported to Calcutta illegally. Forest officals feel that the population of these endangered animal has nose-dived in the region from 11 (Eleven) lakhs in 1990 to a mere 4 (four) lakhs this year (2000) [Source : The Telegraph dt. 21.12.2000]. Such instances on illegal collections of snake skins, animal skins and other parts are very common all over the country. Similar instances of large haul are very often reported in various media.

These records are just the tip of monstrous iceberg of illegal exports which are being carried out surruptituously from this country. Yet, this shows how vulnerable are these floral and faunal species and their parts to the destructive actions of the pilferers.

So whatever rich resources we have in this country are facing a rapid depletion.

Snakes (Reptiles)

Common Name	Scientific Name
Common Kukri snake	*Oligodon venustus*
Common Blind snake	*Typhlops braminus*
Jerdon's Blink snake	*Typhlops jerdoni*
Diard's Blink snake	*Typhlops diardi*

Common Name	Scientific Name
Indian Rock Python	*Python molurus*
Trinket snake	*Elaphe helena*
Ring-tailed Dhaman	*Elaphe contoris*
Indian Rat Snake (Dhaman)	*Ptyas mucosus*
Striped kukri snake	*Oligodon cyclurus*
Indochinese rat snake	*Ptyas korras*
Indian Bronze backed Tree snake	*Ahaetulla tristis*
Twin-spotted wolf snake	*Lycodon jara*
Common wolf snake	*Lycodon aulicus*
Collared Dwarf snake	*Sibynophis collaris*
Chequrered keelback	*Natrix piscator*
Buff-striped keelback	*Natrix S. stolata*
Hooded Tree snake	*Pseudoxenodon macrops*
Grey cat snake	*Bioga gokool*
Mock Viper	*Pasmmodyrastes pulverulentus*
Indian egg eating snake	*Elachistodon westermanni*
Malayan whip snake	*Dryophis prasinus*
Common Green Whip snake	*Dryophis nasutus*
Schneider's water snake	*Enhydris enhydris*
Banded krait	*Bungarus fasciatus*
Common krait	*Bungarus caeruleus*
Lesser Black krait	*Bungarus lividus*
Black krait	*Bungarus niger*
Cobra	*Naja naja*
Russell's viper	*Vipera russelli*
Common Green pit viper	*T. erythruras*
King Cobra	*Ophiophagus hannah*
Red-necked keel back	*Natrix subminiata*
Common kukri snake	*Oligodon arrensis*
Ladder back kukri snake	*O. albocinetus*

Common Name	Scientific Name
Dog faced water snake	*Cerberus rhynchops*
Ring tailed Dhama	*Elaphe contortis*
Olive keelback water snake	*Atretium schistosum*
Dussumier's smooth water snake	*Enhydris dussumicrii*
Siebold's smooth water snake	*Enhydris sieboldii*
Green keelback	*Macropisthodon plumbicolor*
Indian olive keelback water snake	*Macropisthodon plumbicolor plumbicolor*
Dark-bellied marsh snake	*Xenochrophis cerasogaster*
Yellow-spotted keelback water snack	*Xenochrophis flavipunctatus*
Checkered striatus	*X. piscator*
Painted bronze back	*Dendrelaphis pictus*
Shaw's wolf snake	*Lycodon striatus*
Ocellate shield tail	
Diadem snake	*Spalerosophis diadema*
Variegated kukri snake	*Oligodon laeniolatus*
Golden tree or Flying snake	*Chrysopelea ornata*
Sand snake	*Psamophis condararas*
Himalayan pit viper	*Agkistron healaynus*
Saw scaled viper	*Echis carinatus*

Some facts on Tortoises and Turtle (Reptiles)

Indian species have been listed by Dr. I. Das of IUCN. The record pertains to 30 species. The list below show two species of Crocodylidae. Very common species are marked with asterisks(*).

A checklist of fresh water reptiles of India prepared by I. Das, IUCN herpetology specialist is presented below :

Crocodylus palustris	Mugger crocodile
Gavialis gangeticus	Gharial
Curoa amboinensis	Southeast Asian box turtle

**Cyclemys dentata*	Asian leaf turtle
**Geocleyms hamiltonii*	Spotted pond turtle
Hardella thurjii	Crowned river turtle
Kachuga dhongoka	Three stripped roofed turtle
**Kachuga kachuga*	Red-crowned roofed turtle.
Kachuga smithii	Brown roofed turtle
Kachuga synthetensis	Assam roofed turtle
Kachuga tecta	Indian roofed turtle
Kachuga tenoria	Indian tent turtle
Melanochelys trijuga	Indian black turtle
Morenia petersi	Indian eyed turtle
**Aspideretes gangeticus*	Indian softshell turtle
Aspideretes hurum	Indian peacock softshell turtle
Aspideretes leithii	Leith's softshell turtle
**Chitra indica*	Narrow-headed softshell turtle
**Lissemys elegans* (Indian star tortoise)	Point Calimere Wildlife Sanctuary, Mudumalai Wildlife Sanctuary, Kalakkad Wildlife Sanctuary
**Indotestudo elongata* (Elongated tortoise)	Corbett Tiger Rescue Rajaji Wildlife Sanctuary simlipal Tiger Rescue Gorumara National Park
Indotestudo forstenii (Travancore tortoise)	Indira Gandhi Wildlife Sanctuary
Manouria emys (Asian Brown tortoise)	North Cachar Hills, Meghalaya
Aspideretes nigricans (Black softshell turtle)	
Pelochelys cantorii (Asian giant softshell turtle - 5' - 5")	Gata kachan Hooghly, Sundarban, Orissa, Kerala

Geoclamys silvatica (Cochen forest cane turtle)	West Bengal Western Ghat
Melanochelys tricarinata (Tricennate hill turtle)	U.P., Nepal, Kaziranga, Orang
Morenia petusi (Indian eyed turtle)	Brahmaputra, Bihar, Calcutta.
Pyxidea mouhotii (Keeled box) turtle	Cachar

Lizards

Lizards are an inconspicuous, but abundant component of the fauna of India. There are 155 (approx.) species of lizards in India, representing the Geckos, the Garden lizards, the Skinks and the Monitor lizards.

Geckos are nocturnal lizards that live around houses or on trees or on the forest areas. They have amazing running power on smooth surface. These are only reptiles to have developed a true voice.

The Agamids are a major assemblage of medium-sized ground, rock or tree-dwelling lizards. These are diurnal animals.

Skinks are a vast assemblage of mostly small lizards, which are recognized by their long and cylindrical bodies and short limbs. They are found both in the plains and the hills. They are secretive and are found under bark, stones and piles of debris. They are mostly striped, cross-barred or spotted.

Monitors are large sized with mobile head, elongated neck, flattened body with a laterally compressed tail and snake-like forked tongue.

For the present purpose no more information about the existing status of the species could be gathered. A tentative list is as under :

Geckos

Northern House Gecko	Bark gecko
Southern House Gecko	Brook's Gecko
Termite Hill Gecko	Eastern House Gecko
*Rock Gecko	Scaled Gecko
Gliding Gecko	Branded scaled Gecko

*Fan-tail Gecko
Dwarf Gecko
Southern Forest Gecko
Frilled House Gecko

*emareld Gecko

Agamids

Common garden lizards (Blood sucker)

Souther green Calotes

Eastern green or, Jerdons' Calotes

Kashmir agama

Nilgiri salea

*Peninsular rock agama

Forest Calotes

Fan-throated lizard

Draco or flying lizard

Short tailed agama

Blue throated lizard

Cameleons

Indian cameleons

Lacertids

Leschenault's cabita
Jevdon's snake eye

Skinks (Small lizards)

Burrowring skinks
Sand skinks

Monitor Lizard

Common skinks

*Common Indian monitor

Little skinks

Snake Skink

*Yellow monitor

*Water monitor

*Desert monitor
*Agramonitor lizard

Of the lizards, Gecko genus has 54 species, Cnemaspis genus has 11 species, Agamids have 53 species and skinks have 44 species. There are many endemic species and monotypic genera.

Biodiversity in Reptiles and Lizards

The ecological and economic functions of reptiles to the people of India is not as great as the major invertebrate groups

- birds, fishes and mammals. They were so long not being ruthlessly exploited for food or skin though turtles and tortoises were always a covetable item of food for man and were heavily exploited.

During the later part of last century snake and lizard skins and Tortoise/turtle shells and flesh had been heavily exploited to meet human need all over the country.

No authentic data on their number is available. But their biological role is acknowledged and it is feared that the present trend of heavy exploitation for snake and Lizard skins may be the cause for loss of biodiversity of several species such as varanus sp., Python moluras, Ptyas mucosus, Turtles and Tortoises are also threatened due to heavy exploitation.

Several check lists of snakes, turtles and tortoises, lizards, agamids, chameleon, skinks, monitor lizard and lacerids are presented; but these are not to be taken as complete list recorded from India. The lists have been prepared to show the readers about the rich assemblages of faunal species. No data on their individual number can be presented.

The status of balance in nature due to actions and interactions of various species of amphibia, lizards, snakes etc. in the bio.ically and abiotically affected Indian Habitat is very much disputed. It is felt that expansion of settlement, flood, drought, ruthless killing of such species for fun, food and trade have threatened the very existence of all these groups of animals.

3.8 AMPHIBIAN BIODIVERSITY

It is difficult to throw light on the status of various species of amphibia. No numerical count of species have been critically made.

The biological role played by various species cannot be ignored as they have direct bearing on the habitat of agricultural field. Bufo, Ramanella, Limnonectes, Rana, Pailantus, Phacophorous and Ichythyophis are the most common and widely occurring genera. Genera of secondary importance are Amolops, Indiana, Micrixalus and Nictibatracks.

Shrinkage of .aquatic habitat, use of biocides, ruthless killing of species for edible purposes are responsible for depletion of resources which play a very active role in keeping insect fauna under control and they themselves forming food for various reptile and lizard species.

S.N. Banerjee *et al.*, of Rammohan College, Calcutta studied a few species of amphibia as amphibians have been the subject of various investigations because of their transitional phylogenetic position in the animal kingdom, great diversity of modes of life history than any other groups of vertebrates, co-existence of the premitive and advanced type of Karyo type, low dilpoid number with much elongated metaphase chromosomes, ambiguous sex determination mechanisms and a wide range of distribution. They found -

"Certain amphibians, even in habitats that appear to be prestine, are disappearing gradually and the declines have been serious since 1980s..... they may get extinct in the next 30 years." They said, among the vertebrates, amphibians might be the best animal group to be used as bio-indicators of environmental decay.

The importance of aquatic substratum as such is of tremedous importance for the amphibians to survive. Some widespread amphibian genera are *Bufo, Ramanella, Indirana, Limnonectus, Rana, Philautus, Rhacophorus* and *Ichthyophis*.

There are about 1000 species of amphibians in Asia which rely on freshwater at some stage of their larval development. Some adult amphibians may spend all their life near or in the water, especially salamanders and discoglossid frogs. Frogs are pollution indicators.

M. Das *et al.*, (1993) studied amphibian fauna in the district of Sambalpur, Orissa. They identified 10 species viz., *Rana limnocharis, Bufo melanosticutus, Rana cyanophlyctis, Rana tigrina, Microhyla ornata, Ramanella variegata, Uperdon systoma, Polypedates maculatus, Bufo stomaticus* and *Rana variegata*.

The density of occurrence they determined was 8 to 49 numbers per ha.

A checklist of Amphibians of India, Sri Lanka, Bangladesh and Nepal prepared by Indranell Das *et al.*, is reproduced to register the diversity of amphibian in India (209 species) which need aquatic substrata at certain stage of their live to complete their life cycle. (Self communicated by I. Das in August 1997).

The South Asian Herpetofauna
A Conservation Action Plan
Compiled by

Indranell Das and Members of the IUCN/SSC South Ásian Reptile and Amphibian Specialist Group

CHECKLIST OF AMPHIBIANS

PELOBATIDAE

Megophrys boettgeri : India
Megophrys lateralis : India
Megophrys montanta : India
Megophrys parva : India, Nepal
Megophrys robusta : India
Scutlger boulengeri : Nepal
Scutlger nepalensis : Nepal
Scutlger nyingchinensis : India, Nepal
Scutlger occidentalis : India
Scutlger sikinumensis : India, Nepal

BUFONIDAE

Ansonia kamblei : India
Ansonia ornata : India
Ansonia rubrigina : India
Bufo abatus : India
Bufo atukoralet : Sri Lanka
Bufo beddomii : India
Bufo brevirostris : India
Bufo fergusonii : India, Sri Lanka
Bufo himalayanus : India, Nepal
Bufo hololius : India
Bufo kelaartii : Sri Lanka
Bufo Kotagamai : Sri Lanka
Bufo Koynaensis : India
Bufo latestii : India, Pakistan
Bufo melanostictus : Bangladesh, India, Maldives, Nepal, Pakistan, Sri Lanka
Bufo microtympanum : India, Sri Lanka
Bufo olivaceous : Pakistan
Bufo parietans : India
Bufo silentvalleyensis : India
Bufo stomaticus : Bangladesh, India, Nepal, Pakistan, Sri Lanka

Bufo stuarti : India
Bufo surdus : Pakistan
Bufo viridis
Bufo viridis zugmayeri : India, Pakistan
Bufo viridis pseudoraddei : Pakistan
Bufoides meghalayana : India
Pedostibes kempi : India
Pedostibes tuberculosus : India

MICROHYLIDAE

Kaloula baleata
Kaloula baleata ghosh : India
Kaloula pulchra : Bangladesh, India, Nepal
Kaloula taprobanica : India, Sri Lanka
Melanobatrachus indicus : India
Microhyla berdmorei : India
Microhyla chakrapani : India
Microhyla heymonsi : India
Microhyla ornata Bangladehs, India, Nepal, Pakistan, Sri Lanka
Microhyla rubra : Bangladesh, India, Sri Lanka
Microhyla zeylanica : Sri Lanka
Microhyla inornata : India
Ramanella minor : India
Ramanella minor : India
Ramanella montanta : India
Ramanella mormorata : India
Ramanella obscura : Sri Lanka
Ramanella Palmata : Sri Lanka
Ramanella variegata : India
Ramanella variegata : India, Sri Lanka
Uperodon globulosus : Bangladesh, India, Sri Lanka

RANIDAE

Amolops afghanus : India, Nepal
Amolops formosus : Bangladesh, India, Nepal
Amolops nepalicus : Nepal
Amolops monticola : India, Nepal
Amolops senchalensis : India
Chaparana sikimensis : India, Nepal
Euphlyctis cyanophlyctis : Bangladesh, India, Nepal, Pakistan, Sri Lanka

Euphlyctis ghoshi : India
Euphlyctis hexadactylus : Bangladesh, India, Sri Lanka
Hoplobatrachus crassus : Bangladesh, India, Nepal, Sri Lanka
Hoplobatrachus tigerinus : Bangladesh, Bhutan, India, Nepal, Pakistan, Sri Lanka
Indirana beddomii : India
Indirana brachytarsus : India
Indirana diplosticus : India
Indirana gundia : India
Indirana leithii : India
Indirana leptodactyla : India
Indirana phrynoderma : India
Indirana semipalmata : India
Indirana tenuilingua : India
Limnonectes anadamanensis : India
Limnonectes cancrivorus : India
Limnonectes corrugatus : Bangladesh, India
Limnonectes corrugatus : Sri Lanka
Limnonectes doriae : India
Limnonectes greenii : Sri Lanka
Limnonectes keralensis : India
Limnonectes khasiana : Bangladesh, India
Limnonectes kirtisinghei : Sri Lanka
Limnonectes kuhlii : India
Limnonectes laticeps : India
Limnonectes limnocharis : Bangladesh, India, Nepal, Pakistan, Sri Lanka
Limnonectes mawlindipi : India
Limnonectes mawphalangensis : India
Limnonectes murthii : India
Limnonectes nilagiricus : India
Limnonectes parambikulamana : India
Limnonectes rufescens : India
Limnonectes sauriceps : India
Limnonectes shompenonum : India
Limnonectes syhadrensis : India, Nepal, Pakistan
Limnonectes teraiensis : Nepal
Micrixalus fuscus : India
Micrixalus gadgili : India
Micrixalus nudis : India

Micrixalus phyllophila : India
Micrixalus saxicolus : India
Micrixalus silvaticus : India
Micrixalus thampii : India
Nanophrys ceylonensis : Sri Lanka
Nanophrys marmorata : Sri Lanka
Nanophrys guentheri : Sri Lanka
Nanophrys parkeri : Nepal
Nanophrys pleskei : India, Pakistan
Nyctibatrachus aliciae : India
Nyctibatrachus beddomii : India
Nyctibatrachus kempholeyensis : India
Nyctibatrachus deccanensis : India
Nyctibatrachus humayuni : India
Nyctibatrachus modestus : India
Nyctibatrachus major : India
Nyctibatrachus minor : India
Nyctibatrachus sanctipalustris : India
Nyctibatrachus sylvaticus : India
Occidozyga lima : India
Paa hazarensis : India/Pakistan
Paa lleblgii : India, Nepal
Paa minica : India, Nepal
Paa sternosignata : India, Pakistan
Phrynoglossus borealis : India
Rana alticola : India
Rana annandalii : India, Nepal
Rana assamensis : India, Nepal
Rana aurantiaca : India, Sri Lanka
Rana blanfordii : India, Nepal
Rana chalconota : India
Rana curtipes : India
Rana denieli : India, Nepal
Rana erycepeae : Nepal
Rana erythraea : India
Rana garoensis : India
Rana gracilis : Sri Lanka
Rana khare : India
Rana leptoglossa : India
Rana livida : India

Rana malabarica : India
Rana nepalensis : Nepal
Rana nicobariensis : India
Rana nigrovittata : India
Rana pierrei : Nepal
Rana polunini : Nepal
Rana rara : Nepal
Rana rostrandi : Nepal
Rana taipehensis : India, Nepal
Rana temporalis : India, Sri Lanka
Rana vicina : India, Nepal, Pakistan
Taylorana hascheanus : India
Tomopterna breviceps : Bangladesh, India, Nepal, Pakistan, Sri Lanka
Tomopterna dobosonii : India
Tomopterna leucorhynchus : India
Tomopterna rolandae : India, Sri Lanka
Tomopterna swani : Nepal

HYLIDAE

Hyla annectans : India

RHACOPHORIDAE

Chirixalus doriae : India
Chirixalus simus : India
Chirixalus vittahus : India
Philautus adspersus : Sri Lanka
Philautus andersoni : India
Philautus annandalii : Bhutan, India, Nepal
Philautus beddomii : India
Philautus bombayensis : India
Philautus chalazodes : India
Philautus charius : India
Philautus cherrapunjiae : India
Philautus crnri : India
Philautus elegans : India
Philautus femoralis : India, Sri Lanka
Philautus flaviventris : India, Sri Lanka
Philautus garo : India
Philautus glandulosus : India

Philautus hassanensis : India
Philautus hypomelas : Sri Lanka
Philautus kempiae : India
Philautus kottigeharensis : India
Philautus leucorhinus : India, Sri Lanka
Philautus melanensis : India
Philautus namdaphaensis : India
Philautus naralnensis : India
Philautus nasutus : Sri Lanka
Philautus nobeli : India
Philautus parkeri : India
Philautus pulcherimus : India
Philautus shillongensis : India
Philautus shyamrupus : India
Philautus signatus : India
Philautus swamianus : India
Philautus temporalis : India, Sri Lanka
Philautus travancoriacus : India
Philautus variabilis : India, Sri Lanka
Polypedates cruciger : Sri Lanka
Polypedates wques : Sri Lanka
Polypedates insularis : India
Polypedates leucomystax
Polypedates leucomystax teraiensis : Bangladesh, India, Nepal
Polypedates longinasus : Sri Lanka
Polypedates maculatus
Polypedates maculatus maculatus : Bangladesh, India, Nepal
Polypedates maculatus biscutiger : Sri Lanka
Polypedates maculatus himalayensis : India, Nepal
Polypedates zed : Nepal
Rhacophorus appendiculatus : India
Rhacophorus bipunctatus : India
Rhacophorus biscculus : India
Rhacophorus calcadensis : India
Rhacophorus cavirostris : Sri Lanka
Rhacophorus dubius : India
Rhacophorus fergusonianus : Sri Lanka
Rhacophorus jerdonii : Bangladesh, India
Rhacophorus lateralis : India
Rhacophorus malabaricus : India

Rhacophorus macropus : Sri Lanka
Rhacophorus maximus : India, Nepal
Rhacophorus mkicrotympanum : Sri Lanka
Rhacophorus namdaphaensis : India
Rhacophorus naso : India
Rhacophorus pleurostictus : India
Rhacophorus reinwardtii : India
Rhacophorus reticulatus : Sri Lanka
Rhacophorus taeniatus : Bangladesh, India
Rhacophorus tuberculatus : India
Theloderma asper : India
Theloderma moloch : India
Theloderma schmardanum : Sri Lanka

ICHTHYOPHIDAE

Ichythyophis beddomei : India
Ichythyophis bombayensis : India
Ichythyophis glutinosus : Sri Lanka
Ichythyophis longicephalus : India
Ichythyophis malabarensis : India
Ichythyophis orthoplicatus : Sri Lanka
Ichythyophis peninsularis : India
Ichythyophis pseudangularis : Sri Lanka
Ichythyophis sikkimensis : India, Nepal
Ichythyophis subterrestris : India
Ichythyophis tricolor : India
Uraeotyphlus malabaricus :India
Uraeotyphlus menoni : India
Uraeotyphlus narayani : India
Uraeotyphlus oxyurus : India

CAECILIDAE

Gegeneophis carnosus : India
Gegeneophis fulleri : India
Gegeneophis ramaswamii : India
Gegeneophis battersbyi : India

SALAMANDRIDAE

Tylototriton verrucosus : India, Nepal

Himalayan Salamander (*Tylototriton verrucosus*) — **The Tailed Amphibian**

This amphibian is in peril in Darjeeling hills and in north eastern India. In Darjeeling, its habitats are on Gopaldhara, Mirik, Poecheng, Sonada and Chimney which has shrunk due to biotic stress (acidic soil, export, loss of aquatic site). The zoologist feel that the last chance of its survival has vanished as *Cyprinus carpio*, the common carp has been introduced in lakes which destroys egg clutches of this amphibian larvae. The Deptt. of Forests are conscious to save this species from extinction from Darjeeling hills.

3.9 FAUNAL IMPOVERISHMENT

Indepth work on endangered animals of India has been done by Z.S.I. (refer Dr. A.K. Mukherjee's work). Of the detailed work, some 10 distribution maps are presented, which will show how the animals have been pushed back to the wall. Mukherjee's book "Endangered Wildlife of India" a Z.S.I. Publication testify to the utter faunal impoverishment (From Ecology and Biogeography of India 1974).

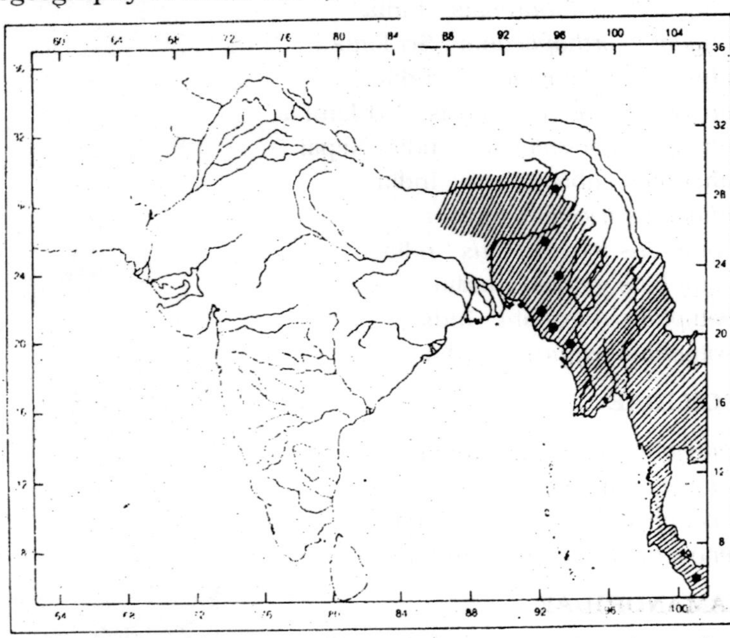

Map. 7. Past (striped area) and the present (black area) distribution of *Rhinoceros sumatrensis*. (Two horned Asiatic Rhinoceros)

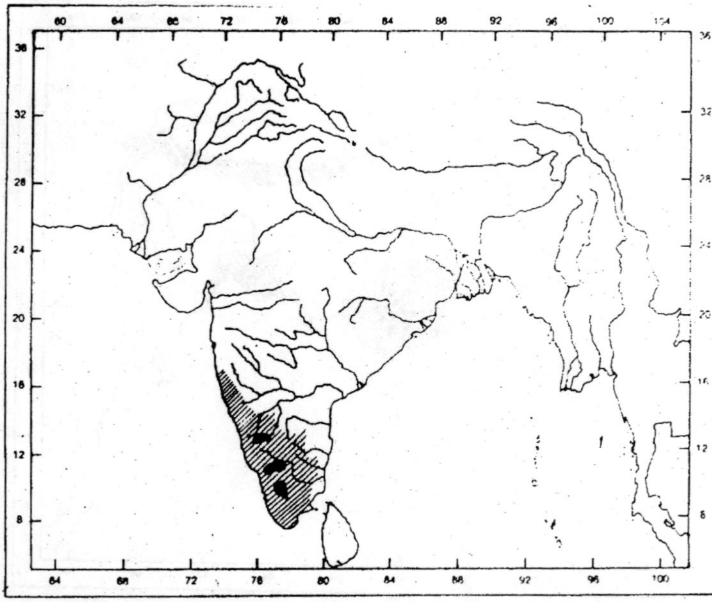

Map 8. Past (striped area) and the present (black area) distribution of *Hemitragus hylocrius* (Nilgiri Thar)

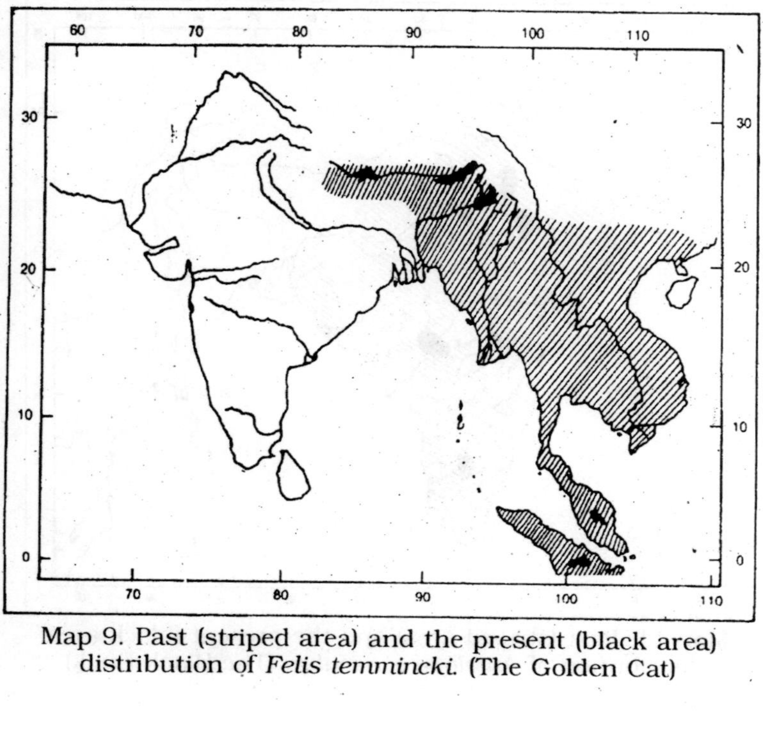

Map 9. Past (striped area) and the present (black area) distribution of *Felis temmincki.* (The Golden Cat)

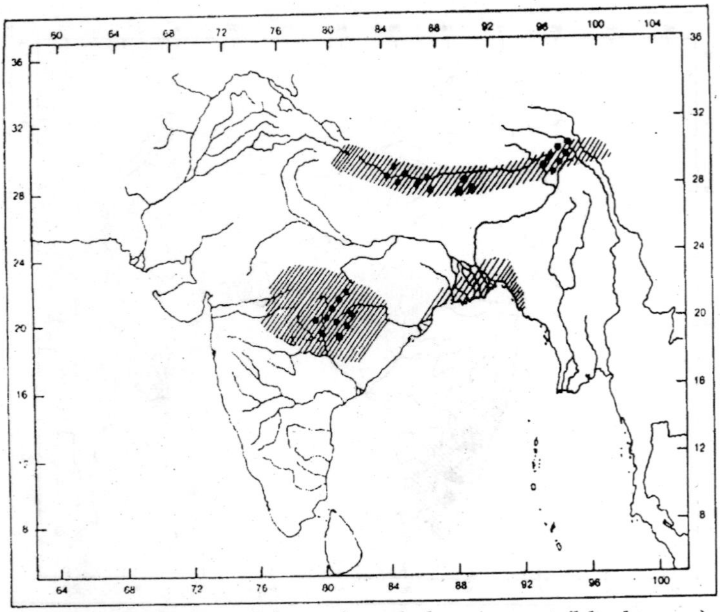

Map 10. Past (striped area) and the present (black area) distribution of *Cervus duvauceli* (Swamp deer)

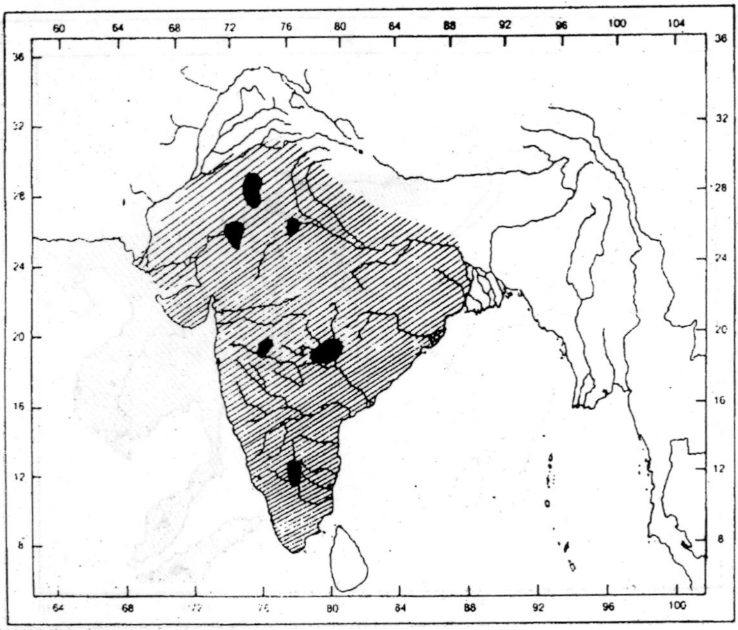

Map 11. Past (striped area) and the present (black area) distribution of *Antelope. cervicubra* (The Black Buck)

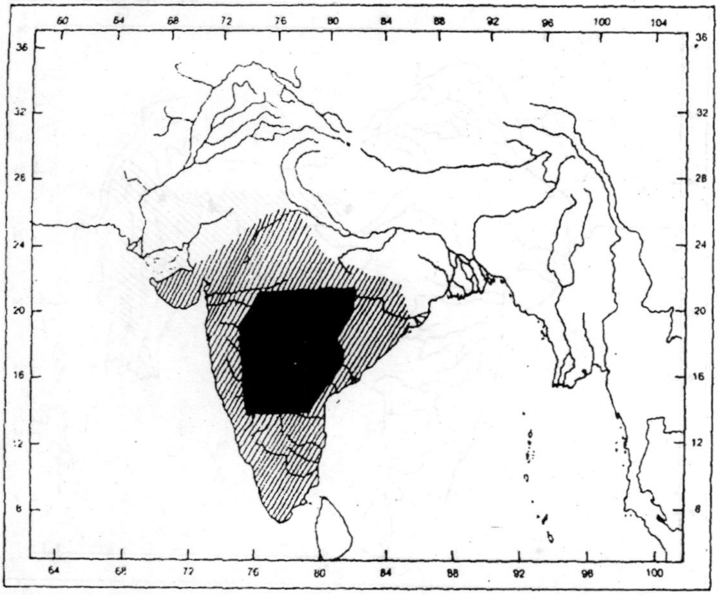

Map 12. Past (striped area) and the present (black area) distribution of *Tetracerus quadricornis*. (Four-horned Antelope)

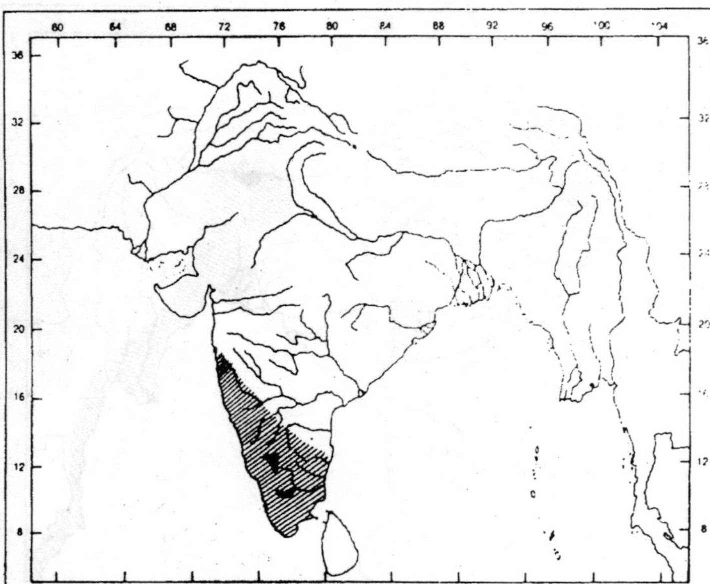

Map 13. Past (striped area) and the present (black area) distribution of *Presbytis johni* (Nilgiri Langur)

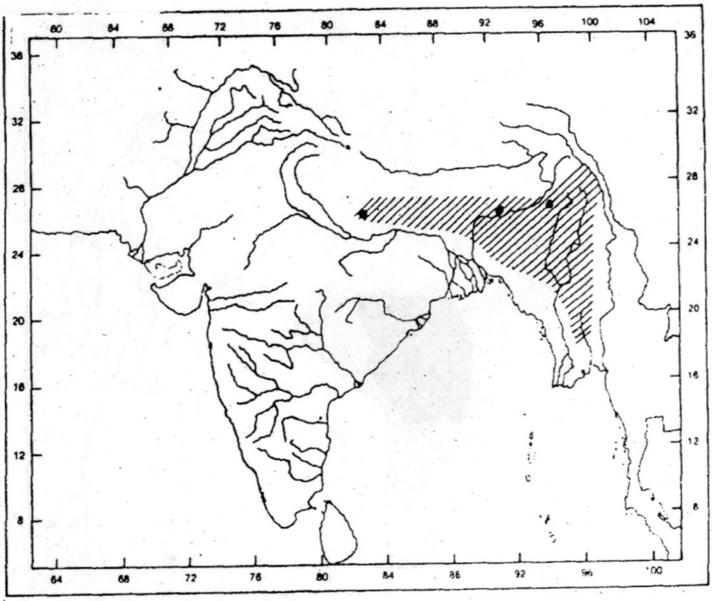

Map 14. Past (striped area) and the present (black area) distribution of *Caprolagus hispidus* (Hispid Hare)

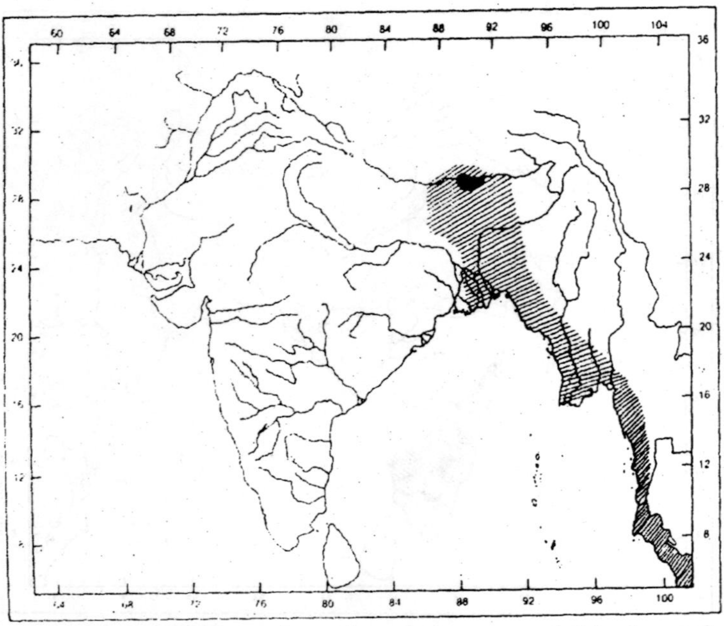

Map 15. Past (striped area) and the present (black area) distribution of *Cairina scutulata*. (White winged wood Duck)

3.10 INVERTEBRATE FAUNAL RESOURCES

Although the present work does not cover the sphere of invertebrate, it has been thought very relevant to reproduce a statement prepared by Z.S.I. in their report on 'Animal Resources of India'. The statement gives broad view of the status of invertebrate fauna. From the figure it may be concluded that the situation is not alarming when compared to the world figure.

A summary of approximate number of species occurring in India and their share in the world fauna are given below :

Table VI. Estimated Number of Species based on available data

Group	India	World	% in India
Protozoa	2577	31250	8.24
Porifera	519	5100	10.18
Siphonophora	118	180	65.56
Sclerectinia	119	7000	1.70
Ctenophora	10	100	10.00
Platyhelminthes	1622	17500	9.27
Turbellaria	47	4000	1.17
Monogenea	295	2500	11.80
Trematodoa	750	6500	11.54
Cestoda	530	4500	11.78
Rotifera	310	2500	12.40
Gastrotricha	88	2500	3.52
Kinorhyncha	10	100	10.00
Nematoda	2350	25000	9.40
Acanthocephala	110	800	13.75
Sipuncula	38	202	18.81
Mollusca	5042	80000	6.30
Echiura	33	127	25.98
Annelida	1093	12620	8.66
Oligochaeta	585	4000	14.27
Polychaeta	428	8000	5.35
Hirudinea	59	500	11.80
Archiannelida	21	120	17.50
Onychophora	1	100	1.00
Arthropoda	57525	952116	6.04
Crustacea	2970	24375	12.88
Anostraca	72	175	41.14
Notostraca	11	15	73.33

Group	India	World	% in India
Conchostraca	27	180	15.00
Cladocera	90	400	22.50
Ostracoda	120	2000	6.00
Copepoda	540	4500	12.00
Branchiura	4	75	5.33
Cirripedia	104	750	15.87
Isopoda	200	4000	5.00
Amphipoda	143	3600	3.97
Decapoda	1535	8500	18.06
Stomatopoda	124	180	68.89
Insecta	50717	839052	6.04
Thysanura	23	1250	1.84
Diplura	16	355	4.50
Protura	20	260	7.69
Collembola	200	5000	4.00
Ephemeroptera	94	2146	4.38
Odonata	491	5500	8.92
Plecoptera	113	2100	5.38
Orthoptera	759	14491	5.24
Phasmida	60	2500	
Dermaptera	320	1800	
Embioptera	33	200	
Blattaria	156	4200	3.71
mantodea	161	2000	8.05
Isoptera	300	2000	15.00
Psocoptera	85	2500	3.40
Phthiraptera	400	3000	13.33
Hemiptera	6500	80000	8.13
Thysanoptera	691	6000	11.51
Neuroptera	315	5000	6.30
Coleoptera	15000	350000	4.28
Strepsiptera	8	300	2.66
Mecoptera	15	350	4.28
Siphonaptera	52	2000	2.60
Diptera	6093	96600	6.31
Lepidoptera	13000	142500	9.12
Trichoptera	812	7000	11.60
Hymenoptera	5000	100000	5.00
Diplopoda	162	7500	2.16
Chilopoda	100	3000	3.33
Xiphosura	2	4	50.00
Arachnida	409	6385	6.40

Group	India	World	% in India
Scorpionida	102	1500	6.80
Pedipalpida	25	85	29.41
Solpugida	15	900	1.67
Opiliones	167	1600	10.43
Pseudoscorpionida	100	2300	4.35
Acari	1915	36800	5.20
Araneae	1250	35000	3.57
Phoronida	3	11	27.27
Bryozoa	170	20000	0.85
Entoprocta	10	60	16.67
Brachiopoda	3	300	1.00
Chaetognatha	30	100	30.00
Echinodermata	765	6226	12.28
Hemichordata	12	118	10.16
Chordata	4894	47674	10.27
Protochordata	116	2173	5.34
Pisces	2546	21723	11.72
Amphibia	204	5145	3.96
Reptilia	428	5375	7.96
Aves	1228	9026	13.61
Mammalis	372	4232	8.79
Grand Total	77452	1211584	6.40

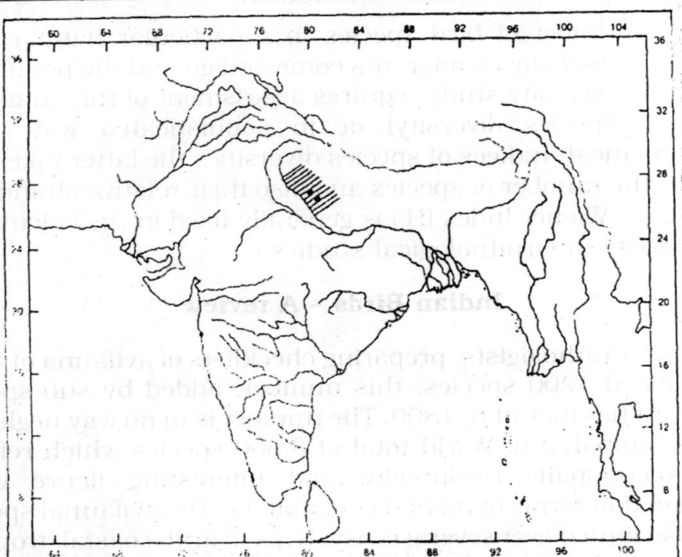

Map 16. Past (striped area) and the present (black area). distribution of Bustard Quail.

3.11 AVIFAUNAL (BIRD) BIODIVERSITY

It is difficult to measure biodiversity status of birds as each species breeds and spends the non-breeding season in congenial habitats of the earth. They occur in specific habitats and each species has a special characteristics, density and abundance. For food a species of bird interact with another species.

It is generally an accepted fact that more diverse ecological communities are more stable; there is a view is that they are more stable because they are complex. To supoort these views factual data cannot be easily produced as research finding is non-existent on this fauna.

The Ornithologists say that birds have a stable population at any region. "The tropical rains forest birds are not only members of extremely diverse communities, but also their population is stable by comparison with temperate forest species" (World Conservation Monitoring Centre, 1992)". They found that two similar species scarcely ever occupy similar niches, but displace each other in such a manner that each takes possession of certain peculiar kinds of food and mode of life in which it has an advantage over its competitor and that two species cannot coexist indefinitiely.

The number of bird species in a particular habitat often change drastically as migrants come and go and the population changes. Diversity study requires assessment of the number of species (Species diversity) or in sophisticated way using mathematical "Indices of species diversity"; the latter takes into accout the number of species and also their relative abundance (Shanon - Wiener Index (H') is generally used in the calculation of diversity in Ornithological studies.)

Indian Birds – A review

The Ornithologists, preparing checklists of avifauna of India have listed 1200 species; this number, added by sub-species and varieties mount to 1800. The number is in no way negligible when compared to World total of 8,600 species which reflects an exceptionally resourceful and interesting figure when evaluated in terms of myriads of habitats. The avifaunal species are distributed over several ecosystems in arid, coastal, tropical, sub-tropical, sub-temperate, temperate and alpine zones, covered by scrub, agriculture fields, mangrove, deciduous, semi- evergreen, evergreen, rains, temperate and alpine forests. The second edition of the 'Fauna of British India' by Blandford and Oates (1889) series on birds enumerated some 2400 forms (species and sub-species) in the then greater India. In their book 'A Synopsis of the birds of India and Pakistan (excluding Mayanmar) S. Ali and D. Repley list about 2061 forms of which over 300 species are Winter visitors chiefly from Palaearctic region of the North. Biogeographically India is put under Oriental region. Primary zones of avifaunal concentrations are Indo-gangetic plains, the Himalayas from Kashmir to Arunachal, Nagaland, Manipur, Mizoram, Tripura and Meghalaya and Peninsular Western-ghat and adjoining hills.

The aves (also some other fauna) of Punjab, Sind, Rajasthan, Kutch resemble those of West Asia, North Africa, Whilst those in the higher Himalayas and the upper Indus Valley resemble those of Central Asia (Palaearctic region). The areas under direct impact of South west monsoon have a very rich avifaunal (also faunal) resources and are of similar nature.

Aves are found in every conceivable habitat - the golden eagle soaring over the snow-clad mountains of Sikkim, the curlew and avocet on mudflat, Baya and Sparrow on farmland, the blue and emerald dove in dense forest, while many are recognizable and many are confusing.

Salim Ali observes "As would be expected, similarly extends to the insect forms dependent upon the plants, which in turn conduce to similarity in the birds' predatory habit upon them. It has, therefore been suggested that all these parallel areas, far-flung though they may be, are perhaps better lumped together in one zoogeographical sub-division". Also,

"....the large race of bird species - this is true of other warm-blooded animals as well - is with rare exceptions, found inhabiting the cooler parts of its distributional range while the smallest inhabits the warmer". (1996)

The Himalayas have acted as barrier between the Orient and Palearctic for the last 10 million years encouraging spaciation. Also the tropical climate of a large number of endemic species, Indonesia supports about 17 per cent of World's bird species.

In 1994, Bird Life International identified over 300 oriental birds species which are globally threatened. Habitat loss and deterioration are the most reasons for threats to wildlife, besides proaching, hunting for food, wildlife trade, pollution and introduced species. The Indian species need a review.

Qualitative survey

Salim Ali's assessment of birds in erstwhile ceylon, Travancore, North-east India, Sikkim and some of other regions depict a realistic picture of total family, genus and species of birds of the areas. Of these areas Sikkim shows a rich assemblage of avifaunal species, as it has tridimentional facets-the altitude making the area most suitable for varied avifaunal species.

Nowhere Dr. Ali or other Ornithologists have provided substantial data on numerical abundance of species. Mention of mere gregariousness of species like sparrow, Weaver bird, white eye, bulbul, swallow, egret, bunting, finch, munia, hill myna, small minivet etc. etc. ...... does not give an idea of the resources of avifauna. Some water-birds show gregarious occurrence in various water bodies, of which open bill stork, pintal duck, seagulls, terns, sand grouse, ducks and few more species are rather common.

Some Factual Assessment of Genera and Species

Sikkim has more than 550 species. The forests, adjacent to Sikkim even in plains record 350 species (Jalpaiguri forests - recorded by E.O. Shebbeare a Forester and others, 1918). The senior author in his survey of avifauna in more than 30 project areas, scattered over Orissa, Bihar, Madhya Pradesh, West Bengal, Assam, Tripura and Uttar Pradesh have recorded more than 250 species in each project area (300 - 500 sq. km.). His assessment of avifauna in the plains of eastern and East-Central India is about 250-300 species which is an index of minimum number of species. A bigger area and observation over longer period is sure to bring to light some more than 400 species at each site. Salim Ali's assessment of birds of Srilanka is 397 species (65, 610 sq. Km.).

In the Indo-gangetic plains of India, the dearth of arboreal birds such as- Finch, Sunbirds, Flower-pecker, Tree Creeper, Wood pecker, Thrush, Chat, Flycatcher, Drongo, Owl, Cuckoo, Parakeet, Hornbills and many other species is owing to the absence of forest and tree covers.

Water bodies

A realistic assessment has been made on aquatic birds of India. There are about 200 species of birds in water bodies of the country (Koller lake of Tamilnadu records 200 species,

Chilka lake records 160 species and Keoladeo-ghana records 360 species).

A critical analysis

From short term qualitative and quantitative study of smaller areas (in plains) it has been possible to record 250-300 species of birds. A long term study is sure to add another 100 to the list.

It is possible to furnish a long list (areawise) of avifauna identified in various protected areas of India. Salim Ali's list cover larger areas such as Sikkim, North-east India, Travancore and Cochin, Satpura hills and Gujarat, Shree Lanka and Other areas.

Various Morphological Forms of Birds

Indian birds cover a wide range of morphological forms and in colouration of feathers, such as -

Tail :- Prominent tail (about 100 sp.), prominent crest (about 40 sp.,) Long tail (about 60 species).

Colouration :- Chiefly black (30 sp.), Chiefly green (37 sp.), Chiefly yellow (25 sp.), Chiefly red (15 sp.), Chiefly blue (20 sp.), White/whitish (14 sp.), Black and white (50 sp.), the rest are greyish.

Species Range of the Sub-continent

(a) *Species range of various rich groups of avifauna (approximate no. of sp.)*

Babblar - 150 sp., Warbler - 90 sp., Fly catcher - 67 sp., Laughing thrush - 60 sp., Finch - 59 sp., Woodpecker - 56 sp., Partridge - 37 sp., Tern - 26 sp.

(b) *Species of less prominent groups*

Parrot - 23 sp., Plover/Lapwing - 24 sp., Dove - 20 sp., Pipit - 18 sp., Sparrow - 14 sp., Munia - 24 sp., Sunbird - 25 sp., Muthatch - 15 sp., Wagtail - 19 sp., Bulbul - 10 sp., Hornbill - 9 sp.

Various species of babblars, Flycatchers, warblers, laughing thrushes and other such rich groups have a great diversity of body forms and foraging styles.

(c) *Common Indian species*

Babblar - 19 sp., Barbet - 12 sp., Bee eater - 5 sp., Bittern - 8sp., Bulbul - 15 sp., Cuckoo - 11 sp., Dove - 12 sp., Duck - 10 sp., Eagle - 16 sp., Fly-catcher - 25 sp., Hawk eagle - 7 sp., Heron - 14 sp., Hornhill - 10 sp., King fisher - 13 sp., Lark - 18 sp., Munia - 10 sp., Myna - 9 sp., Owl - 18 sp., Parakeet - 9 sp., Stork - 10 sp.

These are some statistics which indicate a rich assemblage of avifauna species in this country with a wide range of diversity of forms, characters and numbers

Status

But what about their status? What is the number of individuals of each of these species? No authentic work has been done in this regard. Ruthless killing and trapping of Dove, Duck, Heron, Hornhill, Pigeons, Myna, Parakeet, Stork etc. go on relentlessly. Over-exploitation of trees directly affect the predator avifauna and arboreal species.

In Sikkim State

Salim Ali recorded (1962) 550 species of birds. A rough classification of major birds have been -

Laughing thrush and Babblers - 61 species, Chats and Thrushes - 56 species, Warblers - 60 species, Fly catchers - 27 species, Flinehes - 22 species, Woodpecker - 15 species, Cuckoo - 16 species, Owl - 14 species, Eagles, Hawks & Buzzards - 40 species, Pigeon - 12 species.

Upper Temperate and Alpine Birds of Sikkim (12,000 - 16,000)

The Himalayan Griffon Vulture, the Lammergier or bearded vulture - 15000; the Snow Partridge - 15000', the Sikkim Brew Cock - 16000', the Blood Pheasant - 15000', the Green Sandpiper - 16000', the Common Rubythroat - 15000' and the Blue - fronted Red Start - 15000'.

Lower Temperate Birds. Sikkim (altitude in brackets)

The India Sparrow - hawk, teh Crimson Hornet Pheasant- (12000), the Eastern Solitary Snipe, the Ashy Wood Pigeon - (10000), the Rufous Turtle Dove - (12000) the Himalayan Wad Owl - 12000', the Sikkim Rufous - necked scimmiter. Babbler

- (10000), the Tailed Wren (10000), the plain-coloured Laughing Thrush (12000), the Black-faced Laughing Thrush (12000), the Sikkim Red-headed Laughing Thrush - (12000), the Stripe throated Yahina - (12000) and the Slaty headed Rufous vented Yahina (12000).

Since there is no data on density and frequency of avifaunal species, it is not possible to quantify the status of various species and annual loss or gain in respect of them. With fast dwindling density of forest there is reduction in plant diversity. The count of abundance and relative abundance does not indicate the status as anthropogenic disturbance depletes the flock. This is the experience gainsed at 50 different sites in 8 states recently examined by the authors.

Throughout the World many organizations, especially the International Council for Bird Preservation, have done useful work; the British Trust for Ornithology is another such organisation. India does not have any such powerful organization. Pesticides have affected the egg shells, which do not harden. The abrupt fall of population of Jungle fowl in the vicinity of the Tea Estates and agricultural fields in the district of Jalpaiguri was witnessed in early sixties and thereafter. The dearth of water birds in wetland sites is due to these chemicals and acute anthropogenic activities, leading to shrinkage of aquatic sites.

While we are oblivious of India's avifaunal population in general, the export of some birds was being allowed to the extent of 65,465 of Allexandrin Parakeet, 64,453 Red-breasted Parakeet, 20,186 Slaty headed Parakeet in the recent past.

Some endemic birds face threat both due to single or dual effect of biotic stress and pollution threat; some of the species are :

Narcondom horn bill, Nicobar megapode, Jerden's courser, Bengal florican, Black necked crane, Blood pheasant, Nicobar piegion, Monal Pheasant, the great Indian Bustard, besides migratory Pelicans and Flamingoes and scores of game birds viz., Quails, Partridges, Pheasants etc. Assessment of Population of these birds are of immediate need.

In India 47 species of birds face extinction and many others are threatened. The disappearance of fruit bearing trees such as Ficus sp., Cordia sp., Ehretia sp., Biscofia javanica etc. is definetely causing depletion of bird population. The authors counted 500 numbers of Grackle or Hill Myna in one ficus tree in North Cachar hills and 300 green pigeons on a ficus tree near Alipurduar. Such sights are rare now.

Majority of investigators compile a list of avifauna of a locality which shows species rich area, but the number of individuals of most species is extremely poor.

Number of species of birds

A comparative view.

(i) Pacific Archipelego

Solomon	127 sp.	New Caledonia	77 sp.,
Fizi	54 "	Samoa	33"
Saciely	17 "	Marquesas	11"
Henderson	1"	Alaska	250"

(ii) Medium area (under 500 sp.)

Tasmania	255 sp.	Iraq	354 sp.
Newzealand	256 "	Cyclon	397 "
Serbia	288 "	Israil	400 "
Honglong	289 "	Japan	425 "
Portugal	315 "	W. Australia	436 "
Allurta	317 "	Great Britain	
Macedonia	319 "	Philipine	
Finland	327 "	Ireland	450 "
Norway	333 "	Senegal & Seoraleone	485"
Greece	339 "	Eastern Nigeria	488 "
Afghanistan	341 "	Borneo	554 "
Maine	350 "	Malaya	575 "
Ussuriland	358 "	Burma	953"

(iii) New - Guinea and Tropical Australia 650 Sp.

(iv) Tropical Savanul and Forest Edge.

Eritrea	551 sp.
Ghana	627 "

Cameroon	670 "

(v) North and Central America

Texus	545 "
Mexico	967 "
Surinam	567 "

(vi) North Rodesia 674 :

Sundan	871 "
Angola and Zambia	875 "

(vii) Cango River Basin

Belgian Congo and Ruanda Brundi - 1040 sp.

(viii) Central America

South Mexico and Panam	1190 sp.
Venezuala	1282 sp.
Equator	1357 "
Brazil (Amazan river)	1440 "
(ix) Columbia	1500 sp.
Sri Lanka	397 sp.

Some areas in India

Gujarat	262 sp.
Kerala	400 sp.
Sikkim	500 "
Tripura	341 " (about over 3581 sq. km)
Jalpaiguri duars	350 " (area about 4000 sq. km)

The list shows how rich is the Ornitholigical fauna of India.

Birds protected under Wildlife (Protection) Act, 1972.

In avifaunal population Schedule I of Wildlife (Protection) Act, 1972, mentions 47 species as threatened with extinction, some of which are -

Schedule I, Part III of Wildlife (Protection) Act of 1972

Andaman Teal, Assam bamboo partridge, Bazas, Bengal florican, Black-necked crane, Blood Pheasant, Cheer Pheasant, Eastern White Stork, Forest spotted Owlet, Frogmouth, Great India Bustard, Great Indian hornbill, Hawks, Hooded Crane, Hornbills, Honbare bustard, Humes bar-baked pheasant, Indian pied horn bill, Jerdon's Courser, Lammergeter large falcons, Large Whistling teal, Lesser florican, Mondal pheasant, Mountain quail, Narcondom horn bill, Kerala Forest lerrapin, Nicobar pigcon, Ospray or Fish eating eagle, Peacock pheasant Pea Fowl, Pink headed duck, Scalater's Monal, Siberian White Crane, Tibetan Snow cock, Tragopan pheasants, White bellied Sea Eagle, White-eared pheasant, white spoonbill, White-winged wood duck.

Besides 80 groups of birds have been included in Schedule IV some of which are -

Babblers (Timalinas), Barbets (Capitonidae), Bittern (Ardeidae), Cranes (Anatidae), etc.

Birds of Duars (in district of Jalpaiguri, North Bengal)

E.O. Shebbeare *et al.*, recorded in 1918 more than 350 species of bird in Duars of North Bengal (observed over 4000 Sq. Km.). It was a fantastic study recorded in a notebook, since recovered from debri, which has not been printed. The list is produced here owing to the following reasons -

- The area is small, but species resource is high.
- Birds indicate species of varied habitats.
- More species of arboreal birds recorded.
- Necessity for resurvey the area to review the species status of birds of the area now i.e. after about 82 years.

The same area needs a resurvey of bird species to find out if the conversion of substantial area of mixed forests of large number of species into a few planted trees of commercial value has brought about a change in species pattern. Also expansion of human settlement and area under tea crop by felling of forests has had many impacts on bird species pattern.

Birds of Duars : (Sighted by E.O. Shebbeare in 1918 an unpublished list)

Himalayan Cole Tit	Yeallow-browed Tit
Northern black spotted yellow Tit	Red-headed Tit
Rufuous fronted Tit	Brown-crested Tit
Black-fronted crew Tit Smaller	Red headed crow Tit
Hoary headed crow Tit	Himalayan whirlling Thrush
Jungle Crow	Indian House crow
Green Magpie	Indian Tree Pie
Himalayan Tree Pie	Indian Grey Tit
Green-backed Tit Larger	Red-headed crow Tit
Rufous Georgeted Laughing Thrush	Neck-laced Laughing Thrush
Rufous chinned Laughing Thrush	Striated Babbler
Jungle Babbler	Yellow-eyed Babbler
Mandelli's spotted Babbler	Abbatt's Babbler
Slaty billed short Wing	Chest-nut headed short wing
Nepal Strip throated Yuhina	Indian White eye
Swinhoes White Eye	Yellow-naped Ixulus
White-bellied Heropornis	Common Iora
Ruby-cheek Gold-fronted Heropornis	Orange-bellied Chloropsis
Fairy Blue Bird	Sultan Bird
Fire Cap	Spotted Wing
Yellow Bulbul Bengal	Red-whiskered Bulbul
Velvet-fronted Blue Nuthatch	Cinnamnon-breated Nuthatch
Chestnut-breasted Nutatch	Crow-billed Drongo
Black Drongo	Crow-billed Drongo
ndian Ashy Drongo	Hair-crested Drongo
Lesser Racket-tailed Drongo	Large Racket-tailed Drongo
Turkestan Grasshopper Wasbler	Blyth's Red Warbler

Spotted Bush Wasbler
Brown-necked Tailor bird
Frankin's Wren Warbler
Striated Marsh Warbler
Smoky Willow Warbler
Green Willow Warbler
Blanford's Bush Warbler
Ashy Wren Warbler
Indian Wren Warbler
Grey-headed Shrike
Brown Shrike
Nepal Wood Shrike
Dark-grey Cuckoo Shrike
Large Cuckoo Shrike
Indian Scarlet Minivet
Yellow-throated Minivet
Small Minivet
Maroon Oriole
Indian Grackle
Grey-Headed Myna
Bark Myna
Shooty Fly catcher
European Red-breasted Fly catcher
Slaty-blue Fly
Blue-throated Fly catcher
Brooke's Fly catcher
Grey-headed Fly catcher
Indian Black naped Fly Catcher
White broted fantail Fly catcher
Rufous-bellied Niltava
Common Pied Bush Chat

Indian Tailor bird
Yellow headed fantail Warbler
Large Grass Wabbler
Brown Willow Warbler
Dusky Willow Warbler
Blyth's Crown Willow Wasbler
Yellow-bellied Wren Warbler
Jungle Wren Warbler
Black-headed Shrike
Pale-brown Shrike
Black-backed Pied Shrike
Common Wood Shrike
Black-headed Cuckoo Shrike
Ashy Swallow shrike
Short-bellied Minivet
Rosy Minivet
Indian Black-headed Oriole
Burmese Black-naped Oriole
Common Indian Starting
Common Myna
Pied Myna
Orange Georgetted Fly catcher
Eastern Red-breasted Fly catc
Catcher Little Pied Fly Catcher
Verditer Fly Catcher
Brown Fly catcher
Burmese Paradise Fly Catcher
Yellow bellied Fly Catcher
White-throated fantail Fly catcher
Small Niltava
Indian Bush Chat

Hodgson's Bush Chat	Dark-grey Bush Chat
Himalayan Little Forktail	Leschenauthes Forktail
White-capped Red Start	Daurian Red Start
Hodgson's Red Start	Indian Red Start
Plumbeons Red Start	Indian Blue Throat
Tibet Ruby Start	White-tailed Blue Robbin
Magpie Robbin	The Shyama
Grey-winged Ouzel	Black-throated Ouzel
Pied Ground Thrush	Orange-headed Ground Thrush
Eastern Blue Rock, Thrush	Western Blue Rock Thrush
Chestnut-bellied Roack Thrush	Small-billed Mountain Thrush
Large Brown Thrush	Lesser Brown Thrush
Brown Clipper	Eastern Baya
Large-billed Baya	Chestnut-bellied Munia
Hodgson's Munia	Spotted munia
Indian Red Munia	Common Rose Finch
House Sparrow	Tree Sparrow
Dwarf Bunting	Yellow-breasted Bunting
Chestnut Bunting	Crested Bunting
Indian Sand Martin	The Swallow
Eastern Swallow	Tytler's Swallow
Wire-tailed Swallow	Hodgson's Striated Swallow
Syke's striated Swallow	White Wagtail
Grey Wagtail	Grey-headed Wagtail
Yellow-headed Wagtail	Forest Wagtail
Indian Tree Pipit	Richard's Pipit
Blyth's Pipit	Indian Pipit
Indian Skylark	Ganges Sand Lark
Bengal Bush Lark	Yellow-backed Sunbird
Black-breasted yellow-backed	Sunbird Purple Sunbird
Large Striated Spider-hunter	Litter spider-hunter
Scarlet-backed Flower-pecker	Plain-coloured Flower-pecker

Tickell's Flower-pecker	Fire-breasted Flower-pecker
Blue naped Pitta	Green headed Pitta
Hodgson's Broadbill	Black-naped Woodpecker
Small Himalayan yellow-naped Woodpecker	Fulvours-breasted Pied Woodpecker
Darjeeling Pigmy Woodpecker	Himalayan Pigmy Woodpecker
Northern Rufous Woodpecker	Himalayan Golden backed three-toed Woodpecker
Tckells' Golden Backed Woodpecker	Large Yellow naped Woodpecker
Speckled Piculet	Rufous Piculet
Lineated Barbet	Blue-throated Barbet
Rubby Barbet	Crimson-breasted Barbet (Coppersmith)
Blue-eared Barbet	Indian Roller
Burmese Roller	Broad-billed Roller
Common Indian Bee-eater	Blue-tailed Bee eater
Chestnut headed bee-eater	Blue-bearded Bee-eater
Indian Pied Kingfisher	White-breasted Kingfisher
Beaven's Kingfisher	Three-toed Kingfisher
Brown-headed Stork-billed Kingfisher	Ruddy Kingfisher
Grey Hornbill	Great Hornbill
Indo-burmese	Hornbill Malayan
	Wreathed Hornbill
Rufous-necked	Hornbill European Hoopae
Indian Hoopae	Large White-rumped Swift
Indian crested	Swift White niced spinetail
Himalayan Swiftlet	Horsefield's Nightjar
Franklin's Nightjar	Red-headed tragon
European Cuckoo	Indian Cuckoo
Common Hawk Cuckoo	Drongo Cuckoo
Pied Crested Cuckoo	Red-winged Crested Cuckoo
Indian Planitive Cuckoo	Rufons-bellied Cuckoo

Banded-bay Cuckoo	Viold Cuckoo
Large-billed Malkoha	Common Crow Pheasant
Common Coucal	Lesser Coucal
Indian Koel	Indian Loriquel
Blossom-headed Paraquet	Eastern Blossom-headed Paraquet
Red-breasted paraquet	Large Burmese Paraquet
Grass Owl	Brown Fish Owl
Forest Eagle	Owl Scop's Owl
Spotted Himalayan Scop's	Owl Collared Scop's Owl
Spotted Owlet	Large-barred Scop's Owl
Jungle Owlet	Brown Hawk Owl
Brown or Screech	Owl The Osprey
Cinerious Vulture	Black Vulture
Himalayan Giffon	Himalayan Long-billed vulture
Indian White-backed Vulture	Changeable Hawk Eagle
Hodgson's Hawk Eagle	Crested Serpent Eagle
Pallas Fishing Eagle	Large Grey-headed Fishing Eagle
Hodgson's Fishing Eagle	Rufous-bellied Hawk Eagle
Booted Eagle	Common Pariah Kite
Large Indian Kite	Black-winged Kite
Pale Harrier	Hen Harrier
Pied Harrier	Marsh Harrier
Common Buzzard	Crested Honey Buzzard
The Shikara	Crested Goshawk
Beatra Sparrow-hawk	Black Crested Baza
Blyth's Baza	Peregrne Falcon
Red-legged Merlin	The Kestrel
Red-legged Falconet	Bengal Green Pigeon
Ashy headed Green Pigeon	Orange-breasted Green Pigeon
Thick-billed Green Pigeon	Pin-tailed Green Pigeon
Wedge-tailed Green Pigeon	Green Imperial Pigeon
Hodgson's Imperial Pigeon	Indian Blue-rock Pigeon

Ashy-wood Pigeon	Bronze-winged or Emerald Dove
Indian Rufous turtle Dove	Spotted Dove
Indian Ring Dove Indian	Red-turtle Dove
Burmese Red-turtled Dove	Bar-tailed Cuckoo Dove
Common Pea-fowl	Burmese Jungle-Fowl
Grey-Peacock Pheasant	Black-headed kalij pheasant
Blue-breasted or Rain Quail	Common or Grey Quail
Burmese Bustard Quail	Inlis Bush Quail
Burmese Bustard Quail	Little Button Quail
Blyth's Hill Partridge	Black partridge
Grey Partridge	Swamp Partridge
Blue-breasted Banded Rail	Banded Crake
Brown Crake	White breasted hen
Purple Moore Hen	Bengal Florican
Stone Curlew	Great Stone Plover
Small Indian Pratincole	Bronze-winged Jacana
Pheasant tailed jacana	Red-Wattled Lapwing
Indian Spear winged Plover	Eastern Golden Plover
Grey Plover	Kentish Plover
Little-ringed Plover	Long-billed Ringed Plover
The Ibis-bill	Common Sand Piper
Wood Sand Piper	Green sand Piper
The Green Shank	Little Stint
Lemminck's stint	The Wood Cock
Wood Snipe	Common snipe
Pintail snipe	Painted snipe
Indian River Tern	Black-bellied Tern
White Shafted Ternlet	Indian Little Tern
Eastern White Pelican	Indian Skimmer or Scissor Bill
Large Cormorant	Indian Darter or Snake bird
Indian Shag	Black stork
White-necked Stork	Black-necked Stork

The Adjutant	Lesser Adjutant
Eastern Purple Heron	Great White billed Heron
Dusky Grey Heron	Large Egret
Smaller Egret	Cattle Egret
Pond Heron	Little Green Heron
Night Heron	Chestnut Bittern
Black Bittern	Comb Duck
Brahminy Duck	Whistling Teal
Lesser Whistling	Teal Cotton Teal
The Mallard	The Gadwall
Common Teal	The Pintail
Wigeon	Gargany or Blue-winged Teal
The Shoveller	Red-crested Pochard
The Pochard	White-eyed Duck.

Qualitative and Quantitative assessment of Ornithological Population

A few areas have been chosen for this :

1. Keoladeo - Ghana National Park

It has been stressed that in the event of depletion of wetlands over the major areas, flora and fauna of wetlands will get shelter in the Protected areas of India. A typical example is cited from a study made in the N.P. in 1983. This national park which is primarily dry in winter and summer erupt into a blaze of colourful activities in the monsoon. This park has a spectacular harmony in India and hosts thousands of herons, ibises, spoondbills, storks, cormorants, darters and the number of species in and around the N.P. exceeds 360. Dr. Salim Ali ringed 72,577 birds between the periods 1963 and 1974 of 224 species.

A small N.P. of this size (28 sq. km.) is a typical example and eye opener that the Forest Department must afford complete protection of the P.A.s under them so as to conserve a good population of more than 360 species.

A record of bird census done in 1983 and a list of bird (only English name given) identified in the N.P. is given below :

(Source : The birds of Wetlands by James Hancock, 1994; Important avifaunal species are marked with a 'asterisk').

Keoladeo - Ghana National Park at Bharatpur.
HERONRY CENSUS FOR 1983

Species	No. of Birds	No. of Nest	Nestlings
Large Cormorant	1,494	747	–
Indian Shag	5,555	3,1735	2,038
Little Cormorant	3,165	1,735	408
Darter	591	428	449
Purple Heron	30	15	–
Grey Heron	135	80	24
Cattle Egret	133	96	138
Great White Egret	302	209	332
Intermediate Egret	2,392	1,682	1,559
Little Egret	1,287	1,078	985
Night Heron	245	215	53
Painted Stork	5,666	2,869	49
Open-billed Stork	811	731	1,059
Oriental White Ibis	1,435	968	752
Spoonbill	773	415	30
Pond Heron	1	1	4

This census was carried out by staff at the reserve in the last week of September 1983, and the numbers of nests and nestlings are therefore higher for those species which nested early. A number of species (e.g. Painted Stork) had only just begun to nest, while others (e.g. Open-billed Stork) had nearly completed their nesting. Numbers of birds in relation to numbers of nests vary because many species feed outside the reserve. In 1983 also, pond Herons, which are normally found in large numbers, nested outside the reserve.

Avifaunal Species identified are :

Peregrine Falcon	Grey Headed Lapwing
Oriental Hobby	*Red wattled Lapwing
Kestrel Falcon	*Spur Winged Plover
Black Partridge	Yellow-wattled Lapwing
Grey Partridge	Grey Plover
Common Quail	Dabchick
Jungle Bush-quail	Great Crested Grebe

Common Peafowl
Common Bustard quail
Common Crane
*Sarus Crane
*Siberian While Crane
Demoiselle Crane
Water Rail
Baillon's Crake
Spotted Crake
Ruddy Crake
Brown Crake
White-breasted Water-hen
Water Cock
*Purple Gallinule
Indian Moorhen
*Coot
Pheasant-tailed Jacana
*Bronze-winged Jacana
White-tailed
*Lapwing (Peewit)
Ringed Plover
Little Ringed Plover
Kentish Plover
Lesser Sand Plover
Curlew Bar-headed
Black-tailed Godwit
Spotted Red-shank
Common Red-shank
Marsh Sandpiper
Greenshank
Green sandpiper
Spotted sandpiper
Black necked Grebe
*Rosy Pelican
*Spot billed (Grey)
*Dalmatian Pelican
Large Cormornat
Indian Shag
Little Cormorant
*Indian Darter
*Grey Heron
*Purple Heron
*Pond Heron
Night Heron
*Cattle Egret
*Great White
*Intermediate (Medium)
*Little Egret
Chestnut Bittern
Black Bittern
Lapwing Bittern
Eastern Golden Plover
Spoonbill
Flamingo
Greylag Goose
White-fronted Goose
Goose
Whistling Teal
Ruddy Shelduck
Common Shelduck
Marbled Teal
Pintail
Common teal
Spotbill

Terek sandpiper
Common sandpiper
Turnstone
Pintail Snipe
Fantall Snipe
Jack Snipe
Knot
Little Stint
Temminck's stint
Dunlin
Curlew sandpiper
Broad-billed Sandpiper
Ruff Black-Winged
Red-necked Phalarope
Painted Snipe
Black-winged stilt
Avocet
Stone-curlew
Indian Courser
*Painted Stork
*Asian Open-billed Stork
*White-necked Stork
White Stork
*Black-necked Stork
*Adjatant Stork
*Lesser Adjutant Stork
*Oriental White Ibis
Black Ibis
*Glossy Ibis
White-tailed Sea Eagle
*Pallas's Fish Eagle
King Vulture

*Mallard
Gadwall
Falcated Teal
Wigeon
*Garganey
Shoveler
*Red-Crested Pochard
Common Pochard
White-eyed Pochard
Tufted Duck
*Cotton Teal
Comb Duck
Kite
*Crested Honey Buzzard
Black (Pariah) Kite
Brahminy Kite
Goshawk
Shikra
Sparrow-hawk
Besra Sparrow-hawk
Long-legged Buzzard
Desert Buzzard
White-eyed Buzzard eagle
Bonelli's Hawk-eagle
Imperial Eagle
Tawny Eagle
Steppa Eagle
*Spotted Eagle
*Lesser Spotted Eagle
Collared Scops
Indian Great Horned
Dusky Horned Owl

Indian Long-billed Vulture
Indian White-backed Vulture
Egyptian Vulture
Pale Harrier
*Marsh Harrier
Short-toed Eagle
Crested Serpent Eagle
Osprey
Saker Falcon
Laggar Falcon
Collared Pratincole
Small Indian Prantincole
Herring Gull
Great Black-headed Gull
Brown headed Gull
Whiskered Tern
River Tern
Black-bellied Tern
Little Tern
Common Sand-grouse
Green Pigeon
Blue Rock Pigeon
Eastern Stock Pigeon
Rufous Turtle Dove
Ring Dove
Red Turtle Dove
Spotted Dove
Little Brown Dove
Rose-ringed Parakeet
Pied-crested Cuckoo
Hawk Cuckoo
Koel

Brown Fish Owl
Spotted Owlet
Short-eared Owl
Indian Jungle Nightjar
Syke's Nightjar
Common Indian Nightjar
Franklin's Nightjar
Alpine Swift
House Swift
Palm Swift
*Pied Kingfisher
*Common Kingfisher
*Stork billed Kingfisher
*White billed Kingfisher
Black capped Kingfisher
Blue cheeked Bee-eater
Blue tailed Bee-eater
Common Green Bee-eater
Kashmir Roller
Indian Roller
Hoopoe
Grey Hornbill
Crimson-breasted Barbet
Wryneck
Lesser Golden backed Woodpecker
Yellow-Golden Pied Woodpecker
Pygmy Woodpecker
Indian Pitta
Red-winged Bush-lark
Ashy Crowned Finch-lark
Rufous tailed Finch lark
Short-toed Lark

Sirkeer Cuckoo
Crow Pheasant
Scops Owl Sykes's
Eastern Skylark
Collared Sand Martin
Plain Sand Martin
Dusky Crag Martin Tickell's
Western Swallow
Eastern Swallow
Wire-tailed Swallow
Indian Cliff swallow
Striated Swallow
Grey Shrike
Bay-backed Shrike
Pale brown Shrike
Rufous-backed shrike
Brown Shrike
Golden Oriole
Black headed Oriole
Black Drongo
White-bellied Drongo
Brahminy Mynah
Rosy Pastor
Starling
Pied Mynah
Common Mynah
Bank Mynah
Indian Treeple
House Crow
Jungle Crow
Common Wood Shrike
Black-headed Cuckoo-shrike
Yarkand Crested lark
Crested Lark
Crested Lark
Jungle Babbler
Red-breasted Flycatcher
White-browed Blue Flycatcher
Blue Flycatcher
Blue-throated Flycatcher
Grey-headed Flycatcher
White-browed Fantail Flycatcher
Paradise Flycatcher
Cetti's Warbler
Moustached Warbler
Red-headed Fantail Warbler
Streaked Fantail Warbler
Rufous fronted Long tail Warbler
Indian Wren Warbler
Ashy Long tail Warbler
Tailor Bird
Temminck's Grasshopper Warbler
Thick-billed Warbler
Indian Great Reed Warbler
Blyth's Reed Warbler
Paddy-field Warbler
Blunt-winged Warbler
Booted Warbler
Orphean Warbler
Indian Whitethroat
Lesser Whitethroat
Hume's Whitethroat
Desert Warbler
Chiffchaff

Scarlet Minivet
Short-billed Minivet
Small Minivet
White-bellied Minivet
Common lora
Marshall's lora
Red-whiskered Bulbul
White-browed Bulbul
Yellow-eyed Babbler
Common babbler
Large Grey Babbler
Blue-throat
Magp-robin
Black Redstart
Brown Rockchat
White Wagtail
Grey Wagtail
Large Pied Wagtail
Thick-billed Flowerpecker
Tickell's Flowerpecker
Purple Sunbird
White Eye
Indian House Sparrow
Kashmir House Sparrow
Turkestan House Sparrow
Spanish Sparrow
Yellow throated sparrow
Weaverbird
Black-throated Weaver
Streaked Weaver
Red Munia
White backed Munia
Tytler's Leaf Warbler
Tickell's Leaf Warbler
Grant's Leaf Warbler
Olivaceous Leaf Warbler
Dusky Leaf Warbler
Yellow browed Leaf Warbler
Brooke's Leaf Warbler
Palla's Leaf Warbler
Greenish Willow Warbler
Large Crowned Leaf Warbler
Rubythroat
Forest Wagtail
Yellow-headed wagtail
Black-backed Yellow headed Wagtail
Collared Bushchat
Pied Bushchat
Desert Wheatear
Black Wheatear
Indian Robin
Blue-headed Rock
Blue Rock Thrush
Orange-winged Blackbird
Tickell's Thrush
Grey winged Blackbird
Black throated Thrush
Great Tit
Chestnut-bellied Nuthatch
Spotted Grey Creeper
Olive-backed Tree Pipit
Tree Pipit
Richard's Pipit
Paddy-field pipit

Spotted Munia	Tawny Pipit
Black headed Munia	Blyth's Pipit
Common Rosetinch	Hodgson's Pipit
Black-headed Bunting	Water Pipit
Red-headed Buntin	Black-headed yellow Wagtail
White-capped Bunting	Grey-headed Yellow Wagtail
Grey-headed Bunting	Blue headed Yellow Wagtail
Crested Bunting	

Source : James Hancock (1992)

2. North Bengal, Coochbehar

Quantitative assessment of Ornithofaunal population

The Maharaja of Coochbehar in his Shikar Stories mentioned about the following species of game birds shot in the district early this century which are :

Common teal	- 55% of total bird killed
Garganey teal	- 15% "
Shoveller	- 10% "
White eye	- 6% "
Gadwall	- 6% "
Wigeon	- 3% "
Pintail	- 2% "
Red-eyed pochard	- 2% "
Ruddy shell duck	- 2% "
Spotbill	- 2% "
Baikal teal	- 2% "
Common duck	- 2% "
Mallard	- 1% "

3. North Bengal Alipur duar

	Place	Species	Nos.
(i)	Sagardighi Coochbehar	Lesser Whistling Teal	3100
(ii)	Kaljani Beel Ambari Coochbeahar	Common Pochard	7

		Godwall	115
		Pheasant tailed jacana	105
		Little Cormorant	20
		Little Egret	17
		Pond heron	45
		Lesser whistling teal	90
		Little Greb	18
		Red headed Marlin	1
		Shell drke (Ferruginous duck)	25
(iii)	Birpara Jheel Alipurduar	Little Cormorant	9
		Coot	19
		Pond Heron	12
		Lesser Whistling Teal	86
		Spurwinged Lapwing	7
(iv)	West Pakhri Guri, Barobhisha	Falcated Teal	21
		Bronze Winged Jacana	45
		Shell Drake (Ferruginous duck)	112
		Pond Heron	24
		Gadwall	45
		Large Cormorant	1
		Ruddy Sheldrake	9
		Little Grebe	20
(v)	Badabeel Naraltali	White eyed Pochard	65
		Giant Heron	1
		Coot	20
		Bronze winged Jacana	18
		Grey Headed Lapwing	18
		Little Grebe	14
		Little Cormorant	7
		Spotbill duck	7
(vi)	Satbeki Beel Badachowki mouja Alipurduar	Little Cormorant	35
		Large Cormorant	2
		Coot	10
		Bronze winged Jacana	40
		Indian Moorhen	200
		Pond Heron	150
		Cotton Teal	700
		Asian Openbill Stork	10
		White-breasted water hen	20

	Little Grebe	90
	Pheasant tailed Jacana	85
	Red crested Pochard	7
	Marsh harrier	1
	Grey headed Lapwing	10
	Common Sandpiper	30
	Cattle Egret	120
(vii) Rasik Beek Alipurduar (Recorded on 11.1.97)	Lapwing	10
	Grey headed Lapwing	40
	Indian Moor hen	8
	Coot	64
	Gadwall	24
	White-eyed Pochard	70
	Cotton Teal	50
	Falcated Teal	4
	Little Grebe	220
	Pintail Duck	20
	Lesser adjutant Stork	6
	Purple Heron	4
	Grey Heron	2
	Mallard	6
	Indian Shag	6
	Lesser whistling Teal	364
	Little Cormorant	100
	Median Egret	30
	Little Egret	250
	Cattle Egret	125
	Pond Heron	70

Avifauna of Wetlands

English Name	Scientific Name
A. Water fowl	
Crested Grebe	*Podiceps cristatus*
Cormorant	*P. ruficollis*
Shag	*Phalacrocorax carbo*
Little Cormorant	*P. fuscicollis*
Purple Heron	*P. niger*
	P. purpurea

English Name	Scientific Name
Darter	*Anhinga rufa*
Grey Heron	*Ardea cinerea*
Little Green Heron	*Butorides striatus*
Pond Heron	*Ardeola grayii*
Night Heron	*Nycticorax nycticorax*
Cattle Egret	*Bubulcus ibis*
Large Egret	*Egretta alba*
Intermediate Egret	*E. intermedia*
Little Egret	*E. garzetta*
Chestnut Bittern	*Ixobrychus cinnamomeus*
Bittern	*Botaurus stellaris*
Black Bittern	*Dupetor flavicollis*
White ibis	*Threskiornis melanocephata*
Glossy Ibis	*Plegadis falcinellus*
Spoonbill	*Platalea leucorodia*
Tundra Bean Goose	*Anser fabalis*
Greylag Goose	*A. anser*
Bar headed Goose	*A. indicus*
Brahminy Duck/Ruddy Shelduck	*Tadorna ferruginea*
Common Shelduck	*T. tadorna*
Pintail	*Anas acuta*
Common Teal	*A. crecca*
Mallard	*A. platyrhynchos*
Gadwall	*A. strepera*
Falcated Teal	*A. falcata*
Wigeon	*A. penelope*
Garganey	*A. querquedula*
Shoveller	*A. clypeata*
Spotbill Duck	*Anas poeciolorhyncha*
Greater Scaup	*Aythya marila*
Common Pochard	*A. ferina*

English Name	Scientific Name
Baer's Pochard	*A. baeri*
Tufted Duck	*A. fuligula*
White-eyed Pochard	*A. nyroca*
Red Crested Pochard	*Netta rufina*
White winged Scoter	*Melanitta fusca*
Surf Scoter	*M. perspicilata*
Lesser Whistling Teal	*Dendrocygna javanica*
Large Whistling Teal	*D. bicolor*
Cotton Teal	*Nettapus coromandelianus*
White-winged Woodduck	*Cairina scutulata*
Comb Duck/Nakhta	*Sarkidiorina melanotos*
Sarus Crane	*Grus antigone*
Demoisells Crane	*Anthropoides vigro*
Water Rail	*Rallus aquaticus*
Ruddy Crake	*Amaurornis fuscus*
White-breasted Water hen	*A. phoenicurus*
Water cock	*Gallicrex cinerea*
Purple Moorhen	*Porphyrio porphyrio*
Indian Moorhen	*Gallinula chloropus*
Coot	*Fulica atra*
Pheasant tailed Jacana	*Hydrophasianus chirurgus*
Bronzed winged Jacana	*Metapidius indicus*
B. Waders	
Grey-headed Lapwing	*Vanellus cinereus*
Red-wattled Lapwing	*V. indicus*
Yellow-wattled Lapwing	*V. Malabaricus*
Eastern Golden Plover	*Pluvialis dominica*
Lesser Sand Plover	*Charadrius mongolus*
Little Ringed Plover	*C. dubius*
Plover	*C. hiaticus*
Eastern Whimbrel	*Numenius phaeopus*

English Name	Scientific Name
Curlew	*N. arquata*
Black-tailed Godwit	*Limosa limosa*
Common Red Shank	*Tringa totanus*
Greenshank	*T. nebularia*
Common Sandpiper	*T. hypoleucos*
Wood Sandpiper	*T. glareola*
Green Sandpiper	*T. ochropus*
Terek Sandpiper	*T. terek*
Nordmann's Greenshank	*T. guttifer*
Marsh Sandpiper	*T. stagnatilis*
Solitary Snipe	*Gallinago solitaria*
Pin-tailed Snipe	*G. stenura*
Fantail Snipe	*G. gallinago*
Little Stint	*Calidris munutus*
Temmincks's Stint	*C. temmincid*
Long-toed Stint	*C. subminutus*
Turnstone	*Arenaria Interpres*
Painted Snipe	*Rostratula bengalensis*
Black-Winged Stilt	*Himantopus himantopus*
Avocet	*Recurvirostra avoseta*
Great Stone Plover	*Esacus magnirostris*
Courser	*Cursorius coromandelicus*
Small Pranticole	*Glareola lactea*
Lesser Plea Kingfisher	*Ceryle nudis*
Common Kingfisher	*Alcedo atthis*
Stork billed Kingfisher	*Pelargopsis capensis*
White-breasted Kingfisher	*Halcyon symrnensis*
Grey Wagtail	*Motacilla flava*
Yellow Wagtail	*M. citreola*
Forest Wagtail	*M. caspica*
Pied Wagtail	*M. alba*

English Name	Scientific Name
Large Pied Wagtail	*M. madagascarensis*

C. Scrubland (border areas of wetland)

Rock Pigeon	*Columbia livia*
Ring Dove	*Streptopelia decaocio*
Red Turtle Dove	*S. tranquebarica*
Spotted Dove	*S. chinensis*
Tree Pipit	*Anthus hodgsoni*
Paddy-field Pipit	*Anthus novaesselandae*
Eastern Skylark	*Alauda gulgula*
Sand Lark	*Calandrella cinerea*
Assam Bush lark	*Mirafra assamica*
Ashy Crowned Finch Lark	*Erymopteryx grisea*
Common Myna	*Acridotheres tristis*
Jungle Myna	*A. fiscus*
Pied Myna	*Sturnus contra*

D. Scrub Savannah Land Birds (Swampy areas)

Rose-ringed Parakeet	*Psittacula Krameri*
Red-breasted parakeet	*P. alexandri*
Slaty-headed Parakeet	*P. himalayana*
Common Hawk Cuckoo	*Cuculus varius*
Plaintive Cuckoo	*Cacomantis marulinus*
Crow Pheasant	*Centropus sinensis*
Brown Fish Owl	*Bubo zeylonensis*
Long-tailed Nightjar	*Caprimulgus macrurus*
Blue tailed Bee-eater	*Merops philippinus*
Green Bee-eater	*M. orientalis*
Blue Jay	*Coracias benghalensis*
Black-headed Shrike	*Lanius schach*
Brown Shrike	*L. cristatus*
Black Drongo	*Dicrurus adsimilis*

English Name	Scientific Name
Ashy Drongo	*D. leucophaeus*
House Crow	*Corvus splendens*
Jungle Crow	*C. macrorhynchos*
Redvented Bulbul	*Pycnonotus cafer*
Marsh Spotted Babbler	*Pellorneum palustre*
Abbott's Babbler	*Trichastoma abbotii*
Common Babbler	*Turdoides caudaus*
Striated Babbler	*T. carilei*
Jungle Babbler	*T. striatus*
Nepal Quaker Babbler	*Alcippe nipalensis*
Streaked Fantail Warbler	*Cisticola juncidis*
Streaked Long-tailed Warbler	*Prinia gracilis*
Tawny-flanked Long-tailed Warbler	*P. subflava*
Ashy Wren Warbler	*P. socialis*
Red-breasted Flycatcher	*Muscicapa parva*
Ruby Throat	*Erithacus calliope*
Blue Throat	*E. svecicus*
Tailor Bird	*Orthotomus sutorius*
Maggie Robin	*Copsychus saularis*
Black Redstart	*Phoenicurus ochruros*
Collared Bush Chat	*Saxicolla torquata*
Pied Bush Chat	*S. Caprata*
Baya	*Pioceus philippinus*
House Sparrow	*Passer domesticus*
Spotted Munia	*Lonchura punctulata*
Chestnut Munia	*L. malacca*
White-throated Munia	*L. malabarica*
Red Munia	*Estrilda amandava*

E. Birds of Prey

Black winged Kite	*Elanus caeruleus*

English Name	Scientific Name
Pariah Kite	*Milvus migrans*
Large Pariah Kite	*Milvus migran 'ineatus*
Brahminy Kite	*Haliastur indu* ,
Shikra	*Accipiter badlus*
Changeable Hawk Eagle	*Spizaeteus cirrhatus*
Tawny Eagle	*Aquita rapax*
Large spotted Eagle	*A. clanga*
Lesser Spotted Eagle	*A. pomarina*
Pallah's Fish Eagle	*Haliaeetus leucoryphus*
Grey-headed Fish Eagle	*Ichthyophaga iethyaetus*
White-backed Vulture	*Gyps bengalensis*
Griffon Vulture	*G. fulvus*
Pale Harrier	*Circus macrounus*
Pied Harrier	*C. melenoleucos*
Marsh Harrier	*C. aeruginosus*
Crested Serpent Eagle	*Spilornis cheeta*
Osprey	*Pandion haliatus*
Shaheen falcon	*Falcon peregrinus*
European Kestrel	*Falcon tinnunculus*

F. Other Birds In Wetland Areas

Crested Tree Swift	*Hemiprocne coronata*
Spine-tailed Swift	*Chaetura caudacuta*
Alpine Swift	*Apus melba*
Palm Swift	*Cypsiurus parvus*
Plain Sand Martin	*Riparia paludicola*
Collared Sand Martinq	*R. riparia*
Nepal House Martin	*Delichon nipalensis*
Common Swallow	*Hirundo rustica*
Wire-tailed Swallow	*H. smithi*
Ashy Swallow Shrike	*Artamus fuscus*

Bhutan Birds vis-a-vis B.T.R. (Buxa Tiger Reserve)

It has already been said that Salim Ali *et al.*, have recorded 516 species of birds from Bhutan which lies to the north of B.T.R. of these, 458 are known or are assumed to be the breeding species while 52 species are passage migrants or winter residents. 55 families are represented and the babblers comprise 67 sp. and the Warblers 55. Bhutan is rich in passerine insect eatezrs. BTR has also abundant Babblers and Warblers.

Some birds mentioned by S. Ali *et .al.*, from Samchi, Phuntscholing and areas adjacent to BTR are given below (These are not in the check list) -

East Himalayan	Kestrel Ganges Sand lark
Long billed ringed Plover	Kashmir house Martin
Common Snipe	Orange bellied Chloropsis
Temminck's stint	Himalayan black Bulbul
Small Indian Pratin Cole	Sikkim Spotted Babbler
Pigeon (7 sp. mentioned)	Brown Wren Babbler
Assam great Barbet	Golden headed Babbler
Eastern rapous Wood Pecker	Black throated Babbler
Blue napped pilla Himalayan	Silver eared Mesia.
Streaked spider hunter	Eastern Chestnut
White tailed Blue Robin	Hadgsons Short wing
Crowned leaf Warbler	Golden headed tailor bird
Northern Ashy Grey Wren Warbler	Arunachal White throated Fantail

Darjeeling, Sikkim, Bhutan and north-eastern India together form a very vital zone for ornithological studies. If the forests are protected this richness will sustain.

3.12 ICHTHYOFAUNAL BIODIVERSITY

Expert Aquaculturists are of opinion that it is difficult to measure biodiversity of fish fauna so far as the number of individuals of each species and their species status in diverse habitat besides others factors are concerned. The reasons are:

- Fish systematics have not yet been fully worked out.

– Counting of individuals of each species in water is not possible.

– Most of the species are biotically and abiotically affected due to tremendous impact from both.

Indian Ichthyofaunal resources have a rich diversity presented by innumerable species, genera, besides their morphology, size, shape and biology. They grow in various aquatic habitats. Unlike other commonly recognized vertebrate group fishes have a heterogenous assemblage. They cover a vast array of distantly related vertebrates. Many are closely related to mammals than man. Fishes constitute about half the total number of vertebrates.

The present authors, however, consider another reason more realistic and plausible; The break through in laboratory research in Cuttack in 1957 which enabled fishery Scientists to breed carps and later other fishes in the laboratory by use of pituitory hormone (artificial induction) which revolutionized the fish production in the country by providing millions of fingerlings with minimum of natality. This process has been adopted by fishermen class to produce carps (primarily) which has rapid growth to flood the market at a reasonable low cost. This may be reason that the fishermen has lost interest to pay attention to breeding of very many other species of fish such as Chanda, Koi, Bhola, Magur, Singi, Banspata, Vacha, Bata,

Ichthyofaunal (Fish) loss of Biodiversity

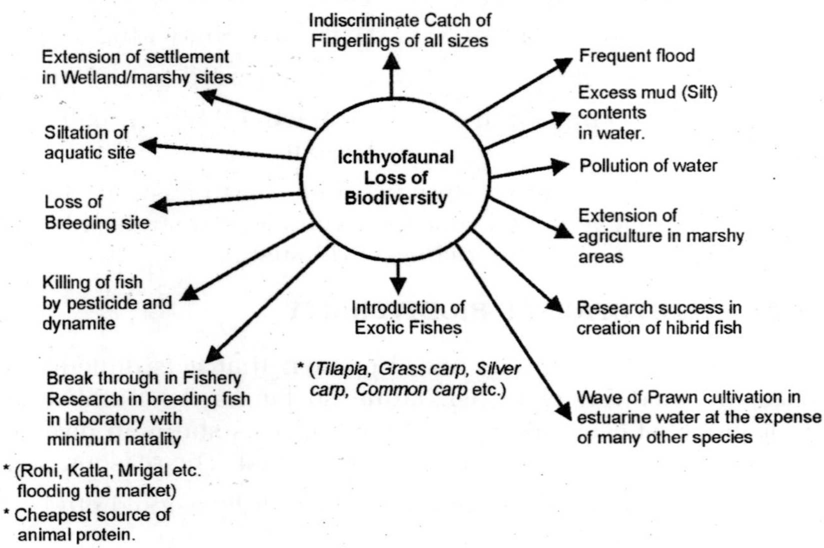

Ban, Tengra, Pholi, Pabda, Punthi etc. Huge quantity of fish of the species Labeo rohita, Catla catla, Cirrhinus mrigala and some exotics like Grass carps, Silver carp, Tilapia flood the market and people buy them at a reasonable low cost.

Introduction of exotic fishes might be another reason for depletion of sporadically occurring fish species as they do not receive adequate attention from fishermen. Salmo gaidneri, S. trulta fario (Trouts) were introduced in the hills in the past; Carpinus carpio, Tilapia mossambica, T. nilotica, Clenopharyngodon idella (Silver carp) and Hypothalmichthys nilotrix (Grass carp) are now mostly produced in fishery farms all over the country, after they were introduced some forty years back.

About the fish biodiversity of India World Bank's Technical Paper 343, 1996 may be referred to. They have produced four statements, three on fishes and one on crabs and shrimps, bringing a comparative picture of 25 Asian countries and also of top ten countries of the world. Mention has been made of

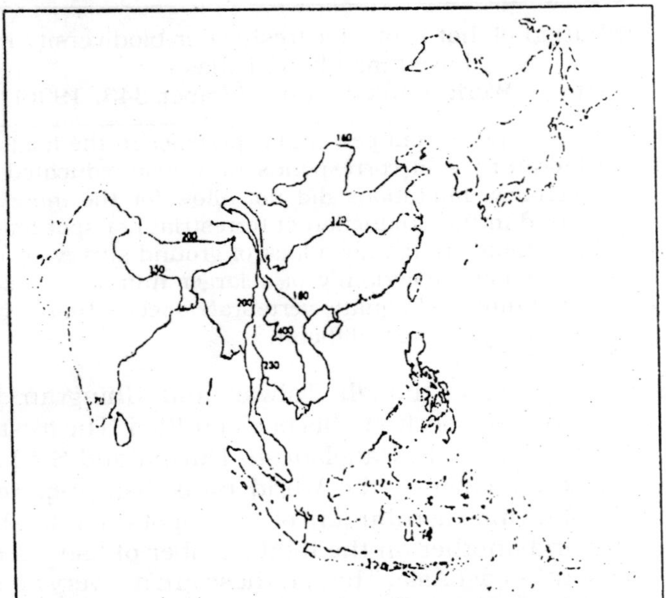

The total number of fish species known from the major river basins of Asia (Source : World Bank Technical paper 343, 1996)

1. Terrestrial organisms have two-dimensional distributions whereas aquatic ones have one-dimensional distribution, i.e. upstream and downstream. In that respect the term 'hot rod' might be more appropriate than hot spot.

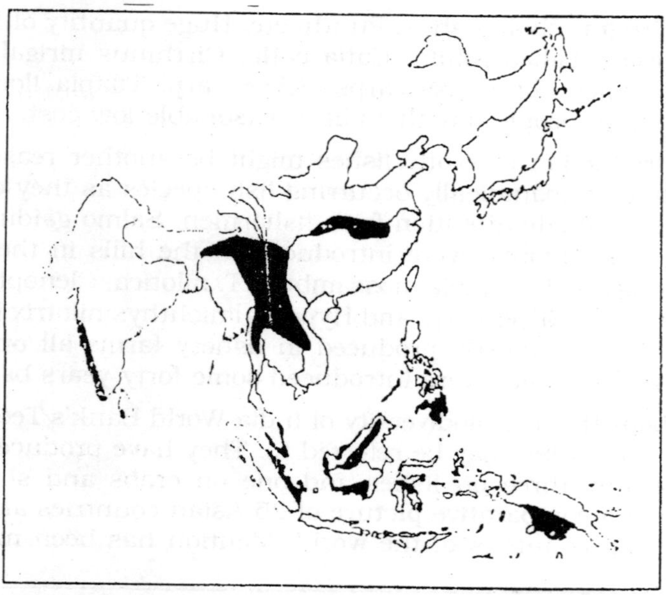

The distribution of 'hot spots' for freshwater biodiversity in Asia, based mainly on fishes
(Source : World Bank Technical Paper 343, 1996)

2. This map is compiled from personal experience in the field, as well as on the literature, and corresponds to a good 'educated guess'. Time and technical limitations did not allow for the more precise procedures used in the production of terrestrial hot spot maps. The technical limitations are mainly a lack of ground surveys over huge areas, the taxonomic uncertainly of a larger number of taxa, and also the larger number of aquatic vertebrate species than terrestrial ones making surveying more difficult.

750 species of fish, though Talwar and Jhingran (1991) mentions about 930 species in his book entitled "The freshwater fishes of India, Pakistan, Bangladesh, Burma and Sri Lanka." Besides these statements, the World Bank has prepared two maps, one showing distribution of 'hot spots' for freshwater biodiversity and another on the total number of fish species in south and south- east Asia. Though these are not very accurate, the hot spot areas marked in India are the western ghats and eastern India. Except the river Mekong, the species of fish in Ganga excels other rivers of South and south-east Asia. The statements and maps are appended hereinafter.

Political units do not, of course, bear much relation to the key determinant in the distribution of freshwater biodiversity, i.e. river basins. Thus the Mekong, Ganges, Kapuas (Kalimantan). and Yangtze basins are the richest in fish species, with the Kapuas the richest in terms of species to length ratio. Lists of biodiversity hot spots of the world have been compiled (Bibby *et al.*, 1992, Groombridge 1992, Dinerstein and Wikramanayake 1993), but again these are based on higher plants, terrestrial vertebrates and butterfiles and ignore aquatic communities. With speciation and distribution patterns in freshwater organisms differing from those of terrestrial ones it is not surprising that many of the terrestrial hot spots, such as oceanic islands, do not contain significant elements of aquatic communities while important regions are entirely missing from existing hot sopt distribution maps (e.g. Lake Inle in Burma, Malili lakes in Sulawesi).

Map shows the distribution of Asian aquatic 'hot spots' based mainly on freshwater fishes, the Mekong basin is arguably the most important hot spot because of its size and the potential and imminent threats to its integrity. The only published estimate of the number of fish species in the Mekong basin in Vietnam, Cambodia, Laos, Thailand and Burma is 298 (Kottelat 1989). This list was based on the scientific literature, confirmed by scientists and supported by voucher material. Allowing for species discovered since, the list may now be around 400 and could be expected to reach 500 if the basin were surveyed thoroughly. This is rather less than the unsubstantiated figures of 1000, 1200 or 1500 fish species circulated by some agencies and experts.

Since much conservation planning is now based on hot spot analyzes (e.g. Johnson, 1996), it is worth indicating that existing national systems of protected areas across Asia have some positive benefits for certain elements of freshwater biodiversity. For example, forested protected areas often include species rich streams, and flood plains may be important as seasonal spawning grounds. There are, however, very few areas which protect the freshwater biodiversity of larger rivers. Exceptions include the Chinese Alligator Reserve in Lawa Yangtze River, and the 400 km, long National Chambal (Gharial) wildlife Sanctuary to conserve the endangered gharial, eleven species of turtles and gangetic dolphin.

In fact out of 750 (may be 950) recorded Indian species about a hundred are commonly or sporadically available in various habitats in the country.. But actually about 50 species are conspicuous in most area and due to biotic impact and the break-through in fish breeding research bringing most of the carps within easy reach of fish eating people, most tasty large fish and a lot of small fish species are gradually being pushed back to the wall.

A checklist of conspicuous and common fishes of India is presented below :

Checklist of Common Fishes of India

Scientific Name	**Indian Name**
Ailia Colla	Kodali
Ambassis nama	Chanda
Ambassis ranga	Ranga Chanda
Amblypharyngodon mola	Murla, Mowrala
Amphipnous cuchia	Bomboch
**Anabas testudineus*	Koil; Koi
Aplochillus pancha	Techoke (Salt lake) Kanakuri
Aspidoparia morar	Peore
Badis badis	Ueda, Bot Koi
Bagarius, bagarius	Goonch, Baghmanch
Barillius barila	Bhola
B. bola	Kulla
B. vagra	Bhola
Botia dario	Gethee, Ranimach
B. rostrata	Ranimach
Brachidanio rerio	Lanpati, Napali
Carassius auratus	Lalmanch
**Catla catla*	Katla
Chaca chaca	Chega, Bhengra
Chagunius chagunio	Kental punti
Chanda nama	Chanda
C. ranga Taka	Chanda
Channa punctatus	Sawli, Leta
**C. striata*	Sawl

[* Asterisks indicate most common fishes in trade in Calcutta]

Scientific Name	Indian Name
C. marulia	Sawl, Gajar
C. gachua	Upal cheng
Chela atper	
C. laubuca	Laupolia
**Cirrhinus mrigata*	Mrigal
C. reba	Rang, Lachu
**Clarias, batrachus*	Magur
Clupiosoma garda	Bachua, Ghaura
Colisa chuna	Khurda, Chuna Kholisha
C. fasciata	Khurda, Kholisha
C. sota	
**Ctenopharyngodon idella*	Grass carp
**Cyprinus carpio* var. cornmunis	Japani rui
**C. carpio* var. specularis	Charpeo
Danio devario	Banspata
Erethistes elongatus	Bot tangra
Esomus danricus	Derwai, Dwarkina
Eutropiichthys vacha	Bachna, Vacha
Gadusia chapra	Khurda, Khoira
Garra gotyla	
Glossogobius giurus	Rajbhuji, Bele
**Heteropeneustes fossilis*	Singh, Singhi
Hypophthalmichthys motitrix	Silver, Carp
**Labeo rohita*	Rohu, Rui
L. calbasu	Kalmanch/Kalbasu
**L. bata*	Bata, Bhangan
L. gonius	Kheri, Ghonia
L. dyocheilus	Khirsa, Naktuti
L. dero	Kursa
L. nandina	Nandina
Laubuca laubuca	Chepkaor

Scientific Name	**Indian Name**
Lepidocephalichthys guntea	Nauria, Gutum
Mastacembelus armatus	Barm, Baine
Mastacembelus pancalus	Pankal (salt lake), Chitra Baine
Muoil parsia	Parse (Salt lake)
M. tade	Bhangan (Salt lake
Mystus seenghala	Singar/Malli/Air
M. cavasius	Tengra
M. vittatus	Tengra, Kalatengra
M. aor	Ghongna
M. bleekeri	Tengra
Nandus nandus	Nandas, Bheda
Nemachilus aurius	Gutum
N. botia	Balibottia
N. corica	Kharka
Notopterus notopterus	Maharmahl. Pholi
N. chitala	Cheetal
Ompok pabda	Pabda, Latl pabda
O. bimaculatus	Pabda, Dhal pabda
Oreochromis mossambicus	Tilapia
O. niloticus	Nilotica
Oryzias melanostigma	Chino (Salt lake)
Osteobrama cotio	Khurri, Gilakani
O. stewartii	Dudcheng
O. striatus	Soal
Oxygaster bacaila	Chilwa, Chela
O. gora	Chal
O. panjabensis	Chal
Pangaslus pangaslus	Pangas
Pellona indica	Khaira/Deb
Pseudeutropius murius	Muribandha
P. garua	Kharula

Scientific Name	Indian Name
Puntlus chagunio	Gulab
P. chola	Tita punthi
P. conchonius	Kanchan putti
P. gonlonotus	Japani punthi
P. phutunio	Punti
P. sarana	Saral Punti
P. sophore	Sarpunthi
P. ticto	Punthi (Tit)
Rasbora daniconius	Darkina
R. elanga	Elenga; Ran darkina
Rhinomugil corsula	Karsola
Rhynchobdella aculeata	Bain
Rita rita	Ghagra, Rita
Scatophagus argus	Pyra chanda (Salt lake)
Sciaena belengeri	Bhola
Setipinna phasa	Phansa
Silonia silondia	Karai/Siland
Tenualosa ilisha	Hilsa
Tetraodon cutcuttia	Tepa, Fatka
Tor tor	Mahasol
Tor putitora	Mahaseer, Mahasol
Trygon narnak	Sankar
Wallagonia attu	Boal
Xenenttodon cancila	Karkla, Karkya

Fish which are available all over the country in huge quantity are *Labeo rohita, Catla catla* and *Cirrhinus mrigala.* Other common species are *Mystus* sp., *Channa* sp., *Puntius* sp. but quantitatively their proportion is far less. In the South India *Labeo fimbriatus, Oxygaster* sp., *Puntius* sp., *Mystus* sp., *Pangasius* sp., *Clarias* sp., *H. fossilis* etc. are in demand. In fact mass production of fish in India is now limited to first three species.

Fishes of Himalayan Aquatic System

In all 164 species of fish belonging to 8 orders, 13 families and 52 genera have been recorded in Himalayan aquatic system (Bhatt). Most of the species are endemic in the region. The commercially important species of fish from sports and food point of view are brown trout (*Salmo truttalario*), rainbow trout (*Salmo gairdneri*), barils (*Barilius bendelisis chedra* and *B. bola*), Masheer (*Tor tor, Tor putilora*), minor carps (*Labeo dero, L. dyocheilus*), snow trout (*Schizothorax* sp.) etc. Aquatic sites are Wular, Dal, Manas, Dal lake, Khajiear, Renuka, Hemkund, Roopkund, Pasigor, Tso, To Morri etc.

Masheer the Covetable Game Fish

Over-exploitation has reduced this beautiful fish to a rare species in India. Masheer has become an endangered species. The population of various species has alarmingly declined in the last few decades in the sub-Himalayan region to the Decean Plateau and Peninsular India.

Gandak Basin

A survey of Gandak Basin has recorded about 56 common species of fish which is listed below. This list gives a present day availability of fish species in the plains of Northern India.

From Ox-bow Lake

1. *Ailia coila*
2. *Amblypharyngodon mola*
3. *Anabas testudineus*
4. *Aspidoparia morar*
5. *Bagarius bagarius*
6. *Barilius bendelisis*
7. *B. bola*
8. *Botia danio*
9. *Botia dayi*
10. *Catla catla*
11. *Chanda ranga*
12. *C. ranga*
13. *Chela laubuca*
14. *C. utrahi*
15. *Cirrhinus mrigala*
16. *C. reba*
17. *Clupisoma garua*
18. *Crossocleilus latius*
19. *Danio rerio*
20. *D. dangila*
21. *Erethistes hara*
22. *Esomus danricus*
23. *Eutropiichthys vacha*
24. *Gadusia chapra*

25. *Glossogobius giurus*	26. *Goniolosa manmina*
27. *Heteropeneustes fossilis*	28. *Labeo bata*
29. *L. gonius*	30. *L. calbasu*
31. *L. rohita*	32. *Lepidocephalichthys guntea*
33. *Mystus aor*	34. *M. seenghala*
35. *M. cavasius*	36. *M. vittatus*
37. *Nandus nandus*	38. *Nangara nangora*
39. *Nemacheilus botia*	40. *Notopterus chitala*
41. *N. notopterus*	42. *Ompok bimaculatus*
43. *Osteobrama cotio*	44. *Oxygaster bacaila*
45. *O. gora*	46. *Puntius ticto*
47. *P. sophore*	48. *P. sarana*
49. *P. chola*	50. *Rita rita*
51. *Rhinomugii corsula*	52. *Sciaena coiter*
53. *Setipinna phasa*	54. *Silondia silondia*
55. *Wallagonia attu*	56. *Xenentodon cancila*

[Source : B.C. Gha in, Bull 63 of CIEFRI, 1989].

Ichthyofauna of North Bengal

Two major contributions, besides many small ones, published in fishery journals, present a reliable picture of the staus of fish in North east India. One was published by Shaw and Shebbeare in 1937 (In the Journal of Asiatic Society) on the fishes of North Bengal and another published by T. Sen in Z.S.I. (Survey paper No. 64, 1985). The former records abour 125 species from a small area and latter 187 species from a much larger area.

Recent Survey of fishes in North Bengal has shown a dismal picture on the occurrence of fish. Hardly 30 to 40 species of fish are commonly found, the balance 70-80 species have a very sporadic or rare occurrence. One can hardly find the species of Garra, Balitora, Bhavania, Psilorhynchus, Parosilorhynchus, Schizothorax, Naemacheiles, Barilius, Barbus, Crossocheilus etc. in the hills. Other species which are sporadic in the plains do not have a very large number of individuals.

An observation on the fish species recorded by Shaw and Shebbeare (selected few species) is presented with comment.

Tenuolosa hilsa	Does not occur
Myslus seenghala	Rare
Danio devario	Rare
Barilus bola	Sporadic
B. vagra	Sporadic
Panio aequipinatus	Sporadic
Chela sp	Sporadic
Notopterus notopterus	Sporadic
Godusia chapra	Rare
Rasbora daniconius	Common
Amblypharyngodon mole	Common
Tor putitora	Sporadic or rare
Puntius sarana	Sporadic
P. ticto	Sporadic
Cirrhinus reba	Sporadic
Labeo bata	Sporadic
L. dero	Sporadic
Osleobrama cotio	Sporadic
Mastacumbelus armalus	Common
Tetraodon cutcutia	Common
Osleobrama colio	Sporadic
O. marulius	Sporadic
Glossogobius giurus	Sporadic
Xenentodon cancrta	Common
Mystus vittalus	Common
M. cavasius	Rare
Wallagonia attu	Rare
Eutropiichthys vacha	Sporadic.

Observation on the Ichthyofaunal resources of North East India

[Based on Work of T.K. Sen of Z.S.I. and M. Singh of CICFRI, Barrackpore]

This area is chosen for biodiversity study as the fauna differs from general Himalayan fish fauna. The species are a mixture of Indogangetic plains, Burmese and Indo Chinese forms. Sinha (1994 - JIFSI) provides a list of 230 giving latest systematic nomenclature. He also mentiones about indiscriminate killing of fish and destruction of habitat. A summary of his paper will be given in this work.

Sen (1985) in his work (ZSI Oce. Paper 64) has listed 187 species of fish of the following genera (No. of species above 2 have been recorded in brackets.

Goniolosa, Gudusia, Hilsa, Notopterus, Chela, Oxygaster salmostoma Barilius (9 sp.), Danio (6 sp.), Esomus, Rasbora (4 sp.), Accrossocheilus hexagonolepis, Aspidoparia, Amblypharya godon, Catla, Cirrhinus, Crossocheilus, Cyprinus, Ctenophrynogodon, Gavra (7 sp.), Hypophthalmichltys, Labeo (8 sp.), Puntius (11 sp.), Schizothorax (3 sp.), Semiplotus, Tor (3 sp.), Psilorhynchus (3 sp.), Balitora, Acanthopthalmus, Aborichthys (4 sp.), Botia (3 sp.), Lepidocephalus (6 sp.), Neomacheilus (12 sp.), Mesonoemacheilus, physoschistura, Somileptes, Aorichthys Batasio, Chandramora, Mystus (6 sp.), Ompok (3 sp.), Silurus, Wallago, Alilia, Clapisoma, Eutropuchthys, Pseudeutropius, Pangasius, Amblyceps, Bagarius, Conta, Erethistoides, Exostoma, Euchiloglanis, Gagata, Glyptothorax (5 sp.), Pseudechensis, Clarias, Heteropneustes, Chaca, Olyra (4 sp.), Aplocheilus, Xenentodon, Channa (6 sp.), Monoplerus, Chanda, Tilapia, Badis, Nandus, Sicamugil, Glossogobius, Anabas, Colisa (3 sp.), Clenops, Mastacembelus (3 sp.), Pillaria, Tetraodon.

The record of 187 species must not be taken that the region is rich in fish density and abundance. Most hill species, such as, Barilius benedelists, Danio horai, Garra lamta, G. nasuta, G. rupicula, G. gotyla, G. kempi, Pulinus shalynius, Schizothorax richardsonii, Psilorhynchus balitory, Balitora bracei, B. macualta, acanthapthalmus sp., Lepidocephalus sp., Noemacheilus sp., Erethistoides sp. are under threat because of lack of or faulty way watershed management and other biotic and abiotic impacts

Notopterus (2 sp.), Ilisha (2 sp.), Gudusia (1 sp.), Goniolosa (1 sp.), Setipina (1 sp.), Angu'la (1 sp.), Moringna (1 sp.), Pisodonopsis (1 sp.), Sa'mo (2 sp.), Chela (2 sp.), Salmostoma (3 sp.), Securicola (1 sp.), Amblypharyngodon (1 sp.), Aspidoparia (2 sp.), Barilus (8 sp.), Bengala (1 sp.), Brachy dania (2 sp.), Danio (5 sp.), Esomus (1 sp.), Parluciosoma (5 sp.), Raimas (2 sp.), Rasbora (1 sp.), Catla (1 sp.), Chagunius (1 sp.), Cirrhinus (2 sp.), Clenopharyngodon (1 sp.), Cyprinion (2 sp.), Cyprinus (2 sp.), Labeo (10 sp.), Neolissochilus (2 sp.), Oreichthys (1 sp.), Osterobrama (4 sp.), Puntius (12 sp.), Tor (4 sp.), Typophthalmichllays (1 sp.), Crossocheilus (1 sp.), Garra (9 sp.), Schizothoraichthys (1 sp.), Schizothorax (1 sp.), Psilorhynchus (3 sp.), Balitora (1 sp.), Aborichthys (3 sp.), Namacheilus (6 sp.), Acanthocobitis (1 sp.), Acanthopthalmus (1 sp.), Acantopsis (1 sp.), Lepidocephalus (6 sp.), Neoeucirrhichthys (1 sp.), Somileptes (1 sp.), Botia (5 sp.), Aorichthys (2 sp.), Batasio (2 sp.), Mystus (7 sp.), Rana (2 sp.), Ompok (3 sp.), Silurus (1 sp.), Ailia (2 sp.), Clupisoma (1 sp.), Pangasius (1 sp.), Amblyciptidae (3 sp.), Bagarius (1 sp.), Conta (1 sp.), Erethistoides (1 sp.), Euchiloglanis (2 sp.), Evostoma (1 sp.), Gagata (1 sp.), Glyptothorax (12 sp.), Hara (2 sp.), Laguria (2 sp.), Nongra (2 sp.), Pseudoecheneis (1 sp.), Sisor (1 sp.), Clarias (1 sp.), Heleropneustes (1 sp.), Chaca (1 sp.), Olyra (2 sp.), Xenentodon (1 sp.), Aplocheilus (1 sp.), Monoptenus (2 sp.), Chanda (1 sp.), Psendoambassis (2 sp.), Johnius (1 sp.), Pama (1 sp.), Badis (1 sp.), Nandus (1 sp.), Oreochromis (1 sp.), Rhinomugil (1 sp.), Sicamgil (1 sp.), Glossogobius (1 sp.), Anabas (1 sp.), Colisa (3 sp.), Ctenops (1 sp.), Channa (6 sp.), Macrregnathus (1 sp.), Maslacembelus (2 sp.), Chaudhuria (2 sp.), Tetraodon (1 sp.).

These 103 genera of fish exhibit both terrestrial and plain water form as well as cold and warm water form.

Sinha (1995) lists the following as threatened Sp.

Raiamas bola
Cyprinion semiplotum
Puntius shalynius
Tor chelynoides
T. putitorsa

T. tor
Schizothoraichlhys progastus
Schizothoran richardsoni
Aborichthys garoensis
Nemacheiluo elongalus
N. reticulofasciatus
Chandhiria indica.

But the present authors feel that true hill stream, semi-torrential form and migratory forms are also threatened. The genera are :

True hill stream form

Garra, Psilorhynelus, Glyptothorox, Enchiloglanis, Exostoma, Psendocheneis.

Migratory form

Anguilla, Schizogyge, Schizothorax, Barilius, Labeo, To, Channa, Mystus, Badis, Anabas, Mastacembelus.

Semi torrential form

Olyra, Pillaia, siturus, Amblyceps, Aborichthys, Bolia, Somileptes Naemacheilus, Crossocheilus, Acanthobilis.

Depletion of Fishery resources

A number of physico-chemical factors affect fish farming which are (i) draining of different presticides from agricultural fields into the water bodies, organic effluents, heavy metals etc. (ii) parasitic infections on fish bodies by protozoans, helminths and arthropods (iii) haemoflagellates including the tripanosomes and trypanoplasms causing heavy mortality in hatcheries. Halder *et al.*, of Zoology Deptt. of Kalyani University obtained haemoglagellales from diverse fish population of Anabas testadineus, Heteropneustes fossilis, Rita rita, trichogaster fasciata and Clarias batracus. They found that fishes of lower age group during their pre-breeding period are more susceptible to trypanosome infection; also that with the increase in global atmospheric temperature, heavy metal toxicity and eutrophication trypanosomal biodiversity might be a threat to fish health in the coming decades. In another investigation Halder *et al.*, found infection of ciliates belonging to the genera Trichodina, Paratrichodina and Tripartiella in 38 edible fishes which are causing biodiversity loss.

Chatterjee *et al.*, of Z.S.I. observed a rich ichthyofaunal (200 species) species at Digha Sea coast but a declining trend in commercial catches due to indiscriminate fishing, disturbance of potential breeding ground.

Z.S.I. in their publication, "Animal Resources of India, (1991)" refers 2546 species under the group pisces (World Sp. 21,723). But a fresh scientifc inventory needs to be made to list actual number of species and genera, 60 monotype genera mark special feature of numerical analysis of fish fauna. The book refers to a severe decline in the fresh water fish fauna which is due to indiscriminate fishing, dynamiting of rivers, construction of dams across rivers, over-utilization of water, cutting down of forest trees, mismanaged farm lands, erosion, pollution bý factory chemicals and competition from introduced species etc. Masheer population has been severely depleted. The following has been listed as (a) endangered and (b) threatened species.

(a) Endangered

Enobarbichthys maculatus
Cyprinion semplotus
Raiamas bola
Tor chelynoides.

(b) Threatened

Notopterus chitala
Neolissochilus hexagonolepis
Labeo fimbrialus
Labeo kontius
Labeo potail
Tor khudree
Tor tor
Tor putitora
Puntius carnaticus
Puntius jerdoni
Gonoproktopterus curmuca
Cirrhinus cirrhosus
Schizothorax richardsonii
Schizothoraichthys progastus
Silonia childreni
Pangasius pangasius
Bagarius bagarius.

It has also been found that fishery has been badly affected by the activities of trawlers from other countries, seeking Tiger prawn and pomfret, destroying large quantities of low-value by-catch. Domestic Shrimp post larvae collectors were also blamed for destroying other fish seed as a result of their activities.

West Bengal has limited ponds, wetland areas and cultivation of fish in paddy fields; also there is poaching and poisoning to a large extent. The urban development has threatened the wetlands of Calcutta (14,000 ha. of wetland has been reduced to 400 ha.). Calcutta gets more than 50% of fish coming from Uttar Pradesh, Tamilnadu, Orissa, Gujarat and Punjab. Common fishes in Calcutta market are Rohi, Mrigal, Tilapia, Catla, Silver carp, common car, Walking catfish, Boal, Climbing perch, shrimp and marine fish (source : Aquaculture Asia April - June - 1998).

The present global picture of inland Shrimp culture fishery is bogged down with environmental issues. Arguments about the farfetchedness of the views put forth by environmentalists highlighting the chemical effects of effluents released from aquaculture firms continue to have a marginal impact on the general belief that the inland shrimp culture activity is harming the aquatic environment. (Fishing Chemes, Oct. 2000).

Ecological degradation of water bodies

Such degradations started a century back when water hyacinth invaded the water bodies. Sunlight cannot penetrate the hyacinth layer so there is no planktonic growth contributing to entrophication in most of the water bodies. Some species such as puntius jerdoni, Bagarius bagarius, Semiplotus semiplotus besides many other have been considered on the verge of extinction in beels of Assam.

"Biodiversity can be an important component in the economic valuation of the wetlands. The change and biodiversity has implications for the food security and livelihood of the population that depends on the beels. The economic value of the biodiversity has not yet been studied. Information on ecological inter-relationships between changes in biodiversity and changes in the primary productivity of the beels is required to establish appropriate policies for *in-situ* and *ex-situ* conservation and for other environmetal related initiatives" (Barman, Nage, 2000).

Biotope improvement is a must for the long-term optimum exploitation of the beels (Jhingran 1979) in terms of both environmental protection and productivity. The resources will dwindle in due course of time if biotope improvement is not made (Yadav, 1987). Formerly, conservation ethics were followed in re-exploitation of beel. Catching and killing of brood fish and Juventiles were prohibited.

CHAPTER – 4

SPECIAL SITE FOR STUDY OF BIODIVERSITY AND ECOLOGY

— Case of floating Island of Laktak Lake, Manipur

A special and unique (habitat) Biodiversity Reserve to study biodiversity

Located at Moirang in the Bishnupur district, 45 km South of Imphal lies the Loktak Lake. Surrounded by hills the lake covers about 300 sq. km. of water spread. It now forms a National Park designated as Keibul Lamjao National Park.

Phoom or Phoomdi locally means extensive patches of weed floating on shallow water. The phoom are formed of intensive dense growth of aquatic weeds and grass. The dense network of root shoot anastomose into a thick floating platform, trapping organic and inorganic debris, layer upon layer, on top of which weeds keep on growing. The phoom is usually 1.5 to 3 metre in thickness, though the thickness depend on their age. The people walk over them and build bamboo hutments. There are several such islands of weeds.

The unique aspects are as below :

- World's only species of deer called Thamin or 'Dancing deer' scientifically known as *Cervus ildi* inhabits this phoom.
- Fishery is a primary source of income of local people as about 2500 tonnes of fish is caught annually.

Thousands of fisher-folk (Nga mees) are engaged in fishing

from these islands. Several of them are permanent dwellers in huts built on phooms. These are floating fishery villages.

Depletion of habitat

The idyllic environs are under extreme abuse of the fisherman. Shifting cultivation in the adjoining hills has resulted in siltation of the lake. Bank erosion is another problem. The water level of the lake has gone-up after construction of Loktak Hydro-electric Project which is also polluting the water. The construction of Ithal barrage has resulting in poor production of *Osteobrama balengeri* which is most tasty fish of the lake : migration from Chenduri river has stopped.

This is a unique habitat and it is not known if any where one such type of habitat occurs.

A check - list of flora and fauna in and around such Phoomdi is reproduced from the work of IUCN, communicated through T. Singh, Conservator of Forests, Manipur.

IMPORTANT FLORAL/FAUNAL SPECIES THAT COLONIZE THE *PHOOMDI*

(A) FLORAL SPECIES

Alpinia galanga

Alternanthera philoxeroides

Argyreia nervosa Burm.f.

Arundo donax

Azolla pinnata

A. nigra Gaertn.

Brachiaria mutica

Carex indica

Coix lacryma-jobi

Colocasia esculenta

Crotolaria juncea

Cymbopogon nardus Linn.

Cynodon dactylon

Cyperus difformis

Cyperus rotundus

C. alata

Desmodium motorium

Echinochloa stagnima

Eichhornia crassipes

Enhydra fluctuans

Erianthus arundinaceus

Erianthus procerus

Hedychium coronarium

Imperata cylindrica

Ipomoea aquatica

Jussiaea repens

Leersia hexandra

Mikania micrantha

Monochoria hastaefolia

Monochoria hastata

Oenanthera javanica

Persicaria chinensis

Phragmites karka

Pistia strotiotes

Polygonum barbatum

P. orientale

P. perfoliata

P. plebejum

P. posumbu Buch. Ham. ex.

Rumex nepalensis

R. Maritimus

R. vesicarius

Saccharum munja

Saccharum spontaneum

Sacciolepis myosuroides

Sagittaria sagittifolia

Salvinia cucullata
Salvinia natans
Setaria pallidefusca
Zizania latifolia

(B) FAUNAL SPECIES

(i) Mammalia

Anoyx cinerea	Clawless otter
Arctonyx collaris	Hog badger
Axis axis	Erxleben Chital/spotted deer
A. porcinus	Zimmermann Hog deer
Canis lupus Linn.	Indian Wolf
Cannomys badius	Bay bamboo rat
Cervus eldi eldi	Brow-antlered deer
Cuon alpinus	Indian wild dog
Felis catus	
F. chaus	Jungle cat
F. sylvestris	
F. temmincki	Golden cat
Funambulus pennati	Five striped squirrel
Golunda ellioti	Indian bush rat
Lutra lutra	Common otter
L. perspicillata	Smooth Indian otter
Manis crassicaudata	Indian pangolin
Melogate moschata	Ferret badger
Muntiacus muntjak	Barking deer
Mus booduga	Common Indian field mouse
Mus musculus Linn.	Personata House mouse
Mustela erminea	Ermine
Paradoxurus hermaphroditus	Common palm civet
Rattus rattus rattus	Common house rat
Ratufa bicolor	Malayan giant squirrel

Sus scrofa	Wild pig
Talpa micrura	Eastern mole
Viverra zibetha	Large Indian civet
Viverricula indica	Small Indian civet
Vulpes bengalensis	Common fox
V. vulpes	Red fox

(ii) Amphibia

Bufo melanostictus	True toad
Hyla arborea	Tree frog
Polypedates leucomystax	Tree frog
Rana tigrina	Indian bull frog
Rana limnocharis	Paddy field from
Rhacophorus leucomystax	Chunam frog

(iii) Reptilia

*Ahaetulla nasuta Lacepede***	Striped keelback
Amphiesma stolata Linnaeus	—
Bungarus caeruleus Schneider	—
B. faciatus	—
Calotes versicor	—
Chrysopelea ornata	—
Pendrelaphis tristris	—
Elphae radiata	Copperhead
E. helena	—
Lycodon auticus	—
Lycodon jara	Yellow striped Wolfsnake
Nabuija sp.	
Naja naja	Indian cobra
Naja sp.	Ishing Kharou
Natrix subminiata	Red necked keel back
Oligodon albocinctus	—
Oligodon arnensis	—

Oligodon cyclurus	—
Oligodon venustus	—
Opheodrys doriae	—
Ptyas korros	Indo-Chinese ratsnake
Ptyas muscosus	Ratsnake
Python molurus	Indian python
Testudo elangans	—
Trimeresurus albolabris	—
T. erythrurus	—
T. gramineus	—
T. monticola	—
Typhlops braminus	—
Varanus salvator	—
Vipera russelli	—
Xenochrophis piscator	—

(iv) Ornithofauna

Acridotheres tristis tristris	Indian myna
Aegithina tiphia	Common iora
Alauda gulgula	Lesser skylark
Amaurornis phoenicurus phoenicurus	White breasted water hen
Anas acuta	Pintail
Anas clypeata	Shoveller
Anas crecca crecca	Common teal
Anas penelope	Wigeon
Anas poecilorhyncha poecilorhyncha	Spotted bill duck
Anas querquedula	Blue winged teal
Anas strepera streperaa	Gadwall
Anser anser rubriostris	Eastern graylag goose
Anser indicus	Bar headed goose
Ardea alba modesta	Eastern large egret

Ardea cinerea	Grey heron
Ardeola bacchus	Chinese pond heron
Ardeola grayii	Indian pond heron
Aythya ferina	Common pochard
Aythya nyroca Gulden	White eyed pochard
*Bambusicola fytchii****	Assamese bamboo partridge
*Batrachostomus hodgsoni hodgsoni****	Hodgson's frogmouth
Buteo buteo japonicus	Buzzard
Bubo falvipes	Tawny fish owl
Bubo nepalensis nepalensis	Forest eagle owl
Bubulcus ibis	Cattle egret
Centropus sinensis intermedius	Crow pheasant
Ceryleudis leucomelanura	Pied kingfisher
Columba pulchricollis	Ashy wood pigeon
Corvus macrohynchos levaillantii	Eastern jungle crow
Coturnix coturnix coturnix	Common grey quail
Coturnix coturnix japonica	Japanese grey quail
Dendrocygna javanica	Lesser whistling teal
Dendrophasa bicincta bicincta	—
Dicrurus annectans	Crow billed drongo
Dicrurus hottentttus hottentottus	Spangled drongo
Dicrurus odsmiles albrictus	North Indian black drongo
Ducula bandia griscicapilla	Grey-headed imperial pigeon
Egretta garzetta	Little egret
Egretta inermedia intermedia	Median egret
Emicurus leschenaulti indicus	Leschenault's forktail
Emicurus schistacens	Slaty backed forktail
Ephippiorhynchus asiaticus	Black necked stork
Excalfactoria chinensis chinensis	—
Falco biarmicus jugger	Lagger falcon
Francolinus francolinus	Black francolin

Francolinus francolinus melanonotus	Assam black partridge
Francolinus pintadaenus phayrei	Burmese francolin
Fulica atra atra	Coot
Gallicrex cinerea cinerea	Water cock
Gallinago gallinago gallinago	Fantail snipe
Gallinago solitaria solitaria	Eastern solitary snipe
Gallus chloropus indica	Indian moorhen
Gallus gallus spadiceus	Burmese red jungle fowl
Gracula reliogiosa intermedia	Northern hill myna
*Grus monacha Temminck****	Hooded crane
Grus sharapii	Burmese sarus
Halicyon smyrnensis	Smyrna kingfisher
Hirundo daurica	Daurian swallow
Macronygia unchall tusalia	Bar-tailed cuckoo dove
Metopidius indicus	Bronze winged jacana
Milvus migrans	Black kite
Milvus migrans govinda	—
Monaracha zurea	Black naped blue fly catcher
Motacilla caspica	Grey wagtail
Motacilla citreola	Yellow headed wagtail
Motacilla maderaspatensis	—
Motacilla alba	White pied wagtail
Muscadivora cenea sylvatica	—
Netta rufina Pallas	Red crested pochard
Nettapus cormandelianus	Cotton pygmy goose
Nycticorax nycticorax nycticorax	Night heron
Oenopopelia tuanquebaria humilis	—
Oriolus oriolus kundoo	—
O. zanthornus zanthornus	—
Passer domesticus indicus	—

P. rutilans intensior	Yunnan cinnamon tree sparrow
Pelecanus onocrotalus	White or rosy pelican
Perdicula manipurensis manipurensis	Manipur painted quail
Picoides cathpharius pyrrhothorax	
Pluvialis dominica fulva	Eastern golden plover
Polyplectron bicalaratum bakeri	Bhutanese peacock pheasant
Porphyrio porphyrio	Indian purple moorhen
Psittacula cyanocephalia	—
Psittacula eupatria	—
P. Krameri	—
Pycnonotus jocosus monticola	Assam red whiskered bulbul
Rhodonessa caryophyllacea	Pink headed duck
Rostratula benghalensis benghalensis	Painted snipe
*Sarcogyps calvus Scopoli****	—
Scolopax rusticola rusticola	Woodcock
Sphenocereus apicaudus	—
Spilornis cheela cheela	Crested serpent eagle
Streptopelia chinensis	Spotted dove
Streptopelia orientalis agricola	—
Streptopelia orientalis meena	Western turtle dove
Streptopelia xanthocycla	—
Sturnus contra superciliaris	Burmese pied myna
Sturnus malabaricus	Grey head myna
Sturnus pagodarum	Black head myna
Syrmaticus humiae humiae	—
Tadornaw tadorna	Common shelduck
Tadorna ferruginea	Ruddy shelduck
Terron apicauda	Pin-tailed green pigeon
Treron bicincta bicincta	Indian orange breated green pigeon

Treron curvirostra nipalensis	Thick-billed green pigeon
Treron pompadora phayrei	Ashy headed green pigeon
Tringa totanus eurhinus	Eastern red shank
Turnix maculatus maculatus	—
Turnix sylvatica dussumier	Little bustard quail
Turnix suscitator blumbipes	Burmese bustard quail
Turmix tanki tanki	Indian yellow legged button quail
Tyto alba	White owl
Tyto alba stertens	Indian barn owl
*Upupa epos longirostris***	Burmese hoopoe
Venellus cinereus	Grey headed lapwing
Venellus indicus atronuchalis	Burmese red-wattled lapwing

CHAPTER - 5

PROTECTED AREAS AND SUSTAINABLE DEVELOPMENT OF INDIAN PROJECTS — A REVIEW

In India the network of protected areas consist of more than 80 National Parks, 441 Wildlife Sanctuaries covering more than 1,50,000 sq. Kms. of the Country. A National Coordination Committee has been set-up to promote effective inter-departmental coordination for the control of illegal trade in Wildlife and Wildlife products in the country. Also an Inter-state committee has been set-up to review the Wildlife (Protected) Act, 1972, and their relevant laws.

The authors intend to discuss in nut-shell various ramifications of Protected Areas (P.A.) and the measures taken by the Central and State Government to protect them. The discussions will also cover the aspects of sustainability of development of the projects briefly.

At the outset it is necessary to define the terms Protected Areas (P.A.) and Sustainable Development (S.D.) which are intimately connected with each other.

Protected Area (P.A.) is defined by the Convention of Biological Diversity as "A geographically defined area which is designated or regulated and managed to achieve specific conservation objectives."

Other organizations viz., The Fourth World Congress on National Parks and Protected Areas (1992), and Global Biodiversity Strategy (1992) also define the 'Protected Area' on same lines.

It's resources need wise maintenance and utilization by accurate inventory and taking protective measures. P.A.s contribute to the conservation of living resources and sustainable development by —

- Maintaining ecological processes and life supporting system,
- Preserving genetic diversity,
- Maintaining environmental stability to decrease intensity of flood, erosion, drought, grazing, fire and conserving water and soil moisture.
- Maintaining productivity of eco-system to conserve water and resources.
- Providing conservation education,
- Providing opportunities for rural development, and
- Providing base for recreation and promoting of tourism.

Various strategies of P.A. remain incomplete unless the projects are sustainable; a sustainable development, however, has a complex definition as conceived by several authorities from the world. *The following is a summary showing the concepts of various authorities/organizations from the world.*

		Various Conceptions	Authorities
(a)	(i)	Eco-centric	Hugh Barton (2000)
	(ii)	Anthropocentric	"" ""
(b)	(i)	People-centred (Social, Environmental and Economic)	W.C.E.D. (1987)
(c)	(i)	Maintenance of environmental quality.	Simon Bell *et al.*, (1999)
	(ii)	Development that improves quality of life.	"" ""
	(iii)	Economic development subject to maintaining the services and quality of natural resources.	"" ""
	(iv)	Sustainability to be viewed in a holistic manner including economic, social and ecological components.	"" ""
	(v)	Envisages imperative necessity of measuring the Biodiversity.	"" ""

(d)		Sustainable Development indicates social, economic, environmental facets.	United Nations (1991)
(e)	(i)	To maintain essential ecological processes & Life support system.	W.W.F., I.U.C.N. & U.N.E.P. (1980)
	(ii)	To preserve genetic diversity	"" ""
	(iii)	To ensure sustainable utilization of species and ecosystme.	"" ""
(f)		"Sustainable Development (is) a development strategy that manages all assets, natural resources, and human resources, as well as financial and physical assets for increasing long-term wealth and Well beings."	Robert Repello (1986)

According to I.U.C.N. (1991) Sustainable Development implies increasing human productivity and the quality of life while keeping within the carrying capacity of supporting ecosystem. Strategies for Sustainable National Development seek to improve and maintain the well-beings of people and ecosystem.

Project Monitoring

Monitoring the projects were confined to look into the following aspects (only a few, mentioned among many).

- Improved P.A. management viz. Planning Organizational strength, Consolidation, Apacity building and Key ecosystem restoration issues.
- Village eco-development, viz. Mobilisation of awareness; Attitude of villagers, Staff, N.G.Os. etc.
- Nature education and awareness campaign for staff N.G.Os and general public.
- Impact monitoring and research to understand the issue and finding solutions.
- Analysis of the existing funding sources, financing mechanism in the project area and state.

- Analysis of the range of funding sources and financing mechanisms, including opportunities for recovering cost of selected services being used by different states and abroad.
- Consulting the Forest Department to make rough projection for post-project budget requirement to maintain and restore Biodiversity in the project area and continuing village eco-development initiative.
- Identification and meeting the key-stakeholders who could pro-actively promote the adaption of new funding sources and funding mechanism.
- Re-commending the strategy on how the bio-diversity conservation programme in the project can achieve financial stability.

Biodiversity vis-a-vis Conservation Projects in India

Since the ratification of the Conservation-on Biological Diversity by India in February 1994, several steps have been taken to meet the commitments under the convention as well as to bring the legislative, administrative and policy regime regarding biological diversity in tune with the Article of the convention to avail of the opportunities created by it. India has since participated in several International meetings concerned are bio-diversity. Twenty-ninth December is celebrated as the International Day for Biological Diversity to commemorate the date of entry into force of the convention on Biological Diversity.

Under the Biodiversity Conservation, many projects are undertaken such as Wetland in States, Lake Chilka, Inventory of flora and fauna of some protected areas, Marine Parks etc. The scheme on biodiversity conservation was initiated to ensure proper coordination among various agencies concerned with issues relating to conservation of biological diversity, and to review, monitor and evolve adequate policy instruments for the same.

Subsequent to the signing of the Convention on Biological Diversity by India during the UNCED Earth Summit (1992) extensive consultations were organized at various levels.

Biodiversity related activities were primarily concerned with very many conservation projects some of which are —

- Environmental Impact Assessment (EIA)
- Wetlands
- Lake conservation
- Biosphere Reserves
- National River conservation
- Eco-development
- National Parks
- Tiger Reserves
- Wetland fauna and several others.
- Impact of Ganga Action Scheme.
- Evaluation of Ganga Action Plan Works.
- Yamuna and Gomti Action Plan.
- Damodar Action Plan.
- National River Conservation Plan (NRCP).
- National Lake Conservation Plan (NLCP).
- National Afforestation and Eco-development Board (NAEB).
- Integrated Afforestation and Eco-development Project Scheme (IAEPS).
- Area Oriented Fuelwood and Fodder Project Scheme (OAFFP).
- Raising of Non-Timber Forest Produces (NTFP) including
- Medicinal Plants.
- Monitoring & Evaluation of Afforestation Activities.

The Wildlife Institute of India has been entrusted with the responsibility of acting as a nodal agency for research programmes under the Conservation of Biodiversity Component of the Forest Research Education and Extension Project involving two sites namely the 'Great Himalayan National Park' in Himachal Pradesh and the 'Kalakad- Mundanthurai Tiger Reserve' in Tamil Nadu.

These are a few very broad outlines of some of many biodiversity projects running in India. In fact the foresters and administrators should have a clear idea about this multifaceted subject, 'Biodiversity', so that they may be fully aware of various imperative modalities of this subject while implementing them.

With these in view the authors have collated various definitions and facets of Biodiversity detailed in Chapter 1.

Wildlife Conservation achievements have been a special issue.

The following have been successfully initiated :

- A net work of protected areas have been established.
- Wildlife Institute of India takes care of training and research in Wildlife.
- Effective measures are taken to stop illegal trade in animals. (On the basis of the report of The Subramaniam Committee).
- A National Coordination Committee (NCC) has been set-up to promote effective inter-departmental coordination for control of illegal trade in Wildlife and Wildlife Protection.
- An inter-State Committee has been set-up to review the Wildlife (Protection) Act, 1972 and other related Laws.
- The Indian Board for Wildlife (IBWL) the apex advisory body in the field of Wildlife Conservation has been reconstituted.
- Assistance was provided to the States for taking-up programme of Ecodevelopment around National Parks and Sanctuaries indicating the higher resources in order to achieve ecologically sustainable economic development of these areas.

Project Elephant

This project was launched in 1991-92.

- for ecological restoration of natural habitat,
- for ECO development,
- Conservation of habitat,
- to mitigate man-elephant conflict,
- to preserve migratory routes.

Research Projects

Some of the projects undertaken are :

- All India Coordinated Research Project on Aero-allergens and Human health.
- Coordinated Research Project on Ethnobiology.

- Inter-Government Panel on Climatic Change (IPCC).
- G.B. Pant Institute of Himalayan Environment and Development.
- Research on Biosphere Research, Wetlands and Mangroves.
- Indian Council of Forestry Research and Education DDN. (ICFRE).

Project Tiger

This project was launched in 1973 and remarkable success was achieved —

- to ensure maintenance of a viable population of tiger in India for scientific, economic, aesthetic, cultural and ecological values.
- to presume for all times, areas of such biological importance as a national heritage for the benefit, education and enjoyment of the people.

There are 23 projects covering 33046 sq. km. all over the Country.

Protected Areas Vis-a-vis the Role of Government

The ministry of Environment and Forests, Govt. of India is the nodal agency for the planning, promotion and coordination of environmental and forestry programmes.

The primary tasks involve —

- Conservation and Survey of Flora and Fauna.
- Conservation and Survey of Forests and Wildlife.
- Prevention and Control of Pollution.
- Afforestation and Regeneration of degraded forests.
- Assessment of Environmental Impact.
- Implementing Environmental and Forestry Programme.
- Promotion of Environmental and Forestry Programme.

National Conservation Strategy and Policy Statement on Environment and Development

The National conservation strategy and Policy statement on Environment and Development 1992, the National Forest Policy

1988 and the Policy statement for abatement of pollution 1992 are the major policy instruments of the Government for dealing with various facets of environment and development in a comprehensive manner.

Biosphere Reserve

The Biosphere Reserve Programme is a pioneering effort at pursuing the increasingly difficult yet urgent task of conserving Ecological diversity under mounting pressure. It is to protect Representative ecosystem.

Wetland (National Lake Conservation Plan)

The following are some of wetlands that have received attention :

Kolleru, Wulan, Chilka, Loktak, Bhoj, Smabhar, Pichola, Sasthamkolla and Ashtamudi, Nalsorabar, Hariki, Uyni, Renuka, Pondam, Chakatratal, Kabar, East Calcutta, Sukna, Deepar Beel among several others.

Legislation and Institutional Support

Legislation

Since the enactment of the Environmental (Protection) Act, 1986, several actions have been taken by the Ministry of Environment and Forests to create a comprehensive legal and institutional infrastructure for safeguarding the environment. These include framing of rules, notification of standards, notification of environmental laboratories, delegation of power, identification of agencies for management of hazardous chemicals, setting-up of Environmental Protection Council in the States etc.

The following were enacted :

- The National Environment Tribunal Act, 1995.
- Amendment of Water (Protection and Control of Pollution) Cess Act, 1977.

Government of India's Silent Revolution for Environmental Conservation

In connection with P.A.'s the G.O.I. has summarised in their achievement in the report. "A Silent Revolution for Environmental Conservation" 1999. Some salient features of this report is presented here in after :

Environmental Action Plan for controlling pollution at the national level.

For this the Ministry prepared a draft on National Environmental Action Plan.

(i) Industrial Pollution

(ii) G.O.I. has listed 1551 industrial units in the highly polluted category.

Vehicular pollution

- Use of low ash coal by Thermal Power Plants.
- Environmental Action Plan for control of pollution at religious places.
- Taj Protection Mission.
- Crackdown on Polluting Industries.
- Pollution control in Delhi.
- Sustainable Development strategies.
- Hazardous substance management.

(iii) Environmental Impact Management

- Monitoring of Development projects
- Constitution of Authorities

(Coastal Zone Authority, Aquaculture Authority, Central Ground Water Authority).

- Restoration of mining areas.

(iv) Conservation of Natural Resources

- Forest Protection

(Joint Management, Fire Control)

- Forest cover increase.
- Compensatory plantation.
- Wildlife

(Policy decision, Ban on trade, Ecodevelopment, Project turtle),

(v) Bio-diversity

- Bio-diversity Act
- Natural Policy and Action Strategy on Bio-diversity
- Bio-diversity network

(vi) Marine Ecosystem

(vii) Regeneration and Development

- National Afforestation and Eco-development Board (NAEB)

(viii) National River Development Programme

New Project Proposals (1998-1999)

- Watershed prioritisation of Barak Basin using GIS and Remote sensing
- Study of the influence of Four River Systems on the land and water system of the Kuttand area using remote sensing data and GIS.
- Grassland mapping in Gujarat, Himachal Pradesh and Tamil Nadu.
- Environmental Hazard Zonation and optimal resource use in the Alokananda valley, Garhwal Himalayas using Remote sensing and GIS techniques.
- Environmental Assessment and Monitoring of Kalrayan Hills, Eastern Ghats, Tamil Nadu, using Remote Sensing and GIS.

– Training and Manpower Development.

– Preparation of Mangrove.

Threatened Waterbirds of Assam

To highlight the destruction of migratory birds in this country a media report is reproduced (From The Telegraph, dt. Dec. 30, 2000) entitled "Killing of Birds" at Jhatinga in Cachan, Assam.

National Forestry Action Programme

In order to operationalise the National Forest Policy, 1988, the Government of India decided to formulate a National Forestry Action Programme and signed a project with UNDP and FAO in June 1993. The objectives of the project is preparation of a National Forestry Action Programme and long term, medium term and short term prospective policies as well as identification and quantification of investment and technical assistance proposals in line with the National Forest Policy, 1988. Implementation of NFAP will contribute to Sustainable Development and use of forest resources and to maintain ecological balance.

Various studies such as Zonal Forest Sector review for each state, 16 subject specific studies and 5 sub-contracts have been carried out through national consultants. 5 International Consultants have submitted their report on Forest Sector Policy, Institution Development, Forest Industry Planning, Resources Economics and Forest Sector Review. On this basis symposium were held and detailed guidelines were issued to States for completion of State Forestry Action Programme.

Peoples' participation

Association of Scheduled Tribes and rural poor in regeneration of degraded forests on Usufruct sharing basis have made a remarkable progress.

The National Forest Policy of 1988 provides for involvement of people for managing forests to achieve the desired objectives of forest conservation, extension of tree cover and meeting the requirements of the people.

The objectives of the Schemes are :

- To improve forest based biomass resource base in degraded forest land to manage it on a sustainable basis for the domestic needs of the identified communities.
- To provide gainful employment and a sustainable economic base to Scheduled tribes and other rural poor in the vicinity of their habitation.
- To involve local Scheduled tribes and other rural poor in protection and development of degraded forests.

Protected Area

A group of conservationists felt that India's wildlife wealth survived so long due to intimate association of various animals with the Hindu-Buddhist culture. Many animals have been associated with Hindu Gods.

Besides India's protected . area (about 4 per cent of geographical area), Nepal's 10.6 per cent of PA and Bhutan's 22 per cent PA often offers a better picture of conservation in Northern India.

Royal Bengal Tiger is the National Animal of both India and Bangladesh. All the 23 Tiger Project Reserves are distributed over various bio-geographical regions of India. Each project area protects Forest types and various resident animals along with Tiger.

But the pressure of growing human population poaching, encroachment, pilferage of forest produce are causing deterioration of the forest and wild life.

India's sincere effort related to protected areas should be concentrated to the 23 Tiger reserves and a few other selected biotal rich P.A.'s like Kaziranga, Loktak, Keoladeo-Ghana, Kolletek, Kanchanjhunga (infact most of the P.A.'s of Sikkim), Sandakphu, Neora valley, Desert areas, Gir and Few others. Efforts have already taken shape to protect some threatened animals like — Jercdon's coursew, Forest Owlet, Lion-tailed Macaque, Nilgiri Thar, Siberian Crane, Greater Ardutant, Gangetic dolphine, Gharial, Mugger are some such animals and birds.

A number of reptiles and amphibians have suffered destruction because of their commercial value. Millions of a frog (six-fingered) besides some other species of amphibia and some reptiles were being exported.

(ix) Some significant Programmes, Projects (Completed and Ongoing)

Diagramatic presentation of some programmes and projects have been made as follows to bring to light some significant ones to mark Govt. of India's commendable effort to protect biodiversity.

RESEARCH PROJECT UNDER BIOSPHERE RESERVE PROGRAMME

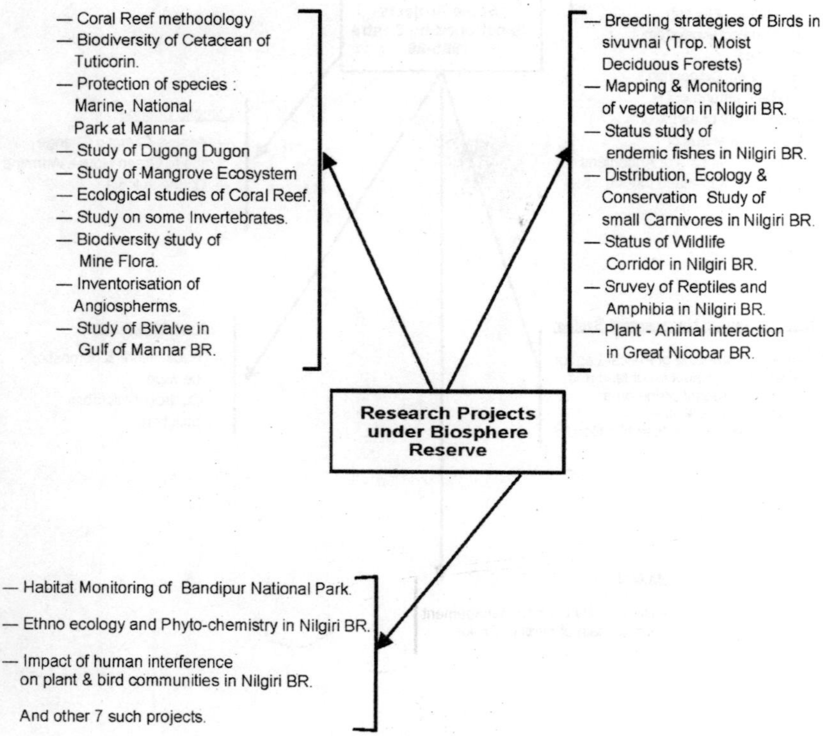

BIODIVERSITY PROJECTS

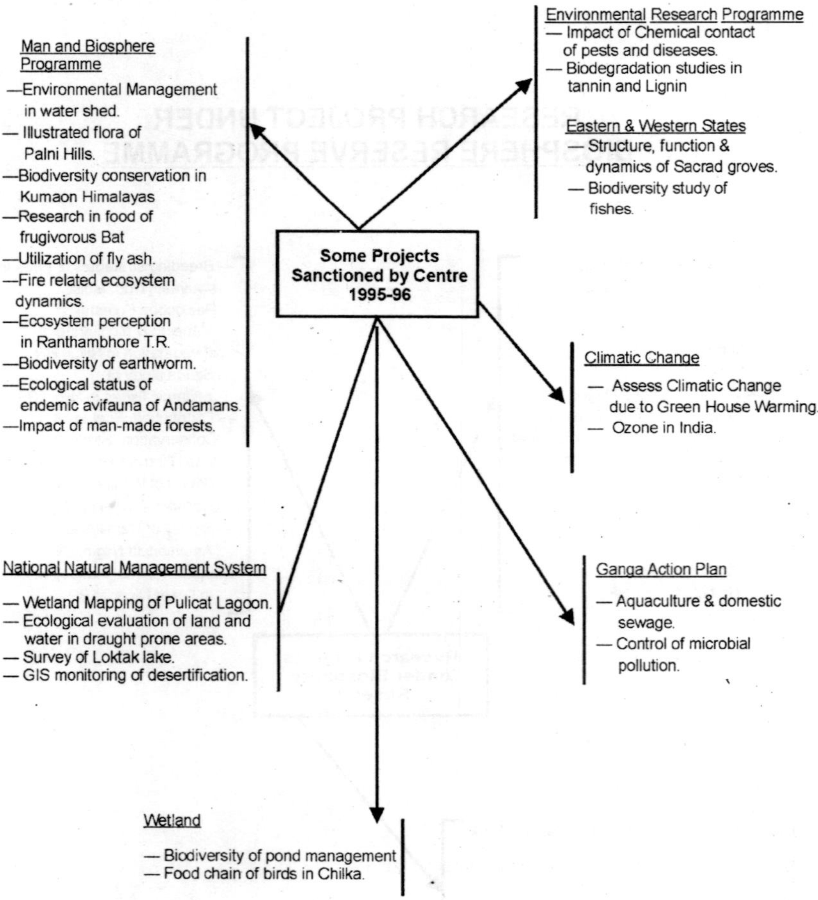

USEFUL PROJECTS COMPLETED

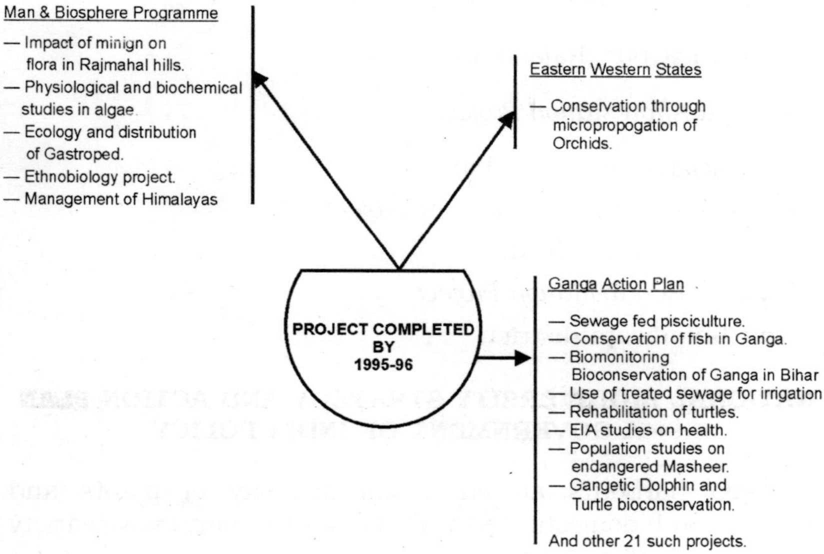

From the facts presented in this chapter it may be valued that the Govt. of India is making one Herculean Effort to protect the floral and faunal resources and put them to sustainable use and for development of the country.

Could the people of India believe that biodiversity loss will be restored in near future?

Externally Aided Forestry Projects

[World - Bank, SIDA, OECF (Japan), ODA (UK), FRG, DDA (UK)]

For sustainable development of various Forestry Projects in India, several World Organisations have come forward with monitory help for implementation of Projects. Some of these are —

– *Forestry Development Projects*

in West Bengal, Maharashtra, Western Ghat, Andhra Pradesh, Himachal Pradesh, Madhya Pradesh, Rajasthan, Gujarat and Orissa.

- *Social Forestry Projects*

  in Orissa and Punjab.

- *Wetland Development Project*

  in Dungarpur (Rajasthan).

- *Aravalli Afforestation Project*

  in Hariana and Rajasthan.

- *Forestry and Ecodevelopment Project*

  in Himachal Pradesh.

- *Bamboo Rehabilitation Project*

  in Chhindwara district.

NATIONAL BIODIVERSITY STRATEGY AND ACTION PLAN — A GOVERNMENT OF INDIA POLICY

India harbours an enormous diversity of plants and animals, both domesticated and wild, and an impressive variety of habitats and ecosystems;

This biological diversity sustains the food, medicinal, clothing, shelter, spiritual, recreational, and other needs of most of India's Population;

This diversity also ensures the essential ecological functions on which life depends, including a steady supply of clean water, nutrient cycling, and soil maintenance;

The diversity is the treasure-house from which future food needs, cures for deadly diseases, and elements for knowledge and technology will be found;

The diversity is seriously threatened by human activities leading to servere shortages of water, fodder, fuel; and natural disasters such as floods, landslides and famines;

There is a need to prepare detailed action plans at state, regional, and sub-state levels to conserve and sustainably use biodiversity, to protect the cultural, scientific, spritual traditions and innovations related to it, and to equitably share the benefits arising from sustainable use;

The participation of women and men from all walks of life, both within and outside the government, rural and urban, is essential for preparing an implementable action plan.

The Value of Biodiversity

The well-being and survival of human populations are dependent on millions of species of plants, animals and microbes. These include species and varieties of crops, livestock and wildlife. This biological diversity is part of our daily life and livelihood and constitutes resources upon which families, communities, nations and future generations depend. Biodiversity has numerous values in agriculture, medicine, food and industry. Biodiversity maintains ecological balance and evolutionary processes, and has spritual, cultural, aesthetic and recreational values. The ecosystem services rendered include pollination, chemical cycling, nutrient cycling, soil maintenance and climate regulation.

The millions of species which constitute biodiversity have their own intrinsic value, as recognized by religions and faiths in India. Traditionally, Indians have celebrated this diversity of life. Conservation and sustainable use of biodiversity is ingrained in the Indian ethos and culture. Even today, millions of people derive their daily sustenance from forests, rivers, grasslands and seas. The practices and beliefs of forest-dwelling tribes, nomadic communities, fisherflok, and farmers remain closely intertwined with the biological diversity they live amidst.

Excessive use coupled with lack of awareness about the actual and potential value of biological resources has led to the loss of biological diversity. Proof of this loss is the extinction of some wild species like the Cheetah, and some domesticated plant varieties and animal breeds. Accelerated loss of biodiversity, largely on account of disempowerment of local communities, commercial exploitation, and fragmentation and degradation of habitats, has become a cause of concern. This loss of biodiversity is being felt through greater water shortage and lack of fodder, fuel, other livelihood resources and so on. And with this, there has been serious erosion in the knowledge base of people, especially the women and men of local communities, and of the cultural values linking people with nature.

The National Biodiversity Strategy and Action Plan (NBSAP)

In 1999, the Ministry of Environment and Forests prepared a National Policy and Macrolevel Action Strategy on Biodiversity through a consultative process. That document was a macro-level statement of policies, gaps and strategies needed for conservation and sustainable use of biological diversity. There is a need to prepare detailed action plans at sub-state, regional, and national levels based on this framework document. Towards this end, the Ministry has accessed funding from the Global Environment Facility (GEF) for preparing the National Biodiversity Strategy and Action Plan (NBSAP).

The NBSAP project envisages the assessment and stocktaking of biodiversity related information at various levels, including distribution of endemic and endangered species and site-specific threats and pressures. Key features of this project include an emphasis on gender sensitive decentralised planning, and the use of interdisciplinary working groups to involve all sectors concerned with biodiversity conservation. These detailed action plans (at sub-state, state, and regional levels) will be consolidated and a national level action plan will be developed.

Government agencies, non-governmental groups and village communities are already contributing towards conservation of biodiversity. However, there is a need to consolidate and coordinate these efforts, to launch new initiatives to plug the gaps in information and action, to put development processes and planning on a more ecologically sound footing, and to promote people's management of their surrounding natural resources.

Such a process of consolidation and new initiatives through a series of plans and strategies at local, state and national levels, cannot be successful without public participation. It is proposed to prepare the NBSAP through a process of widespread consultation and participation across India.

Goals of the NBSAP

To prepare, by early 2002, biodiversity action plans at the following levels —

Local and regional level (a few selected regions in the country, e.g. Karbi-Anglong District in Assam, Gulf of Kachchh in Gujarat, Vidarbha in Maharashtra...);

State level (all of India's States and Union Territories);

Inter-state level for biological regions cutting across states (e.g. Eastern Ghats, Western coasts, Trans-Himalaya);

Thematic level for major topics related to biodiversity; National level, taking into consideration all the above.

Scope of the NBSAP

The term 'biodiversity' is being taken in its holistic sense, to encompass the following levels, including related ecological and evolutionary processes;

Natural ecosystems; e.g., forests, grasslands, wetlands, deserts, mountains, coastal and marine areas.

Wild species and varieties; species of plants, animals, and micro-organisms existing in their natural state, and the genetic variation within each of these species.

Agricultural ecosystems : e.g., farmlands, pastures, capture fisheries, aquaculture.

Domesticated species and varieties; species of crops, livestock (including poultry), captive-breed fish, pets, and micro-organisms in *ex-situ* collections and the genetic variation within each of these species.

It is proposed that the action plan will cover —

- Conservation of biodiversity of all kinds listed above;
- Sustainable use of biological resources, implying their use in such a manner as will not imperil their long-term existence, or will not in other ways threaten biodiversity;
- Social, economic, ethical, cultural, scientific and economic dimensions, including gender relations and equity.

The NBSAP Process

The Ministry of Environment and Forests, Government of India, is overall in charge of the NBSAP process. Its execution is being done by a Technical and Policy Core Group (TPCG) headed by *Kalpavriksh* and Comprising experts from various fields and parts of India, while its administration is being coordinated by the Biotech Consortium India Ltd.

The process of developing NBSAP will be highly participatory in nature, reaching out to a large number of village-level organisations and movements, NGOs, academicians and scientists, government officers from various line agencies, the private sector, the armed forces, politicians and others who have a stake in biodiversity. The NBSAP should be seen as much as a process as a final product. Through such a participatory approach, it will evoke much wider ownership of the process and product and go a long way in ensuring the implementation of what is recommended in the final NBSAP.

[Source : (Ashish Kothari, Kalpavriksh, (Coordinator, NBSAP Technical and Policy Core Group)
C/o Dr. Vibha Ahuja, (Member Secretary NBSAP TPCG)
Biotech Consortium India Ltd. Kundan House, 4th Floor.
16 Nehru Place, New Delhi - 110 019.]

BIBLIOGRAPHY

A. Biodiversity

1. Brodbury, I.K. (1998) : The Biosphere
2. Charles Petal (1995) : Biodiversity loss
3. Das, A.K. (1998) : Faunal diversity
4. Deptt. of Environment (1996) HMSO (London) (1996) : Biodiversity Assessment.
5. Falk D.A. *et al.*, (1996) : Restoring, conservation
6. Furz, B. *et al.*, (1996) : Cultutre and Biodiversity
7. Gastoov, K.S. *et al.*, (1998) : Biodiversity and introduction
8. Huston, M.A. (1994) : Biological diversity
9. Kothari, A. (1997) : Understanding Biodiversity
10. Mater, P. *et al.*, (1995) : Biodiversity loss
11. Malson, L. Hunter (1999) : Maintenance of Biodiversity in forestry
12. National Academy Press (1992) : Conserving Biodiversity
13. Perlman, DL (1997) : Biodiversity : Exploring values
14. Perring, C. *et al.*, (1995) : Biodiversity : Economic & Ecological issue
15. Reitbergen S. (1993) : Tropical Forestry
16. W.N.E.P. (1995) : Global biodiversity
17. World Resources Institute (1993) : Biodiversity prospecting
18. William J.S. (Edited) (1996) : Biodiversity and the land.

B. Medicinal plants :

1. Bakhra, H.K. (1993) Herbs that heads.
2. Basu, B.D. (1997) : Indian medicinal Plants
3. Biswas K.P. & Ghosh A. (1977) : Bharatiya Vanausodhi
4. Bodding P.O. (1986) Studies in Santhal medicine.
5. Calcutta, School of Tropical Medicinal (1959) : Guide to treatment & Pharmacopea
6. Chatterjee A & Pakrashi (1991) : The treatise of Indian medicinal plants.
7. Chaudhuri R.D. (1932) : Herbal Drug Industry
8. Chopra R.N. *et al.*, (1958) : Indigenous Drugs of India
9. Chopra R.N. (1939) : Indigenous drug enquiry.
10. Kirtiker and Basu (1918) : Indian Medicinal Plants
11. Mukhopadhyay G.N. (1974) : History of Indian thrapeuties
12. Mukherjee B. & Chopra R.N. (1954) : A treatese in tropical medicines
13. Nadkarni K.M. (1954) : Indian Materia Medica
14. Sengupta N.N. (1901) : The yurvedic system of medicine.
15. S.S. turtha (1988) : The ayurvedo encyclopaedia
16. Wito monograph (1999) : Selected Medicinal Plants

C. Plants :

1. Chaudhuri, A.B. & Chakraborti, K. (1989) : Sundarbans Mangrove Ecology & Wildlife
2. Chaudhuri, A.B. (1990) : Environment & Resources of Tropical & temperate forests.
   (1991) : Environment of herb-shrub flora.
   (1992) : Mine Environment & Management.
   (1992) : Himalayan ecology as environment.
   (1992) : Plants, Wild life and man.
   (1993) : Forest plants of Eastern India.
   (1993) : Tree & the environment.
   (1995) : Sundarbans
   (1996) : Aquatic ecology
3. Brandis, D. (1906) : Indian Trees
4. Prain, D. (1905) : Bengal plants.

D. Managing Resources

1. David Flint (1999) : Managing Resources

2. OECD (1994) : Environmental indicating
3. World Bank (1994) : World Development report
4. WRI (1994) : Word Resources.

E. Fauna

1. Ali salia (1977) : Field guide to the birds of eastern India.
2. " (1941) : The book of Indian birds.
3. " & D. Repley (1968-74) : Hand book of Indian birds.
4. Baker, E.C.S. (1930-33) : The fauna of British India.
5. Blanford W.T. (1885-1875) : The fauna of British India.
6. Bhamrah H.S. (1990) : Introduced to Reptiles
7. Daniel J.C. (1984) : The book of Indian Reptiles
8. Das A.K. & Sanyal A. ( ) : Faunal diversity of India.
9. Deora, P.J. (1984) : Snakes of India
10. Jayram K.C. (1981) : Fresh water fishes of India
11. Jerden T.C. (1964) : The birds of India.
12. Marten R. (1960) : The world of reptiles
13. Menon A.K. (1978) : Checklist of Fresh water Fishes
14. Merly T.S.N. (1986) : The Snake book of India.
15. Prater S.H. (1928) : The book of Indian animal
16. Shaw G.E. & Shebbeare E.O. (1937) : Fisheries of North Bengal
17. Smith F.A. (1931) : Fauna of British India.
18. Sharma R.C. (1998) : Reptiles.
19. Talwar P.K. & Jhingram A.G. (1991) : Inland Fisheries of India
20. Tikadar B.K. (1992) : Indian Testudines
21. Tikadar B.K. (1992) : Indian Lizards.
22. Whistler H. & Kinnear N.B. (1949) : Popular hand book of Indian birds
23. Wildlife (Protection) Act, (1972) : Govt. Publication.

